Lecture Notes on Data Engineering and Communications Technologies

Volume 115

Series Editor

Fatos Xhafa, Technical University of Catalonia, Barcelona, Spain

The aim of the book series is to present cutting edge engineering approaches to data technologies and communications. It will publish latest advances on the engineering task of building and deploying distributed, scalable and reliable data infrastructures and communication systems.

The series will have a prominent applied focus on data technologies and communications with aim to promote the bridging from fundamental research on data science and networking to data engineering and communications that lead to industry products, business knowledge and standardisation.

Indexed by SCOPUS, INSPEC, EI Compendex.

All books published in the series are submitted for consideration in Web of Science.

More information about this series at https://link.springer.com/bookseries/15362

Roman Oliynykov · Oleksandr Kuznetsov ·
Oleksandr Lemeshko · Tamara Radivilova
Editors

Information Security Technologies in the Decentralized Distributed Networks

Springer

Editors
Roman Oliynykov
Department of Information
and Communication Systems Security
Faculty of Computer Science
V. N. Karazin Kharkiv National University
Kharkiv, Ukraine

Oleksandr Kuznetsov
Department of Information
and Communication Systems Security
Faculty of Computer Science
V. N. Karazin Kharkiv National University
Kharkiv, Ukraine

Oleksandr Lemeshko
Department of Infocommunication
Engineering, Faculty of Infocommunication
Kharkiv National University of Radio
Electronics
Kharkiv, Ukraine

Tamara Radivilova
Department of Infocommunication
Engineering, Faculty of Infocommunication
Kharkiv National University of Radio
Electronics
Kharkiv, Ukraine

ISSN 2367-4512   ISSN 2367-4520 (electronic)
Lecture Notes on Data Engineering and Communications Technologies
ISBN 978-3-030-95160-3   ISBN 978-3-030-95161-0 (eBook)
https://doi.org/10.1007/978-3-030-95161-0

This Springer imprint is published by the registered company Springer Nature Switzerland AG
The registered company address is: Gewerbestrasse 11, 6330 Cham, Switzerland

Preface

With this volume, we analyze challenges and opportunities for infocommunication systems usage, taking in account theory, security, information technologies and communication aspects.

Explicitly, starting with the first chapter "Methods of Ensuring Privacy in a Decentralized Environment" authored by Lyudmila Kovalchuk, Roman Oliynykov, Yuri Bespalov and Mariia Rodinko presented methods and mechanisms used in cryptocurrency blockchains to ensure the anonymity of users. Each of these methods and mechanisms is described in detail with their comparative analysis. It has been determined which of them are most suitable for maintaining personal information privacy. It has analyzed the methods and cryptographic primitives used in blockchain technologies such as the largest by market capitalization and the "most private" digital currencies like Monero, Dash and Zcash. In addition, these cryptocurrencies are the object of close attention and prohibition of many intelligence agencies and financial institutions around the world. This attention also signals on their level of privacy. However, only the Monero cryptocurrency provides privacy for all transactions, and the cryptocurrencies Dash and ZCash allow privacy as the option at the request of the user. The analysis results show how the obtained information can be used to create a network that will operate on a blockchain basis and store, process and transmit private information, such as personal data ensuring its anonymity. It is concluded that to ensure a given level of privacy in the processing, storage, transmission and receipt of confidential personal information, it is possible to use only the appropriate mathematical apparatus, such as ring signatures (such as Monero) or SNARKs, such as Zcash. The example of the most common protocol GROTH16 shows how the information is structured before constructing the SNARK-proof, how the preliminary calculations are performed (the so-called ceremony, or setup), how the proof itself is built and how its correctness is checked. It also shows how to build recursive SNARK-proofs and why and in what situations they are necessary.

The second chapter "Application of Bluetooth, Wi-Fi and GPS Technologies in the Means of Contact Tracking" describes the spread of infectious diseases, including COVID-19, as a complex epidemiological situation exacerbated by powerful transcontinental migration processes that pose a potential threat to human health around the world. It requires a set of tasks aimed at full control of risks and threats to human life. Modern information technologies can be useful for solving a number of scientific and technical problems related to the use of cyberspace for global population monitoring, i.e., when through observation and contacts tracking, it is possible to predict possible negative scenarios and prevent the spread of emergencies and crises, especially in pandemic COVID-19. The high general level of electronic society can keep epidemics and pandemics at the national and international levels under control. Global solutions for building monitoring systems to prevent the spread of infectious diseases already exist and are developing rapidly. This section analyzes, researches and substantiates the possibility of using various information technologies (e.g., Bluetooth, Wi-Fi, GPS, etc.) in contact tracking tools as the main subsystem of global population monitoring. In particular, the principles of construction and the possibility of using these technologies to track contacts are studied. Their advantages and disadvantages, potential attacks on monitoring applications, etc., are analyzed.

The third chapter "Analysis and Research of Threat, Attacker and Security Models of Data Depersonalization in Decentralized Networks" authors focus on relatively new attacks that emerged only after the advent of the blockchain and ways to defend against them. These attacks are aimed at the way the blockchain works and are not attacks on the cryptographic mechanisms that serve it. Some of the blockchain attacks are common, meaning that their basic idea can be applied to any blockchain, regardless of the consensus protocol it uses. Other attacks are special attacks, i.e., they significantly depend on the type of consensus protocol. Authors have considered two main models of general blockchain attacks. In addition to reviewing and comparative analysis of these attacks, the main ways to counter these attacks are also given, which significantly use the features of the operation and maintenance of the blockchain.

The fourth chapter "Cryptographic Transformations in a Decentralized Blockchain Environment" authored by Alexandr Kuznetsov, Dmytro Ivanenko, Nikolay Poluyanenko and Tetiana Kuznetsova describes the distributed decentralized systems built using blockchain technology. The solutions presented in the chapter consider and investigate algorithms and protocols of homomorphic encryption, ring signatures, protocols with zero disclosure, principles of construction of anonymous secure networks, etc.

The authors of the fifth chapter "Statistical and Signature Analysis Methods of Intrusion Detection" describe the existing models and methods of intrusion detection that are mostly aimed at detecting intensive attacks and do not take into account the security of computer system resources and the properties of information flows. This limits the ability to detect anomalies in computer systems and information flows in a timely manner. The latest monitoring and intrusion detection solutions must take into account self-similar and statistical traffic characteristics, deep packet analysis and the time it takes to process the information.

An analysis of properties traffic and data collected at nodes and in the network was performed. Based on the analysis, traffic parameters that will be used as indicators for intrusion detection were selected. A method of intrusion detection based on packet statistical analysis is described and simulated. A comparative analysis of binary classification of fractal time series by machine learning methods is performed. We consider classification by the example of different types of attack detection in traffic implementations. Random forest with regression trees and multilayer perceptron with periodic normalization were chosen as classification methods. The experimental results showed the effectiveness of the proposed methods in detecting attacks and identifying their type. All methods showed high attack detection accuracy values and low false positive values.

The sixth chapter "Criteria and Indicators of Efficiency of Cryptographic Protection Mechanisms" tracking contacts as the main subsystem of population monitoring, and there are practical and theoretical tasks for the mechanisms of cryptographic protection of personal data of users, providing efficiency; objectivity of decision making; depersonalization; generalized processing, reliability and availability of decentralized storage. This section substantiates the criteria and indicators of the effectiveness of cryptographic protection mechanisms. In particular, general theoretical information on the criteria and indicators for assessing the security of symmetric cryptocurrencies, hashing functions, generating pseudorandom sequences, asymmetric cryptographic transformations, etc., is presented. The section discusses general approaches that can be used to determine the criteria and performance indicators of cryptographic protection mechanisms for users' personal data and interaction protocols that provide various security services when tracking contacts with system users.

The main goal of the seventh chapter "Methods of Evaluation and Comparative Research of Cryptographic Conversions" is the security of decentralized systems built on blockchain technology as directly determined by the effectiveness of the applied cryptographic transformations. The classic blockchain technology basically uses several cryptographic primitives: hash function, electronic signature, encryption, etc. In particular, the hash function is used for several purposes as generating an address or user ID; transaction and block hashing to confirm the irrefutable block against network errors as well as the hash value of the previous block is used as a reference in the formation of the next. The use of digital signatures in the blockchain system introduces services of irrefutability and integrity in relation to transactions. In turn, in any blockchain system, there is a pool of transactions, each of which must be signed. This ensures that the transaction will not be altered, the signature of the transaction ensures that it was signed by a person who can then be identified, and, in fact, the signature can help determine "whether the user has the rights to conduct the transaction." Block ciphers are one of the most common cryptographic primitives, which are also used as a structural element of hash functions, message authentication codes, etc. Thus, for the operation of monitoring and tracking applications using blockchain technology, it is necessary to study various cryptographic primitives, in particular, hash functions, electronic signatures, encryption schemes, etc.

The eighth chapter "Cryptographic Mechanisms that Ensure the Efficiency of SNARK-Systems" provides strictly substantiated security estimates of the block cipher HadesMiMC and the hash function Poseidon to the linear and differential attacks. Based on the obtained results, specific parameters for the HadesMiMC encryption algorithm and the Poseidon hash function were determined, which allow them to be used for recurrent SNARK-proofs based on the mnt-4 and mnt-6 triplets. It is also shown how best to select S-blocks for these algorithms, so that this choice was optimal both in terms of security and in terms of computation complexity in SNARK-proofs. It is determined how many rounds are sufficient to ensure the stability of the algorithms for different options. It has determined the number of constraints per bit, for different options, and it is shown that it is significantly less for Pedersen's function, which is now used in the cryptocurrency Zcash. It has shown significant advantages of the considered HadesMiMC block cipher and the corresponding Poseidon hash function in the blockchain as a decentralized environment. These advantages are both a reasonably high level of security against known attacks and a significant increase in performance (up to 15 times) in the construction of appropriate SNARK-proofs, which are an integral part of the mechanisms of anonymity in the blockchain. The obtained results showed that the block encryption algorithm HadesMiMC, based on operations in a prime field, is secure against linear and differential cryptanalysis, with the correct choice of parameters of this algorithm. As a result, the Poseidon hash function, built using the sponge design and the HadesMiMC permutation, will be stable. It was also shown how the parameters of the algorithm should be selected to ensure its specified level of stability. In addition, the number of constraints per bit of information is calculated, and it is shown that the HadesMiMC algorithm and the Poseidon hash function currently have no competitors when used in SNARK-proofs.

The authors of the ninth chapter "Comparative Analysis of Consensus Algorithms Using a Directed Acyclic Graph Instead of a Blockchain, and the Construction of Security Estimates of Spectre Protocol Against Double Spend Attack" have shown the possibility of moving from a blockchain to a more general structure, which has additional capabilities, but remains secure to major attacks. Basically, the new structure is a DAG, which has certain extra features. Various ways of such generalizations and partial analysis of their properties are proposed in the papers. At the same time, the results of such research raise a large number of questions, so the field for research in this area remains extremely large, and the task of scaling the blockchain is very important. Over the last 5–7 years, many consensus protocols have been proposed on DAGs. It should be noted that the use of such protocols, although it leads to a significant increase in the speed of transaction processing, carries certain risks. At present, there is no strict evidence of resistance to major attacks for any of these protocols, such as a double spend attack and a network split attack. The construction of such proofs requires overcoming great analytical difficulties, even if researchers use a very simplified mathematical model of the network. In this section, authors took the first step toward constructing consensus protocol security estimates on DAGs by constructing upper estimates for the probability of a double spend attack on the specter consensus protocol. To increase its security, authors have proposed an additional non-standard algorithm for accepting a transaction by a vendor. They developed methods for calculating the upper estimates of this probability depending on the network parameters and the number of confirmation blocks that the vendor must see before considering the transaction irreversible. For greater clarity, the corresponding numerical results are also given. An important feature of the results obtained is that, in addition to the probability estimate itself, we can also calculate the required number of confirmation blocks sufficient to guarantee the irreversibility of the transaction with a probability as close to 1 as desired.

At the tenth chapter "Models and Methods of Secure Routing and Load Balancing in Infocommunication Networks," authors analyze the existing solutions in the field of network security, the results of which showed that a promising direction of their development is to improve traffic management and routing that should take into account not only network performance parameters but also network security parameters that characterize the efficiency of network intrusion detection systems and analysis of vulnerabilities and risks. A flow-based model of secure routing with load balancing has been developed, in which the modified load balancing conditions, in addition to the bandwidth of the communication link, also consider the probability of its compromise. It has been found that in the process of load balancing, links with a lower probability of compromise are loaded more intensively, unloading more threatening from the point of view of compromise communication links. A model of secure fast rerouting with the protection of structural elements of the network and its bandwidth is proposed, which provides a reactive approach to network security when in case of a probable compromise of the network element all packet flows quickly switch to pre-calculated backup paths that do not pass through the compromised node or link.

The authors of the eleventh chapter "The Methods of Data Comparison in Residue Numeral System" research the information processing in the systems of transmission and processing of digital data, which are built based on a non-positional number system in a residue number system (RNS). The purpose of the research is to develop a method of data comparison, presented in RNS. To make this goal achieved, the existing methods of data comparison in RNS were analyzed, and due to this, a method of data comparison was developed, which allows performing the operation of number comparison in positive and negative numerical ranges. The following methods were used: the methods of analysis and synthesis, as well as the methods of numbers theory. It is recommended to use the presented algorithm, which is based on the developed method, in the practical implementation of the operation of comparing two numbers in RNS. The work presents the implementation of examples of the comparison operation based on the usage of the developed methods.

The twelfth chapter "The Data Control in the System of Residual Classes" discusses the issues related to the development and application of methods for improving the efficiency of operation based on the use of the codes of the system of residual classes (SRC). The research is aimed to present the methods of data operational control in the SRC. The main focus is given to the control methods based on the principle of nullification of numbers in the SRC. It is shown that the resulting redundancy with the introduction of one additional (control) base provides control and correction of errors in the process of performing operations. The algorithms developed and presented in the work allow modeling the data control unit for the practical implementation of the proposed methods. The analysis is performed, the calculated data of the implementation time of the control operation for the considered methods of nullification is obtained, and the conditional amount of the control system equipment is also calculated.

The authors of the last chapter "Traffic Monitoring and Abnormality Detection Methods for Decentralized Distributed Networks" describe modern information systems that are most often created as distributed and include a large number of various devices connected by means of networks. Devices included in this distributed information system have different performance and a set of functions, which are often limited. The development of such distributed systems at the present stage has been transformed into the system of the Internet of Things and further into the Internet of Everything. The widespread development of distributed information systems based on public networks, which penetrate many areas of human activity, creates additional threats to information security and makes them extremely relevant. Traffic monitoring is one of the stages of information security systems functioning. Traffic monitoring results are used to solve the problems of detecting traffic anomalies and detecting attacks and intrusions. Modern detections of attacks and other anomalies are not reliable enough and require further research. The chapter provides the classification of attacks and intrusions, as well as the classification of anomalies and intrusions. The main attention is paid to intrusion detection systems, and a method for detecting network traffic anomalies and intrusions is proposed, which is based on the use of a number of statistical, correlation and informational parameters of network traffic as additional features. The chapter researches a method for detecting traffic anomalies based on machine learning methods. In the end of the chapter, attention is paid to several ways of building and implementing subsystems for detecting anomalies in distributed networks that include devices with limited performance.

Kharkiv, Ukraine

Roman Oliynykov
roliynykov@gmail.com

Oleksandr Kuznetsov
kuznetsov@karazin.ua

Oleksandr Lemeshko
oleksandr.lemeshko@nure.ua

Tamara Radivilova
tamara.radivilova@nure.ua

About This Book

In terms of data centralization, the potential threat to confidentiality and human rights leads to the spread of decentralized technological solutions since the centralized model involves the use of a single database that records identification codes of meetings between users. In the decentralized model, such information is recorded on individual devices, and the role of the central database is limited to the identification of phones by their anonymous identifier, in the case of sending a warning.

However, system developers identify a number of potential attacks that can not only lead to the disclosure of data, but also to the refusal of the population from monitoring systems. Such attacks include trolling attacks and relay attacks that cause a false probability of disease, power consumption attacks that cause distrust of contact monitoring applications, deanonymization attacks that are morally and ethically damaging and others. Thus, a science approach and single secure solution for building a system for monitoring contacts in the world are missing, which requires research in terms of a number of regulatory and technological international requirements. In this book, the methods for building decentralized private databases without a trusted party are proposed, which allow data processing only in encrypted form, guaranteeing the confidentiality of information and availability of the relevant fragment of original or converted data only to an authorized participant using the database, but not all other participants. This book proposes the approaches to the construction of a decentralized private database without any trusted party, which allows to perform the specified processing only in encrypted form, with the guarantee of confidentiality of information and availability of the relevant fragment of original and (or) converted data only to the participant who owns the relevant data fragment, and there is no trust in all other participants, description of protocols of interaction between individual users, healthcare facilities and systems for monitoring and alerting the population about the epidemiological environment and emergencies. This book also describes the study of peer-to-peer secure fault-tolerant routing of confidential messages on the network, development of number of cryptographic protocols, which makes it possible to deploy an effective decentralized registry (blockchain) system based on them, which allows each participant to know whether he had physical contact with an infected person or other potential risks (as well as the need for self-isolation) and the methods for ensuring a given level of network security and quality of service with efficient use of available network resources based on balancing traffic on multiple routes. The proposed methods that prevent or reduce the existing shortcomings of existing systems and the introduction of technologies based on them will timely inform users about the potential risks of each user, while providing the highest level of guarantees for the privacy of each participant.

Contents

Methods of Ensuring Privacy in a Decentralized Environment

Lyudmila Kovalchuk⬥, Roman Oliynykov⬥, Yurii Bespalov⬥, and Mariia Rodinko⬥

Abstract The section discusses all currently available methods and mechanisms used in cryptocurrency blockchains to ensure user privacy. Each of these methods and mechanisms is described in detail, their comparative analysis is performed. It has been determined which of them are most suitable for maintaining the confidentiality of certain private personal information. The methods and cryptographic primitives used in blockchain technologies of such largest market capitalizations and "most private" digital currencies as Monero, Dash and Zcash are analyzed. In addition, these cryptocurrencies are the object of close attention and ban by many intelligence agencies and financial institutions around the world. This attention also indicates the level of their anonymity. However, only the Monero cryptocurrency is anonymous, of course, and the Dash and Zcash cryptocurrencies only allow the choice of the anonymity option at the user's request. The results of the analysis show how the information obtained can be used to create a network that will operate on a blockchain basis and store, process and transmit confidential information, such as personal data, while ensuring its privacy. It is concluded that to ensure a given level of privacy in the processing, storage, transmission and receipt of confidential personal information, it is possible to use only the appropriate mathematical apparatus, such as ring signatures (such as Monero) or SNARKs, such as Zcash. The example of the most common protocol Groth16 shows how the information is structured before constructing the SNARK-proof, how the preliminary calculations are performed (the so-called ceremony, or Setup), how the proof itself is built and how its correctness is verified. It also shows how to build recursive SNARK-proofs and why and in what situations they are necessary.

L. Kovalchuk
National Technical University of Ukraine "Igor Sikorsky Kyiv Polytechnic Institute", 37, Prosp. Peremohy, Kyiv 03056, Ukraine

R. Oliynykov (✉) · M. Rodinko
V. N. Karazin Kharkiv National University, 4 Svobody Sq., Kharkiv 61022, Ukraine

Y. Bespalov
M.M. Bogolyubov Institute for Theoretical Physics of the National Academy of Sciences of Ukraine, 14-b Metrolohichna str., Kyiv 03143, Ukraine

© The Author(s), under exclusive license to Springer Nature Switzerland AG 2022
R. Oliynykov et al. (eds.), *Information Security Technologies in the Decentralized Distributed Networks*, Lecture Notes on Data Engineering and Communications Technologies 115, https://doi.org/10.1007/978-3-030-95161-0_1

1

Keywords SNARK-proof · Anonymity of information · Decentralized system · Cryptocurrency · Blockchain technology · Cryptographic primitive

1   Introduction

One of the most interesting and modern models of decentralized environment that has emerged in the last 10 years is blockchain technology. Although it was originally conceived as a convenient platform for decentralized transaction processing, in the short time of its existence and use, this technology has demonstrated an unexpected number of useful properties. Blockchain technology has long ceased to be seen only as a means of cryptocurrency circulation, and the scope of its application has been really vast, from smart-contracts to electronic document management. This technology is very convenient for storing, processing and sending confidential information. In addition, as it turned out, the blockchain is also convenient to use for the protection of personal data, including to ensure the required level of anonymity of this data and control access to them. It is here that such "classic" and "exotic" technologies as proof without disclosure, distribution of secrets, mathematical methods of providing multilevel access, etc. have received new life.

Since the use of blockchain in the field of cryptocurrency circulation has become the most widespread, the mechanisms of anonymity are the most developed, worked out and researched in this area. Although we are primarily interested in maintaining confidentiality for private information of various types, not just (and not so much) financial, we will still first review and analyze all existing methods to determine the best way to maintain confidentiality.

In this section, we will look at all the currently available methods and mechanisms used in cryptocurrency blockchains to ensure user privacy. We will describe in detail and carefully analyze each of these methods and mechanisms, perform their comparative analysis and then determine which of them are best suited to maintain the confidentiality of certain private personal information.

2   Ways to Ensure Privacy in the Blockchain

The governments of almost all technologically advanced countries are trying to establish total monitoring and control over the circulation of funds of both their citizens and citizens of other countries that are in the field of view of the relevant intelligence services. This is primarily due to the challenges of combating money laundering, tax evasion and the threat of terrorist financing, but there are also many other reasons for the state's interest in this issue. In particular, knowledge of the source of funds for a person may be additional evidence of his espionage in favor of the state. The desire to hide the movement of their finances, such as the use of cash, cryptocurrencies, precious metals, in some countries may automatically arouse suspicion of some illegal activity.

On the other hand, special services need not only to monitor the circulation of other people's money, but also to be able to hide the circulation of their own. And in general, most ordinary people and businesses have a desire, and in countries with immature democratic institutions, often even the need to maintain the anonymity of their finances. And this is not always because they have planned something illegal, but mostly because they themselves do not want to become victims of illegal actions of criminals. Therefore, with the advent of the first cryptocurrency Bitcoin [1] (BTC) in 2009, many people and organizations have hoped that now they will finally get ways to perform fast, convenient, uncontrolled and anonymous transactions.

Indeed, the Bitcoin concept contained many revolutionary ideas and technological solutions at the time, the most important of which was the way to achieve decentralization in conditions of complete distrust. But since the launch of Bitcoin and until now, the algorithm of the network has not changed. In particular, nothing has been done to create and implement additional privacy mechanisms. However, on the basis of the Bitcoin platform, a number of other cryptocurrencies (altcoins) were created, which tried in one way or another to increase both the privacy of users and the inability to track the turnover of their funds.

Next, the main ways to ensure privacy when using different cryptocurrencies and list their advantages and disadvantages will be looked at.

It should be noted at once that the anonymity of the recipient of funds ends when they withdraw cryptocurrency funds to a registered bank card. Payment for various goods and services in cryptocurrency is currently essentially limited.[1] So, the use of cryptocurrency to pay for real goods or services, as a rule, requires its conversion into the currency of the country. Therefore, here ways to preserve the anonymity of the sources from which the cryptocurrency enters the e-wallet of a particular person will be mainly considered.

2.1 Ways to Preserve the Anonymity of the Source of Income

On the one hand, any cryptocurrency provides some "initial" degree of anonymity. If the user creates their own wallet, and does not entrust its creation to a particular exchange, then the creation of a wallet does not require any personal information about the user. Moreover, he can create any number of wallets so that it will be impossible to determine whether they belong to one person or different. Moreover, modern client programs can automatically create new wallet addresses for the user for each transaction to make it harder to track. In this case, the old addresses will also belong only to them and they can use them at will.

---

[1] https://www.euronews.com/next/2021/09/24/bitcoin-ban-these-are-the-countries-where-crypto-is-restricted-or-illegal2.

But, on the other hand, all transactions are stored forever in the blockchain, i.e. information about the transfer of funds between wallets is publicly available and indelible. So if you can establish a connection between a person and a particular wallet, then the information stored in the blockchain, you can determine all the movement of funds to or from this wallet.

In addition, if you create a wallet on the exchange, it is even harder to remain anonymous. Some exchanges require certain personal information about the client, in accordance with the laws of the country to which the exchange belongs.

Satoshi Nakamoto claimed that his cryptocurrency was anonymous. As it became clear later, by this definition he meant the ability of the user not to provide any personal information about himself to receive digital money. If BTC were extracted only by mining, the cryptocurrency would really be anonymous. But due to certain rules set on major exchanges, many BTC addresses can be associated with real people. In addition, due to the ability of the blockchain to store all information about transactions, you can get actual confirmation of how much the owner of one wallet transferred to the owner of another one, and, also, how much money in the specified account.

About two years ago, Bitfury experts said that about 16% of all BTC address owners could reveal their identities.[2] And a few years before that, the CryptoLux development team conducted a study of the anonymity of transactions in the BTC network and proved that it is possible to successfully deanonymize up to 60% of addresses. In their work, they showed the possibility of linking a BTC account to the IP addresses of users, even if they use the Tor network or other similar programs.

The lack of complete confidentiality was one of the shortcomings of the first cryptocurrency. This has prompted many developers to find ways to improve it, as well as to create new altcoins with the desired properties, including increased anonymity. Next, we will look at how this problem is solved, with varying degrees of success, in the most commonly used cryptocurrencies.

2.1.1 Independent Mining of Cryptocurrencies

The easiest way to get cryptocurrencies from an anonymous source is to own them yourself. In this case, the coins in the wallet are taken as if "from the air". That is, either a new, newly created coin appears in the wallet that has not been in any other wallet, or the coin is transferred from the wallet that corresponds to the mining pool.

This method, of course, has disadvantages. First, the accumulation of coins in this way takes a long time. Second, it requires a constant consumption of resources (electricity) and a constant connection to the Internet.

---

[2] https://cryptocurrencyhub.io/top-3-anonymous-cryptocurrencies-f65d5b11394.

2.1.2 Use of Transaction Mixers: Bitcoin Mixers and Cryptocurrency Dash

The general principle of operation of the transaction mixer is as follows. At one stage of the transfer of funds is a so-called collective transaction (or group transaction), in which several payers participate. The coins they send are broken into small "pieces" that are mixed together and sent to recipients according to a certain algorithm. As a result, recipients receive the correct number of coins, but this amount will consist of different pieces sent by different payers. The anonymization mechanism should hide the unambiguous correspondence between the senders and their coins. As a result, for the recipient and for outside observers, the sender of funds becomes uncertain. The uncertainty is that the funds received could, with equal probability, be sent to any of the participants in the collective transaction. Accordingly, the greater the number of players involved in a collective transaction, the less likely it is possible to guess the actual sender.

BTC-mixers[3] and Dash cryptocurrency mixers (previous name—DarkCoin)[4] are arranged on this principle. Modern mixers can serve several types of cryptocurrencies. In addition, they are enhanced by additional features such as:

- mixing not only the addresses of senders, but also the addresses of recipients;
- delay in the time of delivery of coins (i.e. the amount sent not only comes in parts from other addresses, but also these parts arrive at different times);
- deleting transaction information some time after mixing;
- integration with the anonymous Tor network;
- protection against "dirty" money.

Interestingly, some mixers do not serve residents of certain countries (for example, PrivCoin.io does not serve residents of the Netherlands and the United States).

The disadvantage of BTC-mixers can be considered that the user's funds for mixing seem to be transferred to the mixer. That is, mixing takes place, so to speak, on trust. Since the source codes of many mixers are unknown, it is difficult to say what hidden possibilities may be there (for example, a mixer may remember the "path" of funds while mixing, or store some confidential information about the user and potentially provide it to someone later).

Unlike BTC, Dash cryptocurrency does not require an additional mixer function, as this PrivateSend function exists at the protocol level. Therefore, the use of these mixers does not increase the time of the transaction, in contrast to the use of BTC-mixers. The use of the mixing function is not mandatory, the user decides whether he needs the anonymity of transactions. The user can also select some mixing parameters (for example, the number of independent mixes in different groups).

One of the advantages of Dash compared to BTC-mixers is that the funds remain the property of the user during mixing.

---

[3] https://www.europeanbusinessreview.com/top-15-bitcoin-mixers-in-2021/.

[4] https://www.mycryptopedia.com/dash-privatesend/.

Another advantage of Dash is completely open source. The source code of both the client part (wallets) and the network infrastructure (Masternode) is published for general control for the absence of hidden functions. The same applies to anonymization mechanisms. Using the source code, users can compile their own wallets, which will guarantee no undocumented features.

2.1.3   Use of Anonymous Payment Systems

Alternatives to mixers can be special wallets[5] with a high degree of anonymity, such as Electrum.

There are also wallets with a built-in coin shuffle option. In 2014, Cody Wilson, the creator of the 3D-printed pistol and a fan of cryptocurrencies, and programmer Amir Taaki introduced the Dark Wallet project, a plugin for the browser and client for the Ubuntu operating system. Dark Wallet is based on the CoinJoin method, which is also a transaction mixer. The more users of the wallet, the more opportunities for mixing, and, consequently, the greater the degree of anonymity.

Another way to ensure the anonymity of transactions is to use special networks. At Darknetmarkets,[6] you can find instructions on how to maintain anonymity when conducting transactions on the Tor network. To do this, the user needs to register several wallets both in the anonymous network and in the open segment of the Internet, as well as have a Tor browser and mixer services that support Tor.

A very successful example of the implementation of anonymous transfer of cryptocurrencies is the open source payment system Z-Pay.[7] The main feature of this payment system is the issuance and transfer of impersonal payment bills (checks) as payment for goods and services. A check in the Z-Pay system is an anonymous monetary obligation denominated in cryptocurrency or fiat (real currency). The check number is a random combination of 35 digits. When sending such a check number as payment, you do not disclose personal data, and the payment itself is instantaneous. In this case, the payee can withdraw funds in any currency convenient for him. That is, with the use of Z-Pay you can make payments, pay for any services, exchange cryptocurrency and fiat funds without verification.

When converting cryptocurrency funds into fiat funds and withdrawing them to the card via Z-Pay, it looks like a replenishment through a terminal or as a transfer from an individual.

The creators of the Z-Pay.io payment system have also developed a Telegram bot, which functionally completely replicates the service's website. Android and iOS applications have also been created for user convenience.

The disadvantages of this payment system include the commission for the exchange and withdrawal of funds (1–3%). Also, the principle of operation of the system is not completely clear, but what information about the transaction is stored

---

[5] https://electrum.org/#home.

[6] https://darknetmarkets.org/a-simple-guide-to-safely-and-effectively-mixing-bitcoins/.

[7] https://z-pay.io/.

and due to which anonymity is ensured. To a direct question about anonymity, the support service answered "We do not provide your data to third parties", but there was no detailed explanation.

To register in the system you need to enter an email address and phone number, or log in via social network. Therefore, anonymity cannot be called complete.

2.1.4 Purchase and Sale of Cryptocurrencies Through Crypto Machines or Exchanges Without Verification

The easiest way to buy bitcoins is through a crypto machine (a special device for buying cryptocurrencies, similar to an ATM). Most crypto machines, in addition to BTC, also support Ethereum, Bitcoin Cash, Litecoin, Dash, Dogecoin, as well as anonymous Zcash and Monero, and some even more than 40 different cryptocurrencies. The purchase procedure is very simple, but the commission is quite high: 5–6% of the transaction amount from the hryvnia and + 0.001 BTC commission of the Bitcoin network itself.

In Ukraine, crypto machines already exist in Kyiv[8] (although they are constantly migrating around the city, and the companies that serve them are constantly changing) and in Odesa. A crypto machine has recently appeared in Irpin.[9] About a year ago, 8 crypto machines were installed in Ukraine and 26 in Russia.

To use the crypto machine, you need to scan the address of your wallet and deposit cash, at least a certain minimum amount. For example, in Ukraine, the minimum amount a year ago was 500 hryvnias ($20), because a smaller amount will simply be "eaten" by commissions. In Russia it was about $750. It takes no more than 10 min to process a request, after which cryptocurrencies appear in your wallet.

Some "high-level" models of crypto machines, for example, from Genesis Coin company, allow not only to buy cryptocurrency, but also to exchange it for fiat money. Unlike buying a cryptocurrency, selling it through a cryptocurrency ATM takes more time because it takes place in two stages. The user needs to do the following.

1. Indicate the amount of cryptocurrency he plans to sell.
2. Go through the verification procedure (if necessary), which we will discuss below.
3. Scan the QR-code on the screen of the device or on the issued check to transfer the specified amount to the operator's wallet (the time allotted for this procedure is limited).
4. Wait for the confirmation of the transaction by the nodes of the cryptocurrency network (usually two confirmations are enough).
5. Scan the QR-code for cash withdrawal, after which the crypto machine will give out fiat money.

---

[8] https://coinatmradar.com/bitcoin_atm/12265/bitcoin-atm-general-bytes-kyiv-palladium-siti/.

[9] https://coinatmradar.com/bitcoin_atm/12448/bitcoin-atm-general-bytes-irpin-ambar-shop/.

When exchanging cryptocurrency for fiat money, the crypto machine also receives a commission. It should be noted that the commission held by crypto machines tends to decrease due to increasing competition between operators. A few years ago, the commission was 10–12%, and today—5 to 6%.

Let's consider in more detail on the problem of anonymity and possible verification. First, according to the description of the crypto machine protocol, it is associated with some cryptocurrency wallet (see BATM Terminal Management video,[10] "internal wallet" for 2 min 19 s). At the address of this wallet, generally speaking, it is possible to track this crypto machine as a source of replenishment. And although the identity of the person who contributed the funds remains unknown, its geographical location can be identified.

As for the anonymity of the recipient, there are different options. When exchanging cryptocurrency for fiat money through a crypto machine, verification may be required in the following cases:

1. the amount to be exchanged exceeds the set limit (however, in most such cases, operators simply do not allow the exchange of amounts that exceed the limit by setting limits);
2. mandatory verification requirements set in certain countries, regardless of the amount of the transaction.
3. The verification procedure can be performed in various ways, for example, by taking a photo with a certain supporting document in hand, or by scanning fingerprints, or by reading payment card data, or by phone number.

All these requirements are very different for different countries and, moreover, change very quickly. Governments in many countries are unable to determine the law on cryptocurrencies in general and cryptocurrency ATMs in particular, which leads to increased loyalty to cryptocurrency transactions, and, conversely, to the permanent withdrawal of cryptocurrency ATMs and other equipment involved in such transactions. Therefore, the ability to withdraw fiat funds from cryptocurrency ATMs without verification, although it looks very tempting, but before using it you need to carefully study the state of affairs at this time and in this particular country.

Some exchanges also allow transactions with cryptocurrencies without verification, or with limited verification, or with optional verification. Sometimes it is even possible to convert a certain (usually limited) amount of money into fiat currency without verification. In this case, exchanges, as a rule, try not to violate the laws of the country in which they are registered. Therefore, depending on the laws in force, they may also limit the amount that does not require verification, or restrict transactions, for example, not to provide fiat money withdrawal services. A fairly complete list of exchanges that to some extent maintain anonymity and the conditions under which they operate is given in [2].

---

[10] https://www.generalbytes.com/en/products/batmtwo.

2.2 Principles of Functioning of Other Anonymous Cryptocurrencies. Advantages and Disadvantages

Anonymous coins are peer-to-peer (one-to-one) payment systems with their own internal unit of account. Their main goal is to ensure the complete confidentiality of financial transactions with the help of special technologies and cryptographic protocols.

2.2.1 Definitely an Anonymous Cryptocurrency Monero

The developer of the "most important" anonymous cryptocurrency Monero,[11] Ricardo Spani, believes that the digital industry needs financial privacy. Otherwise, the market will be filled with targeted (targeted) advertising, and at worst—will begin crimes against the owners of a significant amount of cryptocurrencies (because of the "traditional" blockchain, everyone can learn the balance of any foreign bitcoin wallet). Monero is so far the only coin with which all transactions are anonymous by default. For all other cryptocurrencies, the anonymity function must be additionally configured (as BTC-mixers) or connected during the transaction (as in the cryptocurrency Dash).

From the Monero public blockchain, you can find out if an account address exists, but you can't check the balance of that account and access the history of transactions associated with it. Monero anonymity is ensured by the following sophisticated cryptographic tools and mechanisms:

- *ring confidential transactions* using range proof and cryptographic commitments (confirmations), which allow to hide the amount of funds sent in the transaction,
- *non-interactive proofs without disclosures*, which allow to verify the transaction without the knowledge of the sender and recipient of funds, without knowing the number of coins and without the use of a trusted party;
- *one-time addresses, and one-time public keys*, which the user must generate using cryptographic mechanisms and which hide the way the funds are moved;
- *ring signatures* that allow you to hide from which address (from a certain group consisting of 11 addresses) funds were sent.

Another undoubted advantage is the open source of this cryptocurrency and the high degree of trust in its developers in the cryptocurrency society.

The disadvantages of Monero include high transaction fees.

---

[11] Nicolas van Saberhagen. CryptoNote v 2.0, 2013 (Monero whitepaper).

2.2.2   The "Most Mathematical" Cryptocurrency Zcash

Zcash cryptocurrency was developed by Zerocoin Electric Coin Company and announced on January 20, 2016. This is an open source cryptocurrency, which is definitely its advantage. It ensures the confidentiality and partial transparency of transactions. This means that all Zcash transactions are published in a public blockchain, but the sender, recipient, and transaction amount remain private.

The level of anonymity of Zcash is noted by many famous companies and individuals. For example, WikiLeaks began accepting charitable contributions with Zcash,[12] and Edward Snowden called this cryptocurrency the most interesting alternative to BTC.

For the anonymity of the transaction, the cryptocurrency Zcash [3] first used a non-disclosure proof of a new type called zk-SNARK [4]. The abbreviation zk-SNARK stands for "zero-knowledge Succinct Non-interactive ARguments for Knowledge", i.e. "non-interactive compressed arguments of knowledge with zero disclosure". The term "non-interactive" means that the party who proves the knowledge of certain information builds this proof independently, without interaction with the verifying party. The phrase "zero-knowledge" means that the party who proves the knowledge of some classified information builds the following arguments for the examiner, which have the following properties:

- on the one hand, completely (using a complex mathematical apparatus) convince the examiner that this information is known to it;
- the probability of convincing the examiner without knowing this information is almost zero;
- on the other hand, proving knowledge of this information does not provide any information about the secret itself.

To understand how such a controversial task can be performed, consider the following example. Alice wants to convince Bob that she has a key (or secret code) from some door. They approach the door, Bob makes sure it's locked, then turns away and Alice opens it. Bob checks that it is really open. This example is, of course, very primitive, but it demonstrates the very idea of proving without disclosure—the party who proves the knowledge of information does not disclose the information itself, but shows that it can do something (in our example—open the door; in the case of zk-SNARK—calculate certain values), which could not do without this knowledge.

Note that Monero also uses proof without disclosure, but of a completely different type, not as universal and not as mathematically complex. The advantage of zk-SNARK is not only its versatility, but also less proof. Non-disclosed "compressed" proofs can be checked in a few milliseconds, and the length of the proof does not depend on the amount of information you have to know, for a total of, for example, 288 bytes for the "Groth16" protocol. Compensation for concise proofs and rapid verification is the so-called SetUp protocol, which replaces the trusted party here, is quite cumbersome and requires the coordinated action of many participants in the

---

[12] https://cointelegraph.com/news/wikileaks-now-accepts-donations-in-the-cryptocurrency-zcash.

protocol. SetUp generates some parameters that should be kept secret. In this case, the SetUp protocol will be stable even if at least one of its participants is honest.

The zk-SNARK protocol allows you to check the correctness and legitimacy of the transaction without disclosing any information about:

- sender;
- transfer amount;
- recipient.

Only timestamps are public.

The disadvantage of the zk-SNARK protocol is the large number of computational costs, and almost all of them relate to the SetUp stage. Research is currently underway to simplify this protocol. Meanwhile, most Zcash cryptocurrency transactions are still open (over 90%).

Due to the complexity of the mathematical apparatus used in Zcash, literally dozens of people around the world understand what its stability and anonymity are based on. This is also one of the reasons why it is not as popular as it objectively could be.

2.3 Attitudes of Governments and Intelligence Services Towards Anonymous Cryptocurrencies

Although anonymous cryptocurrencies were created to protect the owners of funds from fraud and other crimes, they themselves proved to be a convenient tool in the hands of criminals. With their appearance, the number of fraudulent attacks and financial fraud has increased significantly. There are many cases where hackers have been able to hide the movement of stolen funds from cryptocurrency platforms and wallets using anonymous coins.

For example, there is a well-known case of malware distribution called Wannacry. This program blocked all files on users' computers and required a ransom equivalent to $300 of BTC. The virus has infected more than 200,000 computers in nearly 150 countries. One research team was able to track the activity of several BTC addresses associated with Wannacry. But then the hackers exchanged the BTC for Monero, so it became impossible to track the further movement of funds.

Interesting statistics on crimes related to the use of cryptocurrencies for the purchase of illicit products (including drugs) are also given in [5]. The authors claim that about a quarter of the total amount of money for which drugs are purchased is accounted for by cryptocurrency.

Therefore, the response to the use of anonymous cryptocurrencies by governments, intelligence agencies and various organizations is appropriate.

For example, in 2017, Europol officially expressed concern about the growing popularity of the cryptocurrencies Monero and Zcash.

Europol has officially warned Zcash and Monero about the dangers of complicity in cybercrime. The document says that Monero is quickly beginning to attract the attention of cybercriminals. "Transactions cannot be tied to a specific user and/or specific address. All coins remain anonymous," the document reads. According to this agency, anonymous cryptocurrencies can be used to exchange borders between criminal syndicates. Moreover, the organization already seems to have the relevant evidence.

"Zcash has not yet been spotted in any illegal activity. However, given the anonymous nature of this cryptocurrency, which hides the addressee in Darknet, we are closely monitoring it," Europol said.

The Financial Services Agency of Japan (FSA) has also banned anonymous cryptocurrencies and suspended trading on Monero, Zcash and Dash due to the possibility of their use for fraud. The regulator believes that these currencies are used to launder illegally obtained profits. The reason for this decision, in particular, is called the break in January 2018 of the Japanese cryptocurrency exchange CoinCheck. The attackers then managed to steal 523 million NEM tokens, worth about $524 million. Masanori Kusunoki, CPO of Japan Digital Design, claims they were laundered in Darknet. After two months of trying to find the funds, CoinCheck announced the cessation of the search, without giving a reason. As the reaction to this event was the ban on anonymous cryptocurrencies, it can be reasonably assumed that some of them were used to withdraw funds.

Also in June 2018, the representative of the US Secret Service, Robert Novi, asked the US Congress to take the necessary measures to combat the use of anonymous altcoins in criminal activities, including money laundering and terrorist financing.

From September 30, 2019, support for Monero (XMR), Dash (DASH), Zcash (ZEC), Haven (XHV), BitTube (TUBE) and PIVX was stopped by the large South Korean exchange Upbit, and from October 10 by the South Korean platform OKEx. In both cases, the exchange said that the decision was due to the inability to fully implement the recommendations of the FATF (Financial Action Task Force).

In the next article, we will consider the mathematical apparatus needed to build a proof without disclosing knowledge of certain information—the so-called SNARK-proof. Using the example of the most common Groth16 protocol, we will show how information is structured before constructing a SNARK-proof, how preliminary calculations are performed (so-called ceremony, or Setup), how the proof itself is built and how its correctness is checked. We will also show how to build recursive SNARK-proofs and why and in what situations they are necessary.

In addition, the threats that arise from the misuse or misconstruction of SNARK-proof will be analyzed and what needs to be addressed to minimize these threats.

3 Groth16 Protocol for SNARK-Proof as a State-of-the-Art Mathematical Device that Allows to Maintain the Anonymity of Arbitrary Data

3.1 Transition from QAP to R1CS

Let p—a large prime number, F_p—the corresponding prime field. So be it $u_{i,q}$, $v_{i,q}$, $w_{i,q} \in F_p$, where $i = \overline{0, m}$, $q = \overline{1, n}$.

SNARK-proof can be used for the following purpose: to prove (without disclosure) that we know the following $a_0, a_1, ..., a_m \in F_p$ (where $a_0 = 1$, and the first l elements may be well known), that the following system of equations:

$$\sum_{i=0}^{m} a_i u_{i,q} \cdot \sum_{i=0}^{m} a_i v_{i,q} = \sum_{i=0}^{m} a_i w_{i,q}, \quad q = \overline{1, n}. \tag{1}$$

System (1) is called "Rank 1 Constrains System", or R1CS.

We first show that system (1) can be reduced to a single polynomial equation by equivalent transformations. And then we reformulate the goal in terms of polynomial equality.

Let's construct $3(m + 1)$ a polynomial:

$$u_i(x), \ v_i(x), \ w_i(x) \in F_p[X], \ i = \overline{0, m}, \tag{2}$$

such that for some given elements $r_1, ..., r_n$ (usually chosen $r_q = q$, for $q = \overline{1, n}$) these polynomials take the given values, namely:

$$u_i(r_q) = u_{i,q}, \ v_i(r_q) = v_{i,q}, \ w_i(r_q) = w_{i,q}, \ i = \overline{0, m}, \ q = \overline{1, n}. \tag{3}$$

For simplicity, we will immediately assume that $r_q = q$, $q = \overline{1, n}$, therefore, Eq. (3) can be rewritten as

$$u_i(q) = u_{i,q}, \ v_i(q) = v_{i,q}, \ w_i(q) = w_{i,q}, \ i = \overline{0, m}, \ q = \overline{1, n}. \tag{4}$$

Further, to construct polynomials (2) with constraints (4), we use the Lagrange formula, therefore, the powers of all polynomials (2) will not exceed $n - 1$.

We can now say that all transformations have been performed so that (1) holds if and only if the following polynomial equality holds:

$$\sum_{i=0}^{m} a_i u_i(x) \cdot \sum_{i=0}^{m} a_i v_i(x) = \sum_{i=0}^{m} a_i w_i(x), \tag{5}$$

for all $x \in \{1, 2, ..., n\}$.

Thus, our goal is now formulated as follows: to prove that there exist such $a_0, a_1, ..., a_m \in F_p$ that $x \in \{1, 2, ..., n\}$ polynomial equality holds for all (5).

Now define a polynomial

$$g(x) = \sum_{i=0}^{m} a_i u_i(x) \cdot \sum_{i=0}^{m} a_i v_i(x) - \sum_{i=0}^{m} a_i w_i(x). \tag{6}$$

According to (5) and (6), the elements $\{1, 2, ..., n\}$ are the roots of this polynomial $g(x)$. According to Bézout's theorem [6], this is equivalent to a system of requirements

$$g(x) \vdots (x - q), \quad q = \overline{1, n} \tag{7}$$

Further, all polynomials $(x - q)$, $q = \overline{1, n}$, are irreducible, including mutually prime; therefore, according to one of the consequences of Euclid's algorithm, system (7) is executed if and only if

$$g(x) \vdots \prod_{q=1}^{n} (x - q). \tag{8}$$

Define a polynomial $t(x) = \prod_{q=1}^{n} (x - q)$. Now we can write (8) as $g(x) \vdots t(x)$.

Once again, we reformulate our goal: to prove that there are such $a_0, a_1, ..., a_m \in F_p$ that

$$g(x) \vdots t(x), \tag{9}$$

where $g(x) = \sum_{i=0}^{m} a_i u_i(x) \cdot \sum_{i=0}^{m} a_i v_i(x) - \sum_{i=0}^{m} a_i w_i(x)$.

Expression (9) can be rewritten as

$$\sum_{i=0}^{m} a_i u_i(x) \cdot \sum_{i=0}^{m} a_i v_i(x) - \sum_{i=0}^{m} a_i w_i(x) = 0 (\mathrm{mod}\, t(x)),$$

or

$$\sum_{i=0}^{m} a_i u_i(x) \cdot \sum_{i=0}^{m} a_i v_i(x) \equiv \sum_{i=0}^{m} a_i w_i(x) (\mathrm{mod}\, t(x)).$$

So we have converted our Eq. (1) to some other object called **QAP** ([7]). And now our goal is to construct non-interactive non-disclosure for a quadratic arithmetic program (**Constructing Non-Interactive Zero-Knowledge Arguments for Quadratic Arithmetic Program, or NIZK Arguments for QAP**): to prove that we know such $a_0, a_1, \ldots, a_m \in F_p$, that

$$\sum_{i=0}^{m} a_i u_i(x) \cdot \sum_{i=0}^{m} a_i v_i(x) \equiv \sum_{i=0}^{m} a_i w_i(x) (\mathrm{mod}\, t(x)), \tag{10}$$

where $t(x) = \prod_{q=1}^{n} (x - q)$.

The polynomial congruence of the form (10) is precisely the so-called Quadratic Arithmetic Program (QAP).

3.2   Setup Protocol for Groth16

We formulated our goal above using congruence (10). Now let's start building proofs—the so-called NIZK Arguments. Next, we will use the more commonly used term "SNARK" instead of "NIZK", where the abbreviation stands for "Succinct Non-interactive ARguments for Knowledge". The first and most cumbersome and time consuming part of SNARK is the so-called "Setup", or "ceremony". During the ceremony, all interactive preliminary calculations are performed and random variables are generated, which are then needed to construct the proof and verify it. Using a large amount of multi-party computations, participants create some quantities that will then be used by the prover to build the proof and by the verifier to test it. Some of the created values are known only to some participants of the ceremony, some are well known. Next we will give a detailed description with an explanation of each action in the ceremony.

As mentioned earlier, some of the values $a_0, a_1, \ldots, a_m \in F_p$ are known to all participants. Without loss of generality, suppose that these are the first $l + 1$ values $a_0, a_1, \ldots, a_l \in F_p$ (in particular, a possible case $l = 0$), and the values $a_{l+1}, \ldots, a_m \in F_p$ are known only to the prover. Quantities $a_0, a_1, \ldots, a_l \in F_p$ are called "*statements*" and quantities $a_{l+1}, \ldots, a_m \in F_p$ are called "*witnesses.*" The prover must prove that he knows all the witnesses for given statements, such that for given and well-known polynomials $u_i(x)$, $v_i(x)$, $w_i(x)$, $i = \overline{0, m}$, and for a given and known polynomial $t(x) = \prod_{q=1}^{n} (x - q)$, a polynomial congruence holds (10).

In what follows, we will use definitions, terms, and symbols from the previous material and in part from [7], but also, for convenience, we will use some other designations other than [7].

Let the triplet $\mathbf{TR} = (\mathbf{G}_1, \mathbf{G}_2, \mathbf{G}_{TG})$ consists of three groups $\mathbf{G}_1$, $\mathbf{G}_2$, $\mathbf{G}_{TG}$ such that:

- $|\mathbf{G}_1| = |\mathbf{G}_2| = |\mathbf{G}_{TG}| = p$, where p is a large prime defined above;
- pairing reflection $e : \mathbf{G}_1 \times \mathbf{G}_2 \to \mathbf{G}_{TG}$ is a bilinear non-degenerate reflection;
- $G_1 \in \mathbf{G}_1$ is a group $\mathbf{G}_1, G_2 \in \mathbf{G}_2$ generator, is a group generator $\mathbf{G}_2$, and $G_{TG} = e(G_1, G_2) \in \mathbf{G}_{TG}$ is a target group generator $\mathbf{G}_{TG}$.

Also, according to [7], we assume that "there are efficient algorithms for calculating the results of group operations, values of bilinear mapping, checking the membership of an element to a particular group, checking the equality of group elements and selecting generators of these groups."

Remark 1 To date, there is only one type for suitable triplet elements—groups G_1, G_2 and G_{TG}: G_1 is some subgroup of an elliptic curve $E(F_r)$ over a simple field F_r; G_2 is some other subgroup of the same elliptic curve $E(F_{r^k})$, but already over some extension of the field F_{r^k}; and the group is some subgroup of the multiplicative group $F_{r^k}^*$. Subgroups G_1, G_2 should have the following property: $G_1 \cap G_2 = \{O\}$, where O is the neutral element of the curve $E(F_{r^k})$. The mapping $e : G_1 \times G_2 \to G_{TG}$ is the so-called *reflection pairing on elliptic curves*.

Remark 2 Now the question may arise: why do we impose such requirements on the characteristics of the field over which the system (1) is given, and the order of the groups in the triplet? Namely: why the orders of the groups G_1, G_2 and G_{TG}, should be equal to the characteristics of the field F_p, where F_p is the field on which the R1CS system was built?

As we noted earlier, our goal is to construct a SNARK to prove (without disclosure) knowledge of the elements $a_0, a_1, ..., a_m \in F_p$ for which it is performed (10). In SNARK we will need to multiply elements $a_0, a_1, ..., a_m \in F_p$ by elements of groups G_1 and G_2. The results of such scalar multiplications will also be some elements of groups G_1 and G_2, accordingly: for example, if $a \in F_p$ and $G \in G_1$, then $aG \in G_1$. It is necessary that the operations in the field F_p and in the group G_1 were "compatible, not contradictory" in the following sense: we require compliance with some "natural" laws for these operations. In particular, the law of distributivity and the law of neutral elements:

- for arbitrary $a, b \in F_p$ and $G \in G_1$, equality must be observed

$$aG + bG = (a + b)G; \tag{L1}$$

- if $0 \in F_p$ is a neutral element of the field, then

$$0 \cdot G = O, \tag{L2}$$

where O is a neutral element of the group G_1.

Now imagine the following situation. Let for some $a, b \in F_p$ it be fulfilled $a = p - b$. In the field F_p it means, that $a + b = 0$. Therefore, according to (L1) and (L2), the following equations hold:

$$aG_1 + bG_1 = (a + b)G_1 = 0 \cdot G_1 = O.$$

But, according to the consequences of Lagrange's theorem [6], equality

$$(a + b)G_1 = O$$

Performed if and only when $a + b \vdots ord\ G_1$, that means $p \vdots |G_1|$ (as $a + b = p$ and $ord\ G_1 = |G_1|$). But p and $|G_1|$ are prime numbers, so the only possible situation when it is executed $p \vdots |G_1|$—is when $p = |G_1|$.

Let us return to the construction of NIZK (or SNARK) according to [7].

Since groups $\mathbf{G}_1$, $\mathbf{G}_2$ are subgroups of a simple order of groups of points of some elliptic curve, we will use the "+" symbol for operations in these groups (it will be clear from the context in which group we perform which operation). Since $\mathbf{G}_{TG}$ is a subgroup of a multiplicative group of some extension of a simple finite field, we will denote as "*" the operation in the group $\mathbf{G}_{TG}$.

It is shown below how the procedure of the ceremony, construction of the proof and its verification is described in [7].

As you can see, the paper itself provides an extremely brief description of the NIZK procedure, without any explanation. This description raises many questions, such as:

- Who exactly chooses the parameters α, β, γ, δ and the so-called *"toxic waste"* x?
- Are these parameters chosen by the participants? That is, they are known to all participants? Or some? Or none of them at all?
- if the values α, β, γ, δ of "toxic waste" x are unknown, how is it possible to calculate A, B and C?
- if these values are known to someone, he can create a huge number of "simulators" of proof, even without knowledge of any of the elements $a_0, a_1, \ldots, a_m \in F_p$. Indeed, for any pair of elements A, $B \in Z_p$, it can easily calculate the corresponding element $C \in Z_p$ so that it will be executed the last equation in Fig. 1.

$(\sigma, \tau) \leftarrow \mathsf{Setup}(R)$: Pick $\alpha, \beta, \gamma, \delta, x \leftarrow \mathbb{F}^*$. Set $\tau = (\alpha, \beta, \gamma, \delta, x)$ and

$$\sigma = \left(\alpha, \beta, \gamma, \delta, \{x^i\}_{i=0}^{n-1}, \left\{ \frac{\beta u_i(x) + \alpha v_i(x) + w_i(x)}{\gamma} \right\}_{i=0}^{\ell}, \left\{ \frac{\beta u_i(x) + \alpha v_i(x) + w_i(x)}{\delta} \right\}_{i=\ell+1}^{m}, \left\{ \frac{x^i t(x)}{\delta} \right\}_{i=0}^{n-2} \right).$$

$\pi \leftarrow \mathsf{Prove}(R, \sigma, a_1, \ldots, a_m)$: Pick $r, s \leftarrow \mathbb{F}$ and compute a $3 \times (m + 2n + 4)$ matrix Π such that $\pi = \Pi\sigma = (A, B, C)$ where

$$A = \alpha + \sum_{i=0}^{m} a_i u_i(x) + r\delta \qquad B = \beta + \sum_{i=0}^{m} a_i v_i(x) + s\delta$$

$$C = \frac{\sum_{i=\ell+1}^{m} a_i \left(\beta u_i(x) + \alpha v_i(x) + w_i(x) \right) + h(x) t(x)}{\delta} + As + rB - rs\delta.$$

$0/1 \leftarrow \mathsf{Vfy}(R, \sigma, a_1, \ldots, a_\ell)$: Compute a quadratic multi-variate polynomial t such that $t(\sigma, \pi) = 0$ corresponds to the test

$$A \cdot B = \alpha \cdot \beta + \frac{\sum_{i=0}^{\ell} a_i \left(\beta u_i(x) + \alpha v_i(x) + w_i(x) \right)}{\gamma} \cdot \gamma + C \cdot \delta.$$

Accept the proof if the test passes.

Fig. 1 NIZK procedure in Groth16

So, are these elements still unknown? Then who creates them?

Below we will answer all these questions.
The main idea of the ceremony is that all values

$$\alpha, \ \beta, \ \gamma, \ \delta \text{ and "toxic waste" } x \tag{11}$$

are calculated in the course of so-called multi-party computations. None of the participants in the ceremony calculates the values (11) independently, and therefore no one knows these values. Instead, participants calculate the corresponding group elements of the view.

$$\alpha G_1, \ \beta G_1, \ \delta G_1, \ldots \in \mathbf{G}_1 \text{ and } \beta G_2, \ \gamma G_2, \ \delta G_2, \ldots \in \mathbf{G}_2,$$

without knowledge of the quantities themselves (11).

Next we will explain exactly how this happens. First, for simplicity, assume that only three participants participate in the ceremony: $\mathbf{P}_1, \mathbf{P}_2, \mathbf{P}_3$.

Let's start with the simplest algorithm for calculating the value αG_1.

Let all participants $\mathbf{P}_1, \mathbf{P}_2, \mathbf{P}_3$ know the elements $G_1 \in \mathbf{G}_1$ and $G_2 \in \mathbf{G}_2$. They want to create a point together $G_\alpha^{(1)} = \alpha G_1 \in \mathbf{G}_1$ that:

- none of them will know α;
- each of them will know αG_1;
- each of them participates in the calculation αG_1;
- if the behavior of one of the participants is not fair, then all other participants will be able to see it.

The following is the simplest construction algorithm αG_1.

Algorithm 1 *Simplified construction algorithm* $G_\alpha^{(1)} = \alpha G_1$.

1. the participant $\mathbf{P}_1$ generates a random element $\alpha_1 \in F_p^*$, calculates the value $U_1 = \alpha_1 G_1$ and spreads it to the blockchain;
2. the participant $\mathbf{P}_2$ generates a random element $\alpha_2 \in F_p^*$, calculates the value and lays it out in the blockchain;
3. the participant $\mathbf{P}_3$ generates a random element $\alpha_3 \in F_p^*$, calculates the value $U_{123} = \alpha_3 U_{12}$ and lays it out in the blockchain;
4. now put $G_\alpha^{(1)} = U_{123}$;
5. spread $G_\alpha^{(1)}$ in the blockchain.

This algorithm is sufficient only in the "unreal" case, when all participants $\mathbf{P}_1, \mathbf{P}_2, \mathbf{P}_3$ are honest and, moreover, are convinced of each other's honesty. Otherwise, we need to add more appropriate checks after each action of the participants to make sure that each of them really knows the appropriate value of $\alpha_1, \alpha_2, \alpha_3 \in F_p^*$ and uses it in the calculations, and does not state any arbitrary value U_1, U_2 or U_3 instead of correctly calculated (according to Algorithm 1) value. We will show how exactly we can check the correctness of the participants' behavior.

Algorithm 2 *Construction algorithm* $G_\alpha^{(1)} = \alpha G_1$ *with appropriate verifications.*

1. Each participant $\mathbf{P}_i, i = \overline{1,3}$ chooses some random value $\alpha_i, i = \overline{1,3}$, calculates $U_i = \alpha_i G_1$, $T_i = \alpha_i G_2$, $i = \overline{1,3}$, and states the values of U_i and T_i, $i = \overline{1,3}$, in the blockchain.

2. Each participant checks that for the values stated in the blockchain, the following equations are fulfilled:

$$e(U_i, G_2) = e(G_1, T_i), \ i = \overline{1,3}. \tag{12}$$

*** Note that if $\mathbf{P}_i$ is an honest participant, then

$$e(U_i, G_2) = e(\alpha_i G_1, G_2) = e(G_1, \alpha_i G_2) = e(G_1, T_i)$$

and Eq. (12) is satisfied. In other words, equality (12) is a check that the participant $\mathbf{P}_i$ really knows the value α_i and calculated the value U_i and T_i correctly using this value α_i. According to the assumption of the complexity of the CDH problem (Diffie-Hellman computational problem), such a test is quite sufficient. If i and the equation in (12) is not satisfied, then the participant $\mathbf{P}_i$ is dishonest.

3. The participant $\mathbf{P}_2$ calculates $U_{12} = \alpha_2 U_1$ and spreads this value in the blockchain.

4. Each participant checks the implementation of equality:

$$e(U_{12}, G_2) = e(U_1, T_2). \tag{13}$$

*** If (13) is not met, then $\mathbf{P}_2$ is dishonest.

5. The participant $\mathbf{P}_3$ calculates $U_{123} = \alpha_3 U_{12}$ and spreads this value in the blockchain.

6. Each participant checks the implementation of equality:

$$e(U_{123}, G_2) = e(U_{12}, T_3). \tag{14}$$

*** If (14) is not met, then $\mathbf{P}_3$ is dishonest.

7. If (12)–(14) are fulfilled, we will put $G_\alpha^{(1)} = U_{123}$.

*** If at least one of the Eqs. 12–14 is not fulfilled, the participants must remove the dishonest participant from the procedure and repeat it again. In particular, participants can be fined for dishonest behavior.

8. Place $G_\alpha^{(1)}$ in the blockchain.

Note that although Algorithm 2 is still a bit simplistic, it conveys the basic idea of compatible computing in the ceremony, provided there is complete mistrust between the participants.

Using Algorithm 2, each ceremony participant P_j first creates and spreads the following values to the blockchain:

$$\alpha_i G_1, \beta_i G_1, \delta_i G_1, x_i G_1, \beta_i G_2, \gamma_i G_2, \delta_i G_2, x_i G_2.$$

In creating them, it uses the values α_i, β_i, δ_i, x_i, γ_i, which it selectively selects with F_p^* and keeps secret.

Using these values set out in the blockchain, the other participants perform the appropriate steps described in Algorithm 2, and after a certain number of iterations receive the following "simple" values, which are calculated in the ceremony:

$$G_\alpha^{(1)} = \alpha G_1;$$

$$G_\beta^{(1)} = \beta G_1;$$

$$G_\delta^{(1)} = \delta G_1;$$

$\left\{ X_i^{(1)} = x^i G_1 \right\}_{i=0}^{n-1}$ (with additional checks that when constructing $X_{i+1}^{(1)}$ with $X_i^{(1)}$ the participant P_j multiplied $X_i^{(1)}$ by the same value x_j that he used when constructing the value $x G_1$);

- $G_\beta^{(2)} = \beta G_2;$
- $G_\gamma^{(2)} = \gamma G_2;$
- $G_\delta^{(2)} = \delta G_2;$
- $\left\{ X_i^{(2)} = x^i G_2 \right\}_{i=0}^{n-1}.$

(with additional checks that when constructing $X_{i+1}^{(2)}$ with $X_i^{(2)}$ the participant P_j multiplied $X_i^{(2)}$ by the same value x_j, which he used when constructing the value $x G_2$; in addition, you need to check that each participant P_j used in the construction of $x G_2$ the same value of x_j, as when constructing $x G_1$).

Now it is possible to pass to construction of more "difficult" sizes, namely:

$$\left\{ G_{\alpha\beta\gamma,i}^{(1)} = \frac{\beta u_i(x) + \alpha v_i(x) + w_i(x)}{\gamma} G_1 \right\}_{i=0}^{l} \tag{15}$$

and

$$\left\{ G_{\alpha\beta\delta,i}^{(1)} = \frac{\beta u_i(x) + \alpha v_i(x) + w_i(x)}{\delta} G_1 \right\}_{i=l+1}^{m} \tag{16}$$

where x is "toxic waste".

Below we present an algorithm for constructing quantities (15); to calculate the values of (16) uses the same algorithm, with the value δ instead γ.

First, recall that all polynomials $u_i(\cdot)$, $v_i(\cdot)$ and $w_i(\cdot)$, $i = \overline{0, m}$ are known to all participants (in other words, the coefficients of these polynomials are known). Participants also know all the values $\left\{ X_i^{(1)} = x^i G_1 \right\}_{i=0}^{n-1}$ from the previous steps. We will show how, using these quantities, to calculate $u_i(x)G_1$.

Let $u_j(x) = f_{n-1}^{(j)} x^{n-1} + f_{n-2}^{(j)} x^{n-2} + \ldots + f_1^{(j)} x + f_0^{(j)}$, $j = \overline{0, m}$. Then

$$\sum_{i=0}^{m-1} f_i^{(j)} X_i^{(1)} = \sum_{i=0}^{m-1} f_i^{(j)} x^i G_1 = \left(\sum_{i=0}^{m-1} f_i^{(j)} x^i \right) G_1 = u_j(x) G_1, \ j = \overline{0, m} \quad (17)$$

To calculate $\beta u_j(x) G_1$ using the value $u_j(x) G_1$ obtained in (17), participants successively multiply $u_j(x) G_1$ by β_1 (participant $\mathbf{P}_1$), then by β_2 (participant $\mathbf{P}_2$), and finally by β_3 (participant $\mathbf{P}_3$), usually by performing appropriate checks after each multiplication to make sure that the values $\beta_1, \beta_2, \beta_3$ are really the same used in the construction of $G_\beta^{(1)}$ and $G_\beta^{(2)}$.

In this way, participants calculate the values.

$$\beta u_j(x) G_1, \ \alpha v_j(x) G_1, \ \text{and} \ w_j(x) G_1$$

and then add the results, $j = \overline{0, m}$.

Therefore, the participants calculate the sums $\left(\beta u_j(x) + \alpha v_j(x) + w_j(x) \right) G_1$, $j = \overline{0, m}$, and then multiply these sums sequentially by the corresponding elements so that the resulting sum is multiplied by γ^{-1} (for $j = \overline{0, l}$) or by δ^{-1} (for $j = \overline{l+1, m}$). For these successive multiplications, they can use Algorithm 2, but with some corrections: after the participant $\mathbf{P}_i$ has selected the appropriate value γ_i (or δ_i), it calculates γ_i^{-1} mod p (or δ_i^{-1} mod p) and then multiplies the corresponding value by γ_i^{-1} mod p, but not by γ_i (or by δ_i^{-1} mod p, but not by δ_i). And, of course, at each step, all participants perform appropriate checks for the correctness of this multiplication. As a result, the values (15) and (16) are obtained and checked for correctness.

To complete the ceremony, participants must still calculate the values $\left\{ T_{\delta,i}^{(1)} = \frac{x^i t(x)}{\delta} G_1 \right\}_{i=0}^{n-2}$.

Recall that the polynomial

$$t(y) = \prod_{i=1}^{n} (y - i)$$

It is known to all participants, so they also know polynomials $y^i t(y)$ and can find values $\left\{ x^i t(x) G_1 \right\}_{i=0}^{n-2}$ similarly as calculated $u_j(x) G_1$ in (17). Participants then multiply sequentially $t(x)G_1$ by δ_1^{-1} (participant $\mathbf{P}_1$), then by δ_2^{-1} (participant $\mathbf{P}_2$) and finally by δ_3^{-1} (participant $\mathbf{P}_3$), also with appropriate checks after each multiplication, and finally obtain values $\left\{ T_{\delta,i}^{(1)} = \frac{x^i t(x)}{\delta} G_1 \right\}_{i=0}^{n-2}$.

Now all the values that needed to be set during the ceremony are calculated.

For the convenience of further presentation, we will rewrite all the values calculated during the ceremony in our symbols:

$$\sigma_1 G_1 = \left\{ G_\alpha^{(1)}, \; G_\beta^{(1)}, \; G_\delta^{(1)}, \; \left\{ X_i^{(1)} \right\}_{i=0}^{n-1}, \; \left\{ G_{\alpha\beta\gamma,i}^{(1)} \right\}_{i=0}^{l}, \; \left\{ G_{\alpha\beta\delta,i}^{(1)} \right\}_{i=l+1}^{m}, \; \left\{ T_{\delta,i}^{(1)} \right\}_{i=0}^{n-2} \right\}$$
(18)

$$\sigma_2 G_2 = \left\{ G_\beta^{(2)}, \; G_\gamma^{(2)}, \; G_\delta^{(2)}, \; \left\{ X_i^{(2)} \right\}_{i=0}^{n-1} \right\}$$
(19)

where

$$\sigma_1 = \left\{ \alpha, \beta, \delta, \left\{ x^i \right\}_{i=0}^{n-1}, \; \left\{ \frac{\beta u_i(x) + \alpha v_i(x) + w_i(x)}{\gamma} \right\}_{i=0}^{l}, \right.$$
$$\left. \left\{ \frac{\beta u_i(x) + \alpha v_i(x) + w_i(x)}{\delta} \right\}_{i=l+1}^{m}, \; \left\{ \frac{x^i t(x)}{\delta} \right\}_{i=0}^{n-2} \right\},$$
$$\sigma_2 = \left\{ \beta, \gamma, \delta, \left\{ x^i \right\}_{i=0}^{n-1} \right\}$$

3.3 Construction of Proof for QAP

According to the protocol described in [7], selects random variables $r, s \in F_p$ and then calculates three quantities, which are elements of the respective groups and form the so-called proof. These values are as follows:

$$A^{(1)} = AG_1, \; B^{(2)} = BG_2 \quad B^{(2)} = BG_2 \mathrm{Ta} C^{(1)} = C G_1,$$
(20)

where

$$A = \alpha + \sum_{i=0}^{m} a_i u_i(x) + r\delta,$$

$$B = \beta + \sum_{i=0}^{m} a_i v_i(x) + s\delta,$$
(21)

$$C = \frac{\sum\limits_{i=l+1}^{m} a_i(\beta u_i(x) + \alpha v_i(x) + w_i(x))}{\delta} + \frac{h(x)t(x)}{\delta} + As + Br - rs\delta.$$

Note that, like the values σ_1 and σ_2, as well as others, calculated during the ceremony, the values A, B, $C \in F_p$ are also unknown to the participants, including those unknown to the proofreader and the inspector. But knowledge of quantities (18) and (19) is quite sufficient for a proofreader to construct quantities (20). We will now demonstrate how the proof builds the proof.

First, the value is calculated

$$A^{(1)} = AG_1 = \alpha G_1 + \sum_{i=0}^{m} a_i u_i(x)G_1 + r\delta G_1. \tag{22}$$

Note that the first term in (22), αG_1, was calculated during the ceremony and is therefore known to all participants. The third term, $r\delta G_1$, the prover can calculate as

$$r\delta G_1 = rG_\delta^{(1)}, \tag{23}$$

where the value r was chosen by the prover, and the value $G_\delta^{(1)}$ is known to all, as the buda is calculated during the ceremony.

To calculate the second term in (22),

$$A^{(1)}_{a_1 \ldots a_m} = \sum_{i=0}^{m} a_i u_i(x)G_1, \tag{24}$$

the prover first calculates $u_i(x)G_1$, according to the procedure described in (17), and then, using "well-known" values a_i, $i = \overline{0, l}$, and "secret" values known only to him a_i, $i = \overline{l+1, m}$, calculates $A^{(1)}_{a_1 \ldots a_m}$. Then the value $A^{(1)}$ is calculated as the sum of values (22), (24) and (23).

The value

$$B^{(2)} = \beta G_2 + \sum_{i=0}^{m} a_i v_i(x)G_2 + s\delta G_2 \tag{25}$$

is calculated similarly, according to procedure (17), using "well-known" information, a_i, $i = \overline{0, l}$, and "secret" information a_i, $i = \overline{l+1, m}$, known only to the prover. "Secret" information required to calculate is $a_i v_i(x)G_2, i = \overline{0, m}$.

The value $C^{(1)}$ can be written as

$$C^{(1)} = \sum_{i=l+1}^{m} a_i G^{(1)}_{\alpha\beta\delta,i} + \frac{h(x)t(x)}{\delta}G_1 + sA^{(1)} + rB^{(1)} - rsG_\delta^{(1)}, \tag{26}$$

where $B^{(1)} = BG_1$.

The elements $G^{(1)}_{\alpha\beta\delta,i}$, $i = \overline{l+1, m}$, are known to all after the ceremony, and the values a_i, $i = \overline{l+1, m}$ are known only to the prover, so the prover can easily calculate the first term in (26). He can also calculate $sA^{(1)} + rB^{(1)} - rsG^{(1)}_\delta$.

Before the calculation of $\frac{h(x)t(x)}{\delta}G_1$, recall that the polynomial $h(\cdot)$ known to the prover (and only to him). That is, the prover knows all its coefficients h_i, $i = \overline{0, n-2}$. Let's write this polynomial as $h(y) = h_{n-2}y^{n-2} + \ldots + h_1 y + h_0$, $h_i \in F_p$, $i = \overline{0, n-2}$.

The values of $T^{(1)}_{\delta,i} = x^i \frac{t(x)}{\delta}G_1, i = \overline{0, n-2}$, were calculated during the ceremony. Therefore, the prover can calculate the value of $\frac{h(x)t(x)}{\delta}G_1$ as

$$\frac{h(x)t(x)}{\delta}G_1 = \sum_{i=0}^{n-2} h_i x^i \cdot \frac{t(x)}{\delta}G_1 = \sum_{i=0}^{n-2} h_i T^{(1)}_{\delta,i},$$

where the values h_i, $i = \overline{0, n-2}$, are known to the prover (and only him), and the values of $T^{(1)}_{\delta,i}$, $i = \overline{0, n-2}$, were calculated during the ceremony.

Now the prover calculates the value CG_1 as the sum of the above values, and this completes the construction of the proof.

Note that to calculate the proof it is not necessary for the prover to know all the values that are components of the vectors and. It is enough for him to know only some of these quantities. The values required by the prover to form a proof form the so-called *proof key* (PK).

For Groth16 PK is defined as

$$\text{PK} = \left\{ G^{(1)}_\alpha, \; G^{(1)}_\delta, \; \left\{ X^{(1)}_i \right\}_{i=0}^{n-1}, \; \left\{ G^{(1)}_{\alpha\beta\delta,i} \right\}_{i=l+1}^{m}, \; \left\{ T^{(1)}_{\delta,i} \right\}_{i=0}^{n-2}, \; G^{(2)}_\beta, \; G^{(2)}_\delta, \; \left\{ X^{(2)}_i \right\}_{i=0}^{n-1} \right\}.$$

$$(27)$$

3.4 SNARK Proof for QAP

It will be recalled that the examiner knows all system-wide parameters p, G_1, G_1, G_2, G_2, G_T, open quantities a_i, $i = \overline{1, l}$ and polynomials $u_i(\cdot)$, $v_i(\cdot)$, $w_i(\cdot)$, $i = \overline{0, m}$. He also knows the values (18)–(19), calculated during the ceremony, and the values $A^{(1)}$, $B^{(2)}$, $C^{(1)}$, calculated by the prover.

According to the protocol [7], the inspector must calculate the values

$$V_{AB} = e\left(A^{(1)}, B^{(2)}\right)$$

$V_{\alpha\beta} = e\left(G^{(1)}_\alpha, G^{(2)}_\beta\right)$ (this value can be calculated in advance),

$$V_{\alpha\beta\gamma,i} = e\left(G^{(1)}_{\alpha\beta\gamma,i}, G^{(2)}_\gamma\right), \quad i = \overline{0,l}, \tag{28}$$

$$V_{C\delta} = e\left(C^{(1)}, G^{(2)}_\delta\right).$$

Then the examiner checks the equality:

$$V_{AB} = V_{\alpha\beta} \cdot \left(\sum_{i=0}^{l} a_i V_{\alpha\beta\gamma,i}\right) \cdot V_{C\delta}, \tag{29}$$

where multiplication and addition are operations in a finite field F_{r^k}.

Note that all values in the right-hand side of Eq. (28) were calculated during the ceremony. These values are needed to verify the proof, so they form the so-called *proof verification key* (VK):

$$VK = \left\{G^{(1)}_\alpha, G^{(2)}_\beta, G^{(1)}_{\alpha\beta\gamma,i}, \ i = \overline{1,l}, \ G^{(2)}_\gamma, \ G^{(2)}_\delta\right\}.$$

3.5 Correctness of SNARK-proof verification

Correctness means that if the prover's action is correct, according to the protocol (ie is honest), then an honest verifier always recognizes the proof as correct. In other words, an honest prover will always convince an honest examiner of the correctness of his behavior and knowledge of the relevant information. That is, Eq. (29) always holds if the proof was created by an honest prover.

To prove the correctness, it suffices to show that equality (29) holds for the quantities (22), (25) and (26).

According to (28), and also using the property of whiteness of pairing, it is easy to see that the left part (29) can be written as

$$V_{AB} = e\left(A^{(1)}, B^{(2)}\right) = e(AG_1, BG_2)$$

$$= e\left(\left(\alpha + \sum_{i=0}^{m} a_i u_i(x) + r\delta\right)G_1, \left(\beta + \sum_{i=0}^{m} a_i v_i(x) + s\delta\right)G_2\right)$$

$$= e(G_1, G_2)^{\left(\alpha + \sum_{i=0}^{m} a_i u_i(x) + r\delta\right)\left(\beta + \sum_{i=0}^{m} a_i v_i(x) + s\delta\right)} = e(G_1, G_2)^{P_{left}}$$

where

$$P_{left} = \alpha\beta + \alpha\sum_{i=0}^{m} a_i v_i(x) + \alpha s\delta + \beta\sum_{i=0}^{m} a_i u_i(x) + \sum_{i=0}^{m} a_i u_i(x) \cdot \sum_{i=0}^{m} a_i v_i(x)$$

$$+ s\delta \sum_{i=0}^{m} a_i u_i(x) + r\delta\beta + r\delta \sum_{i=0}^{m} a_i v_i(x) + rs\delta^2. \tag{30}$$

The right-hand side of Eq. (29) can be rewritten as

$$V_{\alpha\beta} \cdot \left(\sum_{i=0}^{l} a_i \cdot V_{\alpha\beta\gamma,i} \right) \cdot V_{C\delta} = e(G_1, G_2)^{\alpha\beta} \cdot e(G_1, G_2)^{\sum_{i=0}^{l} a_i(\alpha v_i(x) + \beta u_i(x) + w_i(x))}$$

$$\times e(G_1, G_2)^{\sum_{i=l+1}^{m} a_i(\beta u_i(x) + \alpha v_i(x) + w_i(x)) + h(x)t(x) + As\delta + Br\delta - rs\delta^2}$$

$$= e(G_1, G_2)^{P_{right}},$$

where

$$P_{right} = \alpha\beta + \sum_{i=0}^{l} a_i(\alpha v_i(x) + \beta u_i(x) + w_i(x)) + \sum_{i=l+1}^{m} a_i(\beta u_i(x) + \alpha v_i(x) + w_i(x))$$

$$+ h(x)t(x) + As\delta + Br\delta - rs\delta^2$$

$$= \alpha\beta + \sum_{i=0}^{m} a_i(\alpha v_i(x) + \beta u_i(x) + w_i(x)) + h(x)t(x)$$

$$+ \left(\alpha s\delta + s\delta \sum_{i=0}^{m} a_i u_i(x) + rs\delta^2 \right) + \left(\beta r\delta + r\delta \sum_{i=0}^{m} a_i v_i(x) + rs\delta^2 \right) - rs\delta^2.$$

Now it is enough to prove that $P_{left} = P_{right}$.

Reducing the same terms in P_{left} and P_{right}, we get that it suffices to prove equality $\sum_{i=0}^{m} a_i u_i(x) \cdot \sum_{i=0}^{m} a_i v_i(x) = \sum_{i=0}^{m} a_i w_i(x) + h(x)t(x)$, and this equality is performed according to the construction of QAP.

3.6 Simulation (Forgery) of Proof

As noted earlier, all components of the vectors σ_1 and σ_2 (see Fig. 1) are unknown—both to the prover and to all participants in general. This requirement is significant because the leakage of information, even for some of these values, leads to the possibility of forgery (Proof Simulation), i.e. to the possibility of constructing appropriate values $A^{(1)}$, $B^{(2)}$, $C^{(1)}$ even without knowledge of the information a_i, $i = \overline{l+1, m}$. Thus, if one of the participants somehow learns, for example, the values α, β and δ, he can falsify the proof, which an honest examiner will perceive as correct. Moreover, he will be able to construct a huge number of such proofs as follows.

Algorithm 3 *Forgery of proof*

(1)   choose random A, $B \in F_p$;

(2)   calculate $A^{(1)} = AG_1$, $B^{(2)} = BG_2$;

(3)   calculate $C_1^{(1)} = (AB\delta^{-1}) \cdot G_1$;

(4)   calculate $C_2^{(1)} = (\alpha\beta\delta^{-1}) \cdot G_1$;

(5)   using the values $X_j^{(1)}$, $j = \overline{0, n-1}$ and some techniques described above in the ceremony, calculate

$u_i(x)G_1$, $v_i(x)G_1$, and $w_i(x)G_1$, $i = \overline{0, l}$;

(6)   calculate

$$C_3^{(1)} = \delta^{-1} \sum_{i=0}^{l} a_i (\beta u_i(x)G_1 + \alpha v_i(x)G_1 + w_i(x)G_1)$$

$$= \frac{\sum_{i=0}^{l} a_i (\beta u_i(x) + \alpha v_i(x) + w_i(x))}{\delta} G_1;$$

(7)   calculate $C^{(1)} = C_1^{(1)} + C_2^{(1)} + C_3^{(1)}$;

(8)   send a triplet $A^{(1)}$, $B^{(2)}$, $C^{(1)}$ to the examiner as a constructed proof.

Direct calculations can verify that for this triplet equation (29) holds.

3.7   Recursive SNARKs

The advantage of recursive SNARKs is that we do not have to remember all previous transitions and prove their correctness, but only the last proof. Now we will show how it works.

When constructing each new SNARK proof, we must show that all previous proofs were constructed correctly. After that, we can only store the last of the constructed proofs. If an attacker tries to change some previous information and, accordingly, rework some of the previous SNARK-proofs, then, using the last saved proof, you can detect forgery.

Assume that a transaction can be represented as a system of R1CS constraints over some simple field, such as over a field F_p. So, informally speaking, to create a block of information, we need to build some SNARK-proof that proves the correctness of these transactions. Since we have constructed R1CS over a simple field F_p, we need to construct some elliptic curve E_1, convenient for pairing, which has a subgroup G_1 of order p, to construct the corresponding SNARK-proof. This curve cannot be constructed over the field F_p (if $|E_1| = p$ such a curve is cryptographically weak; and if $|E_1| = hp$, for some $h \geq 2$, such a curve does not exist over the field F_p due to Hasse's theorem).

So we need to build a curve over some other field F_r.

Necessary and sufficient condition for the "convenience of pairing" the curve E_1 with the subgroup G, $|G| = p$, above the field F_r, is the following condition:

$$p \mid r^s - 1$$

for some acceptable s (ideally—for such that $12 \leq s \leq 20$).

So: if we represent transactions as R1CS systems over some simple finite field F_p, then to construct the corresponding SNARK-proof we need to construct some curve E_1, convenient for pairing, over some other simple finite field F_r, where $r \neq p$, and such that $G < E_1$, $|G| = p$ and $p \mid r^s - 1$.

Next, we build a SNARK-proof for the corresponding R1CS over a simple field F_p, ie we calculate a triplet $\left(A^{(1)}, B^{(2)}, C^{(1)}\right)$, where $A^{(1)}, B^{(2)}, C^{(1)}$ are some points of the curve E_1, but, generally speaking, over some extension of the field F_r. Next, we record the obtained proof as a new corresponding R1CS system over F_r. To prove the knowledge of the relevant information about this new system, we, similarly to the previous considerations, must construct another curve E_2, convenient for pairing, over some other simple field (not F_r) F_q, such that

$$r \mid q^l - 1$$

for some appropriate l.

We would also like to add some new constraint systems that describe the new transactions to this R1CS system above F_r, but we can't do that! Because R1CS for transactions can only be created over a field F_p that is not equal to a field F_p! So, we need to develop a device that allows us to "return" to the field F_p.

To solve this problem, we could use some curve E_2 over a field F_q that $q = p$, if, of course, such a curve exists. To do this, we also need to fulfill the condition $r \mid p^l - 1$. If we can construct such an elliptic curve, we can use it both to construct a SNARK-proof for R1CS over F_r and to "go back" to the field F_p, because the coordinates of the curve E_2 will be elements of some extension of the field F_p. Therefore, after constructing such a SNARK-proof (using a curve E_2), we will create some new R1CS system over the field F_p, and we can now add new systems to this system that correspond to the new transactions.

So, to build recursive SNARK-proofs, we need the following two triplets, $\mathbf{TR} = (\mathbf{G}_1, \mathbf{G}_2, \mathbf{G}_{TG})$ and $\mathbf{TR'} = \left(\mathbf{G}_1', \mathbf{G}_2', \mathbf{G}_{TG}'\right)$, with the following properties:

- triplet $\mathbf{TR} = (\mathbf{G}_1, \mathbf{G}_2, \mathbf{G}_{TG})$: $\mathbf{G}_1$ is some subgroup of some elliptic curve $E_1(F_r)$ over some simple field F_r; $\mathbf{G}_2$ is some subgroup of the same elliptic curve, $E_1(F_{r^k})$ but already over some extension F_{r^s} of the field F_r; $\mathbf{G}_{TG}$ is some subgroup of the multiplicative group $F_{r^s}^*$ of the field F_{r^s}, and $p \mid r^s - 1$ and $|\mathbf{G}_1| = |\mathbf{G}_2| = |\mathbf{G}_{TG}| = p$;

- triplet $\mathbf{TR'} = \left(\mathbf{G}_1', \mathbf{G}_2', \mathbf{G}_{TG}'\right)$: $\mathbf{G}_1'$ is some subgroup of some elliptic curve $E_2(F_p)$ over some simple field F_p; $\mathbf{G}_2'$ is some subgroup of the same elliptic curve $E_2(F_{p^s})$, but already over some extension F_{p^l} of the field F_p; the group

G'_{TG} is some subgroup of the multiplicative group $F^*_{p^l}$ of the field F_{p^l}, and thus $r \mid p^l - 1$ and $\left| G'_1 \right| = \left| G'_2 \right| = \left| G'_{TG} \right| = r$.

Note that triplets $\mathbf{TR} = (\mathbf{G}_1, \mathbf{G}_2, \mathbf{G}_{TG})$ and $\mathbf{TR'} = \left(\mathbf{G'_1}, \mathbf{G'_2}, \mathbf{G'_{TG}} \right)$ must be different because:

$$\mathbf{G}_{TG} < F^*_{r^k} \Rightarrow |\mathbf{G}_{TG}| \mid r^k - 1 \Rightarrow |\mathbf{G}_{TG}| \neq r$$

The only triplets known today that can be used to construct recursive SNARK proofs are the so-called MNT-4 and MNT-6 triplets [8], which are used in the Coda protocol (new name—MINA) [9]. For these triplets, the degree of expansion is equal to $s = 4$ and $k = 6$. Because these degrees are so small, to achieve an acceptable level of resilience (primarily to MOV attack and to the problem of discrete logarithm in a multiplicative finite field group) we had to choose very large fields F_p and F_r for which the bit length of the bit characteristic is $|p| = |r| = 753$.

In our works, we also considered the possibility of constructing more "convenient" cycles for recursive SNARK-proofs, studied the necessary and sufficient conditions for the existence of appropriate cycles over a given field. In particular, an algorithm was developed that easily (even on a personal laptop) allows you to build analogs of triplets MNT-4 and MNT over different fields, as well as arguments about the "unity" of this format of cycles. Several hypotheses have also been put forward regarding the existence of such and similar cycles, which are planned to be strictly proved in the future.

4 Summary

1. Regarding the preservation of anonymity, confidentiality and privacy of personal information during its storage, processing and transmission. We will discuss how the analyzed information can be used to create a network that will operate on a blockchain basis and store, process and transmit confidential information, while maintaining its anonymity.

It should be noted that the use of so-called mixers, such as those using BTC or Dash, is not possible when processing information, because personal information cannot be broken into small particles, such as coins, and mix the same ones, because personal information is very personal. Further, the use of various specific payment systems, as well as cryptocurrencies, will not solve this problem, as the processing of personal information, its transmission and receipt, due to its specificity and personalization, cannot be compared with the processing and receipt of impersonal cryptocurrencies. In addition, when processing personal information you need to provide different levels of access to it (for example, the user—full access to his personal information, his doctors—according to specialization, the employer—a certificate of permits and restrictions determined by health status candidate for the position, etc.).

This leads to a clear conclusion: to ensure a given level of anonymity when processing, storing, forwarding and receiving confidential personal information, it is **possible to use only the appropriate mathematical apparatus, such as ring signatures (such as Monero) or SNARKs, such as Zcash**.

The main advantage of such methods of ensuring anonymity is that:

- the mathematical apparatus used on these platforms to ensure anonymity is fully mathematically sound and sufficiently tested, which provides a sufficient level of trust in it;
- this device has been tested and improved for almost 10 years, resulting in increased transparency and ease of use;
- there are numerous scientific papers that explain the principles of this mathematical apparatus and justify its stability, while all scientific and software is completely open and constantly discussed in various forums;
- the information that can be processed using this mathematical apparatus can be any and have an arbitrary structure, without any restrictions, due to the universality of the apparatus;
- to ensure the impossibility of substituting information, it is not necessary to store all information on local devices with all its transformations, but it is enough to store only the last option together with proving the correctness of the transition between previous states;
- the proof of the correctness of the transition between states is relatively short, its length does not depend on the amount of information it confirms, ie with increasing amount of information, the amount of memory required to protect it from forgery does not change.

2.   **Regarding the preservation of anonymity and private information when using cryptocurrencies.** Today, anonymous cryptocurrencies such as Monero, Dash and Zcash are the largest private digital currencies by market capitalization. In addition, these cryptocurrencies are the object of close attention and ban by many intelligence agencies and financial institutions around the world. This attention suggests that they are indeed anonymous. However, only the Monero cryptocurrency is anonymous, of course, and the Dash and Zcash cryptocurrencies only allow the choice of the privacy option at the user's request.

It should be noted that the question of the possibility of using these currencies, as well as other means of payment for anonymous financing of persons in different countries, significantly depends on which country it is. This is primarily determined by the country's legal framework for cryptocurrencies in general and anonymous cryptocurrencies in particular. Ukraine, for example, made a step to the legislation of cryptocurrencies [10]. Therefore, final conclusions can be drawn after examining the law of a particular country or after examining cases that may have been heard in the courts of that country and what decisions have been made on them. And it should be borne in mind that the attitude of countries to cryptocurrency is changing very dynamically, sometimes from one extreme to another.

1. Transfer cryptocurrency to fiat funds through a crypto machine that does not require verification—in this case, both the sender and the recipient remain anonymous. One problem is that there are very few such cryptocurrencies, but those that exist now may not exist tomorrow (for example, due to changes in legislation or the government's attitude to cryptocurrency).
2. Withdraw funds to your bank card—in this case, the sender remains anonymous. But in countries where anonymous cryptocurrencies are banned, such a path can be difficult.
3. Transfer the anonymous cryptocurrency to BTC and then withdraw the BTC to your bank card (if there is no complete ban on cryptocurrency in the host country)—in this case, the sender also remains anonymous.

The section considered the mathematical apparatus needed to build a proof without disclosing knowledge of certain information—the so-called SNARK-proof. The example of the most common Groth16 protocol showed how information is structured before constructing a SNARK-proof, how preliminary calculations are performed (so-called ceremony, or Setup), how the proof itself is built and how its correctness is checked. It also shows how to build recursive SNARK-proofs and why and in what situations they are necessary.

In addition, the threats that arise from the incorrect use or incorrect construction of the SNARK-proof were analyzed and what should be considered to minimize these threats.

References

1. Nakamoto S (2009) Bitcoin: A Peer-to-Peer Electronic Cash System/Satoshi Nakamoto (9 c). https://bitcoin.org/bitcoin.pdf
2. Best Crypto Exchanges Without KYC Verification in 2021! https://bitshills.com/best-non-kyc-crypto-exchanges/
3. Sasson EB, Chiesa A, Garman C, Green M, Miers I, Tromer E, Virza M (2014, May) Zerocash: Decentralized anonymous payments from bitcoin. In: 2014 IEEE Symposium on security and privacy. IEEE, pp 459–474
4. Ben-Sasson E, Chiesa A, Tromer E, Virza M (2014b) Succinct non-interactive zero knowledge fora von Neumann architecture. In: Fu K, Jung J (eds) USENIX 2014, pp 781–796. San Jose, CA, USA. USENIX Association
5. Ivantsov SV, Sidorenko EL, Spasennikov BA, Berezkin YM, Sukhodolov YA (2019) Cryptocurrency-related crimes: key criminological trends. Vserossiiskii kriminologicheskii zhurnal. Russian J Criminol 13(1): 85–93. https://doi.org/10.17150/2500-4255.2019.13(1). 85-93. (In Russian)
6. Lidl R, Niederreiter H (1997) Finite fields, no 20. Cambridge university press
7. Groth J. On the size of pairing-based non-interactive arguments. Cryptology ePrint archive: Report 2016/260, pp 1–25. https://eprint.iacr.org/2016/260.pdf

8. Miyaji A, Nakabayashi M, Takano S (2001) New explicit conditions of elliptic curve traces for FR-reduction. IEICE Trans Fundamentals Electron Commun Comput Sci 84(5): 1234–1243. https://pdfs.semanticscholar.org/7bb2/1519af1bfc52fe820e4e69802cab2fb8630b.pdf?_ga=2.252642121.1568939187.1585419833-2102199312.1585419833

9. Bonneau J, Meckler I, Rao V, Shapiro E. Coda: Decentralized cryptocurrency at scale. Cryptology ePrint Archive: Report 2020/352, pp 1–47. https://eprint.iacr.org/2020/352.pdf

10. Ukraine becomes the latest country to legalise Bitcoin and cryptocurrencies. https://www.euronews.com/next/2021/09/09/ukraine-becomes-the-latest-country-to-legalise-bitcoin-and-cryptocurrencies

Application of Bluetooth, Wi-Fi and GPS Technologies in the Means of Contact Tracking

Alexandr Kuznetsov⦿, Nikolay Poluyanenko⦿, Anastasiia Kiian⦿, and Olena Poliakova⦿

Abstract The spread of infectious diseases, including COVID-19, is a complex epidemiological situation that is exacerbated by strong transcontinental migration processes that pose a potential threat to human health worldwide. This requires the implementation of a set of tasks aimed at full control over the risks and threats to human life. Modern information technology can be useful in solving a number of scientific and technical problems related to the use of cyberspace for global population monitoring, ie when monitoring and tracking contacts can predict possible adverse scenarios and prevent the spread of emergencies and crises, especially in the COVID-19 pandemic. A high overall level of e-society can keep epidemics and pandemics at the national and international levels under control. Global solutions to build monitoring systems to prevent the spread of infectious diseases already exist and are evolving rapidly. This section analyzes, explores and substantiates the possibility of using various information technologies (eg, Bluetooth, Wi-Fi, GPS, etc.) in contact tracking tools as the main subsystem of global population monitoring. In particular, the principles of construction and the possibility of using these technologies to track contacts are studied. Their advantages and disadvantages, potential attacks on monitoring programs, etc. are analyzed.

Keywords Blockchain technology · Monitoring systems · Contact tracking applications · Decentralized systems · Wi-Fi and GPS technologies

A. Kuznetsov (✉) · N. Poluyanenko · O. Poliakova
V. N. Karazin Kharkiv National University, Svobody sq., 4, Kharkiv 61022, Ukraine
e-mail: kuznetsov@karazin.ua

N. Poluyanenko
e-mail: gorbenkou@iit.kharkov.ua

O. Poliakova
e-mail: olena.polyakova@nure.ua

A. Kuznetsov · N. Poluyanenko · A. Kiian
JSC "Institute of Information Technologies", Bakulin St., 12, Kharkiv 61166, Ukraine

© The Author(s), under exclusive license to Springer Nature Switzerland AG 2022
R. Oliynykov et al. (eds.), *Information Security Technologies in the Decentralized Distributed Networks*, Lecture Notes on Data Engineering and Communications Technologies 115, https://doi.org/10.1007/978-3-030-95161-0_2

1 Introduction

The spread of infectious diseases, including COVID-19, a complex epidemiological situation exacerbated by powerful transcontinental migration processes that pose a potential threat to human health around the world, requires a set of tasks aimed at full control of risks and threats to human life. Modern information technologies can be useful for solving a number of scientific and technical problems related to the use of cyberspace for global population monitoring, ie when through observation and tracking of contacts it is possible to predict possible negative scenarios and prevent the spread of emergencies and crises, in particular pandemic COVID-19. The high general level of electronic society can keep the spread of epidemics and pandemics at the national and international levels under control. Global solutions for building monitoring systems to prevent the spread of infectious diseases already exist and are developing rapidly.

This section analyzes, studies and substantiates the possibility of using various information technologies (eg, Bluetooth, Wi-Fi, GPS, etc.) in contact tracking tools as the main subsystem of global population monitoring. In particular, the principles of construction and the possibility of using these technologies to track contacts are studied, their advantages and disadvantages, potential attacks on monitoring applications, etc. are analyzed.

2 Functional Purpose of Bluetooth, Wi-Fi and GPS Technologies in the Context of Implementation of Applications for Tracking Contacts

In order to reduce the spread of COVID-19, leading countries use all possible tools: from social constraints to technical applications. In particular, the development and implementation of mobile applications to prevent COVID-19 is ongoing. Today there are three categories of such applications:

- digital contact tracking: applications that help maintain social distance, contact tracking, and isolate suspected or confirmed cases;
- symptom tracking: programs that help conduct surveys and track symptoms;
- immunity certification: applications that establish and certify immunity, for which there will be a long time for development.

Applications of the last two types are autonomous and do not require the use of additional mechanisms, unlike applications of the first type.

Summarizing the information on the operation of existing contact tracking applications, it should be noted that in order to function properly, three additional technologies are involved: Bluetooth, GPS and Wi-Fi or mobile Internet [1].

GPS technology is based on the simple navigation principle of marker objects. A marker object is a landmark whose coordinates are precisely known. Satellites in

Earth orbit play the role of marker objects in GPS. They rotate quickly, but their location is constantly monitored, and each navigator has a receiver tuned to the desired frequency. Satellites send signals that encode a large amount of information, including the exact time. Accurate time data is one of the most important for determining geographical coordinates: based on the difference between the recoil and reception of the radio signal, the satellites calculate the distance between themselves and the navigator.

Bluetooth is known to be one of the most popular and widely used wireless technologies. Specifically, contact tracking applications use Bluetooth Low Energy (BLE), which is part of the Bluetooth 4.0 specification, which also includes Classic Bluetooth and Classic Bluetooth and Bluetooth High Speed Protocols. Compared to classic Bluetooth, BLE is designed to use less power while maintaining a similar communication range. BLE is a technology that is always disabled and transmits only short amounts of data when needed. This significantly reduces power consumption, making it ideal for use in cases where a permanent long-term connection with low data rates is required. This gives the BLE an advantage over the usual modification of Bluetooth in terms of distribution of applications among the population, as high power consumption can lead to the abandonment of the use of applications. Moreover, the number of users who use the application directly affects the effectiveness of the idea of monitoring contacts [2].

In turn, Wi-Fi is a wireless technology used by computers, phones, tablets and many other devices.

The functional purpose of these technologies is as follows.

1. **GPS**. Use similar technology in applications to track users' locations and build real-time terrain maps. This has a number of advantages. First, provide users with information about outbreaks in certain areas, which can warn many people from visiting such places. Second, inform users about the potential risk of infection to reduce the number of cases.

2. **Bluetooth**. When users contact for more than 15 min at a distance of less than 1.5 m, their devices exchange identifiers (the principle of formation of identifiers depends on the specific implementation of the application) using BLE technology. In the future, these identifiers are meeting tags that can be used to reproduce all contacts in the event of illness and to inform users that they may be at risk.

3. **WiFi or Mobile Internet**. Internet connection is required to transmit to the central server user IDs in case of their disease for further analysis and sending messages to users who have been in contact with patients for 14–21 days, depending on the specific implementation [3].

2.1   Advantages and Disadvantages of Using Bluetooth and GPS to Implement Contact Monitoring Applications

Since the beginning of the development of mechanisms for contact tracking, two main areas of development have been identified depending on the technology used: applications based on Bluetooth and GPS. As mentioned above, GPS solutions operate on the principle of fixing the location of users, followed by tracking the history of locations and building, on this basis, a map, the probable spread of the disease in the area. However, GPS technology does not allow you to correctly capture the location of the device in all cases. For example, there is low accuracy in moving and underground environments, as well as limited vertical accuracy, when the majority of residents or employees of one high-rise building are fixed within the error, which is especially relevant in urban conditions and can lead to false suspicion.

Applications that use Bluetooth technology, in turn, make a statement of the fact of meeting devices that were at a distance of less than 2 m for a certain period of time. In the case of Bluetooth technology, the indoor recording range is up to 10 m, but the RSSI (power level indicator) follows the inverse square law and decreases rapidly with the distance required for calibration for maximum efficiency.

Shortly after the release of the applications, global experience has shown that GPS technology has not become so widespread, compared to its alternative, due to inaccurate data and user concerns about the transfer of their location. For this reason, the possibility of implementing contact tracking based on Bluetooth, namely its modification of low-power Bluetooth (BLE) in order to minimize the cost of batteries of end-user devices [4].

3   Analysis and Research of the Use of Bluetooth, Wi-Fi and GPS Technologies in Terms of Implementation of Application Building Protocols for Tracking Contacts

Currently, there are two vectors of research in the areas of implementation of Bluetooth solutions: centralized and decentralized. The centralized model involves the use of a single database that records the identification codes of meetings between users. In the decentralized model, such information is recorded on individual devices, and the role of the central database is limited to the identification of phones by their anonymous identifier in the case of sending a warning. The general principle of operation of applications is as follows (Fig. 1) [5].

Fully decentralized models are DP-3 T and Exposure Notification and semi-centralized—PEPP-PT, on which most of the functioning government applications are built. In the following we will consider the main points concerning the mentioned higher protocols and we will make their initial comparison.

A and B—Devices exchange special codes during a meeting.

B is diagnosed with the disease. He fixes it in the application.

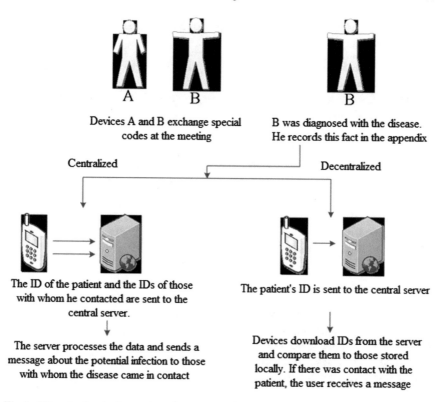

Fig. 1 General scheme of operation of contact tracking applications

Centralized Decentralized.

To the central server the ID of the patient and identifiers of those with whom he was in contact are sent.

The server processes the data and sends a message about the potential infection to those with whom the patient has been in contact.

To the central server the ID of the patient is sent.

Devices download IDs from the server and compare them to those stored locally. If there was contact with the patient, the user receives a message.

3.1 Exposure Notification Protocol

Exposure Notification—a mechanism developed by the joint efforts of Apple and Google, first unveiled on April 10, 2020. It is implemented at the operating system level, which allows for more efficient background work, in contrast to other alternative technologies. The joint development allowed to ensure compatibility between devices that run on Android and iOS and make up the vast majority of the global market. After installing the technology, the principle of its operation on the device is as follows [6].

1. The device generates a secret 32-byte key Tracing Key. This key does not leave the device and is stored on it.
2. Daily Tracing Key—a secret 16-byte key generated daily. It is generated using the HKDF function as follows:

$$Daily\ Tracing\ Key$$
$$= HKDF(Tracing\ Key, (UTF - 8$$
$$("CT - Daily\ Tracing\ Key")\|D_i), 16),$$

D_i—the number of the day, calculated according to the formula

$$unix\_timestamp\ div\ (60 \cdot 60 \cdot 24)$$

where

- div—it is an integer division operation;
- symbol ‖ represents concatenation;
- NULL is a salt that is not used in this case;
- 16—number of bytes at the output. These keys are then passed on if the user is diagnosed with the disease

3. Rolling Proximity ID—16-byte identifier generated every 10 min. Similar identifiers are used for exchange between contacting devices. It is generated using the truncated HMAC function (first 16 bytes):

$$Rolling\ Proximity\ ID$$
$$= Truncate(HMAC(Daily\ Tracing\ Key,$$
$$(UTF8("CT - Rolling\ Proximity")\|T_i)), 16),$$

T_i—time slot number, calculated as

$$unix\_timestamp\ mod\ (60 \cdot 60 \cdot 24))\ div(60 \cdot 10),$$

where:

- div—integer division;
- mod—the remainder of the division.

4. During contact, the devices store each other's Rolling Proximity IDs locally.
5. In case of infection, the user downloads the last 14 daily keys to the server, where other users download them at regular intervals

6. After downloading, the device calculates the Rolling Proximity ID of each contact and compares the calculated values with the values stored locally.
7. If a match is found, the user will receive a message about contact with an infected person.

Thus, low-power Bluetooth is used to log meetings and then send messages to neighboring devices. Each terminal device locally records and stores contact IDs for the last 14 days, and in case of diagnosing the disease sends its IDs to the central server, which are then processed by each device to establish the fact of previous contact with the patient. Developers also aim to use the received signal strength (RSSI) of beacon messages as a source for intimacy. RSSI and other signal metadata will also be encrypted to counter deanonymization attacks.

All digital contact tracking protocols, with the exception of the first-party Google/Apple protocol, have performance degradation on iOS devices. These problems occur when the device is locked or the program is not in the foreground. This is a limitation of the operating system, which results from how iOS manages its battery life and resource priority. The Android application does not have these problems, as it can ask the operating system to disable battery optimization and because Android is more allowed for background services [7].

3.2 PEPP-PT Protocol

Pan-European Privacy-Preserving Proximity Tracing (PEPP-PT) is an open protocol designed to track contacts, which is also based on low-power Bluetooth technology. Unlike its two presented alternatives, PEPP-PT uses a central server to process user contacts and send messages about possible infection.

On the one hand, this approach reduces confidentiality, but on the other hand, it allows health authorities to monitor. The general functioning of applications based on the PEPP-PT protocol can be represented as follows [8].

1. The user passes by a combination of a call confirming the work and CAPTCHA. The validation algorithm is a script defined in RFC7914 and popularized in various blockchain systems, such as Dogecoin and Litecoin.
2. After the user registers in the program, the server issues a unique 128-bit pseudo-random identifier (PUID). It will be marked as inactive until the application resolves the PoW problem with input parameters, cost factor 2, and block size 8. Upon completion, OAuth2 credentials are issued to the client to authenticate all future requests.
3. In case of contact between two users, they must exchange their identification data and register them.
4. In order to prevent clients from being tracked in time by using static identifiers, clients exchange time-dependent temporary identifiers issued by the central

server. In order to generate these temporary identifiers, the central server generates a global secret key, which is used to calculate all temporary identifiers in a short period of time.

5. For each user according to the algorithm represented by the AES encryption algorithm, the ephemeral Bluetooth identifier (EBID) is calculated. These EBIDs are used by customers as temporary IDs in exchange. EBID is obtained in a pre-dated batch in case of poor Internet access.

6. Users constantly broadcast their EBID under the PEPP-PT Bluetooth service ID, as well as scan other clients. If another client is found, they exchange and record EBID, as well as meeting metadata such as signal strength and timestamp.

7. If a user is diagnosed with the disease, he is asked to upload a list of his contacts to the central server. In case of consent, the health authority issues a key authorizing the download. The user then transmits the contact log via HTTPS to the reporting server.

8. After the server receives the contact log, each entry is run through a proximity check algorithm to reduce the likelihood of false positives. The received contact list is confirmed manually, and they, along with a random sample of other users, are sent a message containing a random number and a hash message. This message is used to wake the client and make him check the server for new reports. If the client is on the list of confirmed users, the server will confirm a potential infection for the client, which in turn will alert the user. The reason for the random selection of users is justified by preventing the disclosure to eavesdroppers of information about who is at risk of infection while listening to communication between the client and server.

In summary, it should be noted that despite the lack of real names when registering users, the server processes pseudo-anonymous identifiers, and there is a risk of re-identification. The PEPP-PT protocol is the most criticized of all alternatives by leading scientists. A letter criticizing the approach was published in late April, stating that decisions to restore public information should be rejected.

3.3   DP-3 T Protocol

Decentralized Privacy-Preserving Proximity Tracing (DP-3 T) is an open protocol developed by scientists who emerged from the PEPP-PT initiative due to differences of opinion on the ultimate functioning of the mechanism. It was the DP-3 T that inspired Apple and Google to create their own technology. The operation of the protocol is to perform the following actions.

1. At the beginning, the application generates a secret 32-byte key that does not leave the device.

2. Each day, the application generates a new day key resulting from the hash function of the previous day key SHA-256. In case the user is diagnosed with the disease, the day keys are transferred to the server.

3. Additionally, the application generates temporary EphID's (ephemeral ID) in the amount of 1440, ie one ID per minute of the day. They are used for sharing under the contact of two users. Temporary IDs are obtained from the day key as follows: an array of 23,040 zero bytes is encrypted by the AES CTR algorithm with the HMAC key ("broadcast key" ‖ Daily Key), where the symbol "‖" indicates the concatenation of the salt ("broadcast key") and the daily key (Daily Key).

4. During a contact, user applications exchange each other's temporary IDs and the value used to calculate the distance between devices (RSSI).

5. In case of diagnosing the disease, the infected user downloads the last 14 day keys to the server. Other devices periodically download new keys from the server and generate a corresponding temporary ID, and then check them for compliance with what is stored locally on their devices.

At the time of April 2020, after active research, the governments of Switzerland, Austria, Estonia, Finland and Germany announced that they were adopting DP-3 T as the basis for their national contact tracking applications [9].

Let's summarize the considered information on alternatives of mechanisms of tracking of contacts in the form of the (Table 1).

Table 1 Comparison of solutions for the implementation of contact tracking

Criteria\protocol	PEPP-PT	DP-3 T	Exposure notification
1. Centralization	Semicentralized	Decentralized	Decentralized
2. Using a central server	• Generating id • Receiving a set of contacts from the diseased person • Send a message to the contact of the diseased person	Obtaining IDs of the disease in the last 14 days Provides the ability for devices to load IDs that are sick	Obtaining IDs of the disease in the last 14 days Provides the ability for devices to load IDs that are sick
3. Actions in case of diagnosis of the disease	Confirmation that the IDs of the person you were contacting are sent to the server	Confirmation of sending daily IDs generated within 14 days	Confirmation of sending daily IDs generated within 14 days
4. To establish contact with the diseased	User receives a message from the server about a contact with the disease	The device downloads the IDs of the sick and compares them to their contacts	The device downloads the IDs of the sick and compares them to their contacts
5. Algorithms used	RFC7914 AES	Ed25519 SHA-256 HMAC-SHA256	AES-128

4   Example and Comparison of Implementations of Contact Monitoring Applications Adopted at the Level of World Powers

4.1   COVIDSafe (Singapore)

The application was released on April 26, 2020. It is based on the BlueTrace protocol developed by the Government of Singapore. The app extends traditional contact tracking by automatically tracking meetings between users, and later allows state or local health authorities to warn the user that they have been within 1.5 m of an infected patient for 15 min or more.

Accompanying the presentation of the application, Interior Minister Peter Dutton announced new legislation that would prohibit forcing anyone to transfer data from the program, even if they registered and then received positive results for virus testing. A decision entitled Determination of Biosafety 2020 was adopted, and on 6 May 2020, an amendment to confidentiality (health contact information) was amended to codify it. Legislation regulates how the data collected by the application will be stored, submitted and processed [10].

To register COVIDSafe, you need an Australian mobile number, i.e. foreigners in Australia need a local SIM card. However, there is an obvious drawback. Norfolk Island is an outer territory of Australia, but it uses a different country code +672 instead of +61. This means that Norfolk Island numbers cannot be used to register with COVIDSafe. The Australian government is reportedly working to remedy the situation. The problem was also perceived by the fact that the source application runs on the Amazon Web Services (AWS) platform, meaning the US government could potentially seize the data of Australian citizens. The data is now stored within Australia at the AWS Sydney data center. During a public hearing on COVIDSafe, Randall Bruges, CEO of the Digital Transformation Agency, explained that the decision to use AWS over purely Australian cloud providers was made on the basis of familiarity, scalability and availability of resources within AWS.

4.2   Stopp Corona (Austria)

The Austrian Red Cross began to develop the application relatively early, while many other countries have just begun to discuss concepts. The Stopp Corona app is an application in which participants' mobile phones exchange data with each other through central infrastructure and then store them locally in a mobile phone. To detect contacts between the participating devices, each of these devices attempts to perform so-called handshakes with other devices in their environment. The user can choose whether to perform a handshake automatically or by himself. The purpose of the handshake is to transfer the public key to another device. However, due to the limitations in the amount of data that can be transmitted using Bluetooth Low Energy (BLE), the Central Cloud Service is also used for this purpose in addition to BLE [11].

In a simplified way, the existing architecture can be divided into two parts: on the one hand, the infrastructure necessary for the exchange of handshakes (or public keys) in the case of contact, and on the other—the infrastructure necessary to send reports of infection to the affected contacts.

If the program registers a contact, i.e. another mobile phone with an application installed in the immediate vicinity, the corresponding public key must be exchanged using a handshake. Two different methods are used for handshakes. First, a handshake can be made in which a public (static) key is exchanged through the Google Near Cloud infrastructure via the messaging API. To do this, the public key is stored on Google servers, and the token is exchanged based on Bluetooth, with which this public key can be searched and downloaded. Secondly, you can carry out automatic handshakes. In this case, the public key is exchanged by the p2pkit infrastructure of the Uepaa service provider. Functionally, it's an analogue of Google's variant. In addition to handing over public keys, both iOS and Android apps are currently sending messages to the Red Cross server independently of each other, which is a handshake. This message contains the originally generated (static) identifier of the corresponding client, as well as timestamps rounded to the current hour.

In order to deliver an infection message, a notice of suspicion, or to recall a message, it is sent encrypted to a Red Cross server using Wi-Fi or mobile Internet. The server acts only as a contact point for all programs to store new encrypted messages or periodically receive new encrypted messages. To reduce the number of messages to download, the total number of encrypted messages is reduced to 1/256 of all encrypted messages based on the prefix (first byte of the SHA256 public key hash) of the own public key. This prefix must also be passed to the server along with the encrypted message during the alert.

In general, it should be noted that the architecture of the system is not inherently unprotected. However, this is not desirable in terms of scalability and dependence on central servers. Due to BLE limitations, it is doubtful whether an existing architecture can be implemented jointly without referring to the cloud service. To exchange handshakes between devices, you would need to establish a connection between devices that cannot be implemented under BLE-based iOS.

4.3 NHS (UK)

The NHSX Technology and Research Group has developed an NHS app for researchers from the University of Oxford and uses developers from technology companies such as VMWare. They have also been in contact with Apple and Google, although they use different technologies from these technology companies, but have rejected their approach. The technology, developed by the NHS, will allow smart-phones to track all other devices they have been in contact with for 28 days using Bluetooth signals.

Users who download the app to their phone can voluntarily refuse to write down details of their symptoms when they start feeling unwell. If a person later reports that they are positive for coronavirus, the app will send messages to people who have been in close contact with them over the past 28 days based on their anonymous IDs. The application recommends that those people isolate themselves in case they succumb to the disease. Those who have been in contact will not know the identity of the person who may have suffered coronavirus [12].

If a person then passes the test and tests negative results, he can be released from self-isolation by means of a message through the application.

After that, people with whom they have been in close contact will be sent a message asking to be isolated. This data will not be associated with people's names, but will use anonymous identifiers associated with the device, so phone owners will not know who may have transmitted the virus. However, they will be asked to share the first digits of their zip code.

As mentioned earlier, the NHS application uses Bluetooth signals to check contacts. This means you don't need to connect to mobile data. The app does not currently track GPS signals. The NHS has said it may introduce a location moni-toring system in the future to collect useful pandemic data, but this will be a voluntary factor.

In the new version, NHS has made the transition to the use of Exposure Notification. Currently, the application is used to [13]:

- getting advice about the coronavirus—get information about the coronavirus and find out what to do if the user thinks they have it.
- order repeat prescriptions—review available medications, request a new repeat prescription, and choose a pharmacy where to send prescriptions.
- making appointments—search, book, and cancel appointments in your surgery. View details of your upcoming and past appointments.
- getting health advice—seek information and NHS information on hundreds of conditions and treatments. The user can also answer questions to get instant advice and medical care nearby.
- view your medical record—Securely access your GP's medical record to see information such as your allergies, as well as current and past medications. If the surgical care provided access to a detailed medical record, the user can also see information such as test results and consultation details.
- registration as an organ donor—choose whether to donate some or all of your organs and check your registered decision.
- Finding out how the NHS uses your data—Choose whether the data from your medical records will be transferred for research and planning.

4.4 HaMagen (Israel)

The non-profit organization Start-Up Nation Central has compiled a directory of about 70 Israeli technology companies developing responses to a new virus that has infected more than 4000 people in the country. One of the applications being developed is Hamagen. It was developed by the joint efforts of developers from the Ministry of Health, commercial companies and volunteers from various organizations and the developer community in Israel. Using geolocation technology, the application informs users about any points of contact with known cases of COVID-19. The application is available in five languages, Hamagen has already downloaded more than a million users [14].

Once a set period of time (now once an hour) the application downloads a file with an anonymous list of places visited by diagnosed COVID-19 patients (patients who have been examined by the Ministry of Health and have undergone an epidemiological study with various tools available to the Ministry) from the cloud of the Ministry of Health (including dates and times), and then the program redirects these places to the user's location (including date and time), which are stored in the end device.

Location and time information refers to the device, not the cloud. The user's location is not sent to the Ministry of Health unless the user has authorized the Ministry of Health to obtain this information to help identify people who have been exposed to radiation and who need to be isolated as soon as possible. If the program detects that there is a possibility that you were in the same place and at the same time as the diagnosed patient, you will receive a message from the program indicating the location and time.

Upon receipt of a report of a potential infection, it is recommended that precautions be taken and that this information be verified on the Ministry of Health's website, where these lists are published, as well as maps of the locations of tested patients. If you have any doubts about the accuracy of the information provided in the annex, you can contact the Ministry of Health hotline at * 5400 or the hotline.

Thus, the information is stored as follows:

- location information is stored only on the user's device and is not transmitted;
- information is stored through the SQLite database, accessible only to the program.

HaMagen requires authorization to access the location as well as access to the Internet (for mobile surfing or WIFI) from your device. Internet access is required to download a constantly updated data file containing the history of diagnosed coronavirus patients from the Ministry of Health's cloud (ie the findings of epidemiological studies conducted by the Ministry of Health with a diagnosis of patients).

The application compares the locations of each diagnosed patient with the user's locations for the 14 days preceding each patient's diagnostic data. Therefore, the program needs access to user addresses.

Information stored in the application:

- location history for the last two weeks only (dates, hours, and locations) according to tracking services (and currently does not apply to program installation dates);

- the history of wireless networks (WIFI) that you have encountered only in the last two weeks;
- cross-references to places with diagnosed patients (if any) - only for the last two weeks.

Instances passing diagnosed files:

- The file is transferred from the Ministry of Health to a specific cloud via Azure (cloud services managed by Microsoft) using a secure virtual safe (CyberArk);
- the Ministry of Health communicates with Azure via the ExpressRoute service via special communication lines (not via the Internet);
- The file is stored in the cloud using the Blob Storage service, and the application downloads the file from the cloud.

Information security experts approve information security services for cloud and cloud communications of all parties involved.

User-confirmed location information will be stored in the Ministry of Health's epidemiological investigation system for 7 years, including information marked as "inappropriate" during the epidemiological investigation.

The Ministry of Health places great emphasis on the confidentiality of information. Accordingly, any information transmitted to the Ministry of Health will be transmitted through an encrypted channel and stored on the Ministry's servers in accordance with all information security and confidentiality procedures and protocols applicable to the Israeli health care system and in accordance with the law.

The Ministry of Health conducts regular inspections of maintenance measures to ensure information security and protection of confidentiality, and updates them if necessary.

4.5 Aarogya Setu (India)

This is a tracking application that uses the features of the smartphone GPS and Bluetooth to track coronavirus infection. The application is available for Android and iOS mobile operating systems. Using Bluetooth, he tries to determine the risk if someone is nearby (within six feet of a person) infected with COVID-19, by scanning through a database of known cases across India. Using location information, it determines whether a location belongs to one of the infected areas based on available data. This application is an updated version of a previous program called Korona Kawach (now discontinued), which was previously released by the Government of India.

The appendix shows how many positive cases of COVID-19 are likely within a radius of 500 m, 1 km, 2 km, 5 km and 10 km from the user. The application is built on a platform that can provide an application programming interface (API) so that other computer applications, mobile applications and web services can use the functions and data available in the Aarogya settings [15].

In the three days since its launch, Aarogya Setu has crossed five million downloads, making it one of the most popular government applications. It became the world's fastest mobile app to beat Pokemon Go, with more than 50 million installs, 13 days after launching in India on April 2, 2020.

Rahul Gandhi, the leader of the Congress party, called the Aarogya Setu program a "complex surveillance system" after the government announced that downloading the program would be mandatory for both government and private officials. Following this, others expressed similar concerns about the Aarogya Setu program. The Minister of Electronics and Information Technology responded to these problems by claiming that Gandhi's claims were erroneous and that the application was evaluated worldwide.

The developers of the application stated that obtaining location data is a documented feature of the program, not a disadvantage, as the program is designed to track the spread of the virus-infected population. On May 6, Robert Batiste wrote a tweet telling how many people got sick and infected in the Prime Minister's Office, the Indian Parliament and the Ministry of the Interior. The Economic Times noted that the program's provisions state that the user "agrees and acknowledges that the Government of India is not responsible for any unauthorized access to or alteration of your information." In response, several software developers called for the release of the source code.

4.6   AsistenciaCovid19 (Spain)

In Spain, on April 6, 2020, the Ministry of Economic Affairs and Digital Transformation announced that its application ("AsistenciaCOVID-19") had been launched, allowing people to assess the situation on their own. AsistenciaCOVID-19 requires citizens to enter personal data such as names, telephone numbers, dates of birth, gender, addresses, postal codes and e-mail. AsistenciaCOVID-19 is expected to be updated to provide further analysis of the collected data.

The Ministry stressed that AsistenciaCOVID-19 could, with the consent of users, geolocate data using GPS technology so that the relevant authorities could visualize the display of infections and conduct geospatial analysis, trying to identify high-risk areas. The Ministry then emphasizes that the application guarantees the security of users' personal data [16].

Extending the need for security measures to protect the program, Martinez, Grass and Garcia Del Poyo noted that as set out in the First Additional Regulation of 5 December 2018 on the protection of personal data and the guarantee of digital rights, Spanish public administrations (as well as private third parties provide services to public administrations) are obliged to establish specific security measures in the context of their personal data processing activities. These security measures are provided by a legal instrument specifically aimed at public administrations, the National Security Framework ("ENS"), which defines specific security measures to be applied to the use of electronic means by public administrations in the provision of public services to citizens (e.g. encryption, electronic signatures, backup, erasure procedures).

The Spanish government guarantees the protection and security of the data collected by the application. With the help of the mobile phone's GPS geolocation system and with the explicit permission of the user, the autonomous community in which he is located will be checked in order to personalize the answers on the basis of the relevant legislation.

All the data collected by the program is necessary to be able to offer relevant advice. The data will in no case be used to verify compliance with deterrence measures. Only medical staff and authorized bodies will allow access to the data.

Personal data will be kept during the duration of the health crisis and, once completed, will be kept anonymous and processed for statistical, scientific or public policy purposes for a maximum of two years.

4.7   StayHomeSafe (Hong Kong)

From March 13, 2020, visitors to Hong Kong, placed under mandatory quarantine, will be provided with a carrier device on the bracelet, each with a unique QR code, which links to an application called StayHomeSafe, which you need to download to your phone and add. The application is activated and calibrated by locating (for example, an apartment) and displaying unique signals around the area (WiFi, broadband, etc.). If these signals are different, it is a sign that the person is no longer in such a physical environment and may violate quarantine. This is a warning to both the individual and the government [16].

The Privacy Officer stated that he was satisfied that StayHomeSafe did not collect personal data other than users' phone numbers at check-in. The Chief Information Officer also confirmed that StayHomeSafe has undergone a security and privacy assessment and pre-launch audit, and all data collected is stored in the government's private cloud and protected by various levels of protection to ensure information security. Unfortunately, these reports are not publicly available, the scope and degree of evaluation and audit have not been determined, and the persons conducting the evaluation and audit have not been named.

This is not to criticize the approach to implementing StayHomeSafe, nor to challenge the basis on which it was done. COVID-19 is a very serious health challenge. There is no reason to doubt the statements of the Hong Kong government. StayHomeSafe is one of a number of technological decisions made by governments around the world to combat the current pandemic.

4.8   NZ COVID Tracer (New Zealand)

NZ COVID Tracer app was developed for the Ministry of Health by Rush Digital and relies in part on Amazon's web services platform. AWS is part of the NZ Cloud Services Agreement, All of Government (AOG), which was formed in 2017 after a robust procurement process. AWS services and infrastructure have been reviewed as part of the procurement process and are regularly reviewed based on third-party trust systems.

Any information that an NZ COVID Tracer user has decided to share to track contacts is encrypted before it is sent to the Ministry through the AWS cloud services platform. The Ministry retains control over the decryption keys. The Ministry is actively considering how other technologies can complement NZ COVID Tracer and help improve the speed and efficiency of contact tracking.

The NZ COVID Tracer app works on Android 7.0 or higher devices, as well as on Apple iOS 12 or higher devices. This allows users to scan their QR codes in companies, public buildings and other organizations to track where they were for contact tracking purposes. People can also register their contact tracking data on the official NZ COVID Tracer website. Information about people's movements will be stored for 31 days by the Ministry of Health, which uses Amazon's web services in Sydney to store data before being automatically deleted [17].

It is possible to delete all location data recorded by NZ COVID Tracer, but a separate location cannot be deleted independently. Any organization that has a business or community that has visitors or staff visiting the site and also has a New Zealand number can generate a QR code poster for each of their premises through the MBIE Business Connect service. Each QR code poster contains a code that can be scanned, based on the New Zealand Business Number (NZBN) and global location number. QR code posters should be displayed where staff and/or customers enter the premises.

Two-factor authentication (2FA) provides an additional level of security and makes it difficult to access user information to someone else. The contact tracking app supports 2FA through the use of one-time passwords (TOTP). If a user wants to set up 2FA in a contact tracking app, you first need to download a trusted authenticator app and install it on your device.

4.9   VirusRadar (Hungary)

VirusRadar mobile application to prevent the spread of Covid-19 that Nextsense has developed and transferred to the state of Hungary in support of the fight against Covid-19. Nextsense is a leader in the digital industry, according to a recently published ranking in the Hungarian market, an international software developer.

The Android app works on Android devices version 5.0 or later. The device must support low-energy Bluetooth technology (BLE). The iOS app runs on iOS devices, requires version 11.0 or later. The device must support low-energy Bluetooth technology (BLE). If Android user: VirusRadar can run in the background and use other applications during.

In the case of iOS: VirusRadar works in active mode in the foreground. Users are encouraged to keep VirusRadar during meetings and during community visits and turn the device face down. This activates the power saving mode and the display becomes dark [18].

On iOS devices, you can activate power saving mode by turning your phone down or putting it upside down in your pocket. In this way, VirusRadar can control your environment by storing battery energy. Virus Radar does not require an Internet connection, but you must submit an Internet connection. Other Bluetooth enabled devices, such as wireless speakers, wireless headphones, smartphones and car headsets, do not interfere and can be used as usual. Uninstalling the program will also delete all locally stored data. The user can no longer share the stored data. In this case, the program must be reinstalled.

If you change your mobile phone while saving your mobile phone number, you just need to download VirusRadar to your new device and register with your mobile phone number.

VirusRadar uses the intensity measurement value of the received signal strength (RSSI) from the signal strength between phones. Calibrated RSSI values are used to estimate the estimated distance between users, while the duration of such a connection is fixed by the mobile application itself. VirusRadar detects the distance between mobile devices / applications that exchange data (code aliases). The software exchanges Bluetooth signals with other phones nearby that work with VirusRadar and have bluetooth enabled. The application does not collect or store location data (GPS coordinates).

Only some Android devices require geolocation permission, as Android/Google requires this for applications that require low-power Bluetooth access. VirusRadar does not collect or use GPS location information for any Android device. If the app is hacked or your phone is stolen, attackers will not be able to access any personal information.

VirusRadar user contacts are stored only on the device. If a VirusRadar user become infected with COVID-19, they can voluntarily decide whether to send data stored on the device to epidemiologists. The data does not include mobile phone numbers, but only codes that can be deciphered only by epidemiologists.

4.10   eRouška (Czech Republic)

The application was created by volunteers working under the auspices of the Ministry of Health of the Czech Republic. Data protection is justified by a number of independent experts, including scientists. The application connects via Bluetooth to other phones with eRouška in close proximity and stores their IDs (Fig. 19). If later one of the eRouška users is found to be infected with coronavirus, the epidemiological model will assess whether you have been in contact with it long enough to be warned and sent for further examination. If the user's test confirms infection, they are contacted by a sanitary station to find out who they have been in risky contact with in the past five days to limit further spread [19].

The system will provide access to the phone numbers of authorized hygienic services for persons with whom it was infected in critical contact (closer than 2 m more than 15 min in a total of 24 h) so that they can discuss the next step—voluntary quarantine and priority testing. Consequently, the app helps to contact potentially risky contacts faster and, based on face-to-face interviews, assess whether they are sick.

The eRouška application works on the principle of measuring signal strength between two devices. Technically, it is not excluded that an erroneous detection will be recorded—for example, close contact through a thin wall resembles a more remote contact without interference. These errors, although exceptional, cannot be completely avoided. In addition to the distance (up to 2 m), the length of contact is also relevant for assessing contact as risky—so a simple stop at traffic lights for 1–2 min is insignificant.

The phone number will use a hygiene station to contact users as soon as possible if, based on information from eRouch, an infected person suspects that they have been in contact with him. Therefore, the user registers the phone number when you first start eRouška. The created eRouch ID can only be associated with the user's phone number. In a protected system, only designated employees of hygienic stations have this right, and only those IDs that, based on data sent from infected users, are evaluated as those at risk of coronavirus infection. Unlike other programs, eRouška does not track or collect location information, but only anonymously finds out which other users of the application were convergence. The application does not require an Internet connection to work properly, so it works, for example, in the subway. It is impossible to know from eRouška data when the user has moved and where. ERouška does not record geographic location, but recognizes other eRouška devices in the area using Bluetooth technology. This information is anonymized. The information entered looks something like this: 31.3.2020 from 12:15 to 13:15 next to the user was a program with ID 29,091.

In addition, eRouška does not send information anywhere on its own, it remains stored on the device. Data can be sent to the Ministry of Health or its subordinate hygienic station only when the call center employee contacts the user on the basis of a positive test and asks for his consent to the processing of data collected by the application. Based on this request, the collected data is sent directly from the program to the centralized database of the hygiene station.

You can view the collected data at any time in the app on the My Data tab. The eRouška application does not collect or store data from GPS, but the Android operating system also includes some positioning BLE services that eRouška needs for its operation. Therefore, the user's consent to access the location data is required. In iOS, this consent is not required. Without Bluetooth enabled, it is impossible to detect the proximity of other devices with the eRouška application installed. In addition, we recommend that you allow all devices to see in your phone/tablet's Bluetooth settings. User privacy is the number one priority, and Bluetooth technology offers the most balanced recording accuracy ratio and minimal privacy. Neither GPS technology nor operational triangulation of transmitters can provide the necessary data with the necessary accuracy. One reason is that their functionality is very limited in buildings, garages or subways.

The entire eRouška application system, including a supportive website, has been developed in full compliance with gdpr. The program code is freely available (Android, iOS) and has been audited by independent educational institutions.

4.11   COVID Tracker Ireland (Ireland)

COVID Tracker Ireland is a digital contact tracking program released by the Irish government and the health executive on July 7, 2020 to prevent the spread of COVID-19 in Ireland. The app uses ENS and Bluetooth technology to determine if the user was close. contact with someone for more than 15 min who has shown a positive result on COVID-19. On July 8, the app covered a million registered users within 36 h of launch, which is more than 30% of the population of Ireland and more than a quarter of all On November 21, the app was downloaded by more than 2,190,000 people. The COVID Tracker program development process began on March 22, 2020, when the Head of Health Service (HSE) contacted the Waterford NearForm technical company to build contact tracing. application for Ireland using existing Bluetooth technology in smartphones to support contact tracking.

The app requires users to enable the exposure notification service and will only be available to those with Android 6.0 Marshmallow or later phones, or iOS 13.5 or later [20].

The €850,000 project was attended by representatives of the Department of Health, the Institute for Economic and Social Research (ESRI), An Garda Siocana, the Irish Army and the Irish Science Foundation. NearForm initially worked on a centralized application that would group user data together to study power structures, but in May 2020, the development team contacted Apple and Google to provide beta access to the Exposure Notification System (ENS), developed by two companies, allowing the COVID Tracker app to guarantee user anonymity and ensure that any data transfer to contact inspectors would only take place with the consent of users. Within three months of development, the team had a secure, proven and reliable contact tracking application that worked and was ready for deployment on a national scale.

Stop COVID NI and Scottish Protect Scotland in Northern Ireland are based on the same NearForm technology and interact with it. NearForm made the source code of the program freely available under an MIT license. In September 2020, NearForm deployed localized versions of the COVID Alert program in the US states of Delaware, New Jersey, New York and Pennsylvania.

On October 19, the COVID Tracker app became one of the first waves of national programs related to other European Union countries, after contact tracking programs from Italy and Germany.

The COVID Tracker app uses Bluetooth Low Energy and an exposure notification system from Google and Apple to create anonymous IDs for registration:

- Any phone users who have close contact with this app also have the app installed.
- Distance between users' phone and phone number of other app users. The length of time that phone users are near the phone of other app users.

Every two hours, the app downloads a list of anonymous codes that HSE users shared with other people who use the app and tested positive for COVID-19. If the user is more than 2 m away for more than 15 min with any of these phones, that user will receive a notification on their phone [2].

The Health Executive says the COVID Tracker program protects the privacy of all users and has been designed to protect users' privacy.

On July 18, 2020, researchers from Trinity College Dublin published a report stinging that user privacy is not properly protected in the COVID tracking program if it is used with Google Play services. The report shows how Google Play Apps used on Android devices send extremely sensitive personal data to Google servers every 20 min, including users' IMEI, hardware serial number, SIM card serial number, handset phone number, and Gmail address, potentially allowing you to specify an IP address. based on phone location tracking. Researchers from Trinity College described the data transfers as extremely disturbing in terms of privacy, while the Irish Civil Liberties Council called them completely opaque to users.

4.12 Apturi Covid (Latvia)

Latvia became the first country to launch a contact tracking program using the recently available Exposure Notification API from Apple and Google. The application is designed to increase the response time to the possible impact of COVID-19 to reduce the spread and ultimately destroy the virus. Starting May 29, an app called "Apturi Covid" is available for iOS and Android for all two million residents of Latvia, and the code is available for adaptation in other countries and communities.

"Apturi Covid" is a voluntary, decentralized, fully GDPR-compliant application. It uses Bluetooth technology to identify other nearby devices that also have the app installed. Individuals will only receive notifications if they are 2 m (6 feet) away from a person for more than 15 min who has tested positive for COVID-19. They will receive a notification asking for self-isolation to prevent the further spread of the disease. Due to the decentralized and encrypted nature of the program, it is consistent with EU data protection regulations. The presence of the recently released API from Google and Apple is a fundamental element for ensuring the widespread use of the application [21].

Latvia is experiencing one of the lowest cases of per capita infection in Europe, while applying some of the lowest travel restrictions. This was achieved by combining early restrictions on social distancing and manual contact tracking. The application is a continuation of this approach to the fight against coronavirus and is expected to become a valuable and economical tool used by the Latvian Center for Disease Control.

It is based on a methodology developed by European scientists, including from Latvia, and on a new Bluetooth signal exchange algorithm developed by Google and Apple. Its use is voluntary. The application uses Bluetooth to anonymously detect smartphones nearby (at a distance of ~2 m, there are more than 15 min), on which this application is also installed. This information is stored only on the user's device and is automatically deleted after 14 days.

The SPKC contacts those individuals who have been confirmed to be infected with COVID-19—regardless of whether they use this program or not. The SPKC receives this information from health care institutions, not from the program. If the patient uses this program, the SPKC will send them the code. After entering the program, an anonymous notification will be sent to detected contacts who risked getting infected. Sender identification is not disclosed to recipients. Similarly, an infected person does not know who the recipients are.

When the contact detection feature is activated, the user confirms his/her consent to activating covid-19 impact notifications (exposure notifications): this allows you to exchange contact keys and check the contact risk of the person suffering from the disease. Every time the Application is running, activation of the mentioned consent is checked: if it is not activated, a corresponding message is displayed [22].

The following services must be activated for proper functioning. On Android devices:

- Contact notifications for Covid-19 developed by Google are activated;
- Bluetooth function is activated;
- Device Location Services feature is activated: It's required to detect any nearby Bluetooth devices, but the app doesn't have access or uses device location tracking.

On iOS devices:

- Activation of the covid-19 contact notification developed by Apple
- The user must activate Bluetooth manually because iOS does not allow access to the Bluetooth module and the application does not have the right to activate it.
- The user is asked to enter his phone number in the CDPC Application to contact the User and provide advice on further actions if there is a suspicion of exposure. In this case, the phone number is checked. The user may decide not to enter their phone number or enter a phone number at any other time and start the verification process.

For each User confirming participation in the contact tracking program, a unique Temporary Exposure Key (hereinafter referred to as the FOLDER) is created once a day in a secure phone area that is not available for applications. If the user confirms the presence of the disease, the keys collected during the last 14 days, as well as the start time of their validity, are sent to the CDPC server. A moving proximity identifier (hereinafter referred to as RPI) and an encryption key that protects metadata are created every 10–15 min. RPI and encoded metadata required to quantify the risk of infection using BTLE broadcast packets are periodically transferred from the device.

Transmission does not have a noticeable impact on the power consumption of the phone's battery. Other nearby devices receive these packages, and if the receiving device is registered for a contact tracking program, it stores the received RPI in its secure memory. This information is not available for the application.

If the user is sick, he/she is asked to transfer the TEK keys collected in the last 14 days to the CDPC, which will publish them to a public server. After that, TEK can no longer be obtained from protected memory. In addition, it is not possible to obtain the current day's TEK to protect your privacy. Other devices regularly load these keys from the CDPC server. The application transmits a set of keys to the exposure notification API, which checks the protected memory of the device for any contact with the specified keys. If a contact is found, the Exposure Message API quantifys the risk in the device according to the settings you set.

For each recorded contact, the API returns its Exposure Report if the user has been in contact with an infected person. The aforementioned Impact Report contains the following data: the day (without time) when exposure occurred; Exposure duration: 5, 10, 15, 20, 25, 30 (if longer, rounded up to 30); BTLE signal interference level, which allows you to set the approximate distance; quantitative risk of infection.

4.13 Immuni (Italy)

Immuni is a technological solution that focuses on the application for smartphones, available for iOS and Android.

The application is equipped with an irradiation system that helps prevent potentially positive users on SARS-CoV-2 at an early stage. This system keeps track of contacts between Immuni users even when they are completely unfamiliar. When a user submits a positive result to SARS-CoV-2, the program uses this system to notify other users at risk. The system is based on Bluetooth Low Energy and does not use any geolocation data, including GPS data. So, although the application knows that there has been contact with an infected user, how long it lasted, and can estimate the distance that divided the two users, he cannot say where the contact took place, as well as to reveal the identities of those involved. The app then recommends to at-risk users what they should do. Recommendations may include self-isolation (which helps minimize the spread of the disease) and going to the GP (so that the user can get the most appropriate help and reduce the likelihood of development of severe complications) [23].

As already noted, immuni radiation notification system uses Bluetooth Low Energy. This has some advantages over a location-based solution:

- The system is more accurate. In many contexts, geolocation has an accuracy of tens of meters. Meanwhile, Low Energy Bluetooth signals allow you to record contact that occurs within a radius of several meters from the user. These are cases of contact relating to the transmission of SARS-CoV-2.

- The battery is used more efficiently. Bluetooth Low Energy excels when it comes to energy efficiency. This is important because it is reasonable to expect that the speed of uninstalling the program will correlate with battery consumption.
- Geolocation data is not required. Thanks to Bluetooth Low Energy, the contact is tracked without tracking the location of users. This can lead to the application being more welcomed by the public and can contribute to wider implementation by increasing the usefulness of Immuni.

To implement its irradiation features, Immuni uses Exposure Notification from Apple and Google. This allows Immuni to overcome certain technical limitations, therefore, to be more reliable than it would be possible.

After installation and adjustment on the device (device_A), the program generates a temporary exposure key. This key is randomly generated and changed daily. The app is also starting to transmit a Bluetooth Low Energy signal. The signal contains a moving intimacy identifier (ID_A1, provided as fixed in this example for simplicity). This is generated from the current temporary irradiation key. When another device (device_B) that runs the program receives this signal, it ID_A1 write it locally to its memory. At the same time, the device_A writes the identifier device_B (ID_B1 is also assumed to be fixed in this example).

If later device_A detects a positive result on SARS-CoV-2, following the protocol defined by the National Health Service, it has the ability to upload temporary exposure keys to the Immuni server, from which Immuni can receive moving intimacy identifiers recently broadcast by device_A (including ID_A1). Periodically device_B checks new keys uploaded to the server for a local ID list. ID_A1 be a coincidence. The app not device_B user that he may be at risk and provides advice on what to do next (e.g. isolation and calling a GP). In practice, when it comes to determining whether a user is threatened by device_B, identifying that he was in close proximity to the user device_A is not enough. Immuni assesses this risk based on the duration of impact and the distance between the two devices. This is assessed based on the weakening of the BLE signal received by device B. The longer the impact and the closer the contact, the higher the risk of transmission of the virus. Contact that lasts only a couple of minutes and occurs at a distance of several meters is usually considered a low risk. The risk model may develop over time when more information about COVID-19 becomes available [24].

To ensure that only users who actually test positive for SARS-CoV-2 upload their keys to the server, the download procedure can only be performed with the assistance of an authenticated medical operator. The operator asks the user to provide the code generated by the application and enters it into the back-office tool. Downloading can only succeed if the code used by the data authentication application matches the code that the health operator enters into the system.

The only epidemiological data that Immuni collects relates to the user's impact on infected users. Data includes:

- Day when exposure occurred
- Exposure duration

- Signal attenuation information used to estimate the distance between devices of two users during exposure

The application can send epidemiological information to the server only after loading temporary impact keys. When the health operator informs the user about its positivity to the SARS-CoV-2 test, any available epidemiological information for the previous 14 days will also be downloaded. Downloading data must be initiated by the user and approved by the healthcare operator. To help protect your privacy, your data is downloaded without requiring you to authenticate yourself (for example, without a phone number or email confirmation). Moreover, traffic analysis prevents fictitious downloads.

4.14   STAYAWAY COVID (Portugal)

STAYAWAY COVID is an application that works on iOS and Android mobile phones and aims to support the country in its efforts to prevent the spread of COVID-19. By monitoring the last contacts of users, the software allows you to simply and safely know if they have not been exposed to the disease.

Stayaway COVID is completely voluntary and free of charge, and never requires access to personal or personal data [25].

If a person who has a positive COVID-19 test, users who have been in close contact will receive a notification no later than the next day.

Notifications will appear in the user application, namely those whose devices were next to the mobile phone of a person diagnosed with COVID-19 - less than two meters and more than 15 min in a row. In addition, according to the current rules, only users who have had a risky effect on a person diagnosed with COVID-19 are notified two days before the first symptoms appear; in the case of asymptomatic persons, the period is two days before doing the COVID-19 test. This risk period is determined during the creation of the Legitimization Code and will never exceed 14 days.

After inserting code into the application, STAYAWAY COVID deactivas the tracing; therefore, users who may have been at risk after this date will not receive notifications. Tracking will restart only after the user diagnosed with COVID-19 resumes and reinstalls STAYAWAY COVID.

In a family environment, and although people often face risk, it is usually not so common for their mobile phones to be nearby for the aforementioned amount or time, or less than two meters apart, without any obstacles. Therefore, messages among close family members may be less frequent than expected or, in some cases, absent. However, as people tend to become closer and more familiar in these environments, the need for digital contact tracking tends to be less important.

Using the Exposure Message feature of the operating system does not require or use the GPS interface. The request made by Android is incorrect and is a cause for concern in all applications using EN. Google is aware of the problem, but adds that "this is a limitation of the operating system itself." In Android, the Bluetooth Low

Energy service (BLE) is connected to the Location Services and must be activated to use it [26].

Yes, the program requires Bluetooth. However, the app uses Bluetooth Low Energy (BLE), which, as the name implies, requires less power than bluetooth, which is commonly used to connect the device to the car's speakers, headphones or stereo system. The application requires periodic access to the official and public server. This can be done via Wi-Fi or mobile data transfer.

The assessment is based on the most up-to-date scientific knowledge in accordance with the recommendations of health authorities. To date, staying with a person diagnosed with COVID-19 for about 15 min at a distance of less than two meters can significantly increase the likelihood of infection. If these precautions are reviewed by the health authorities, the application will be adjusted accordingly.

STAYAWAY COVID issues a local warning, i.e. the program will warn you by informing you that you have been in contact with someone diagnosed with COVID-19. No external organization knows the identity of the user or his mobile phone number; therefore, these organizations cannot send messages either by text message or by any alternative means. In addition, no external organization has the information necessary to assess the risk of infection of the user, since the data is stored exclusively on the user's mobile phone.

The STAYAWAY COVID app was developed exclusively for iOS, from Apple, and Android, from Google, operating systems and depends specifically on the API notifications about the exposition of the two companies. Thus, the program is only compatible with versions of Android that are equal to or greater than version 6.0 (Marshmallow) and iOS equal to or greater than version 13.5.

The expected behavior of the program should remain active even if you restart the device. However, it is recommended that you open the program to ensure that notifications are received in a timely manner.

The software receives notifications even when running in the background or even when the device is restarted. However, in specific cases, restarting the device may prevent the application from running in the background. Therefore, it is worth restarting the program after restarting the phone, thus making sure that the program is active.

4.15 Koronavilkku (Finland)

Koronavilkku is a program developed by the Finnish Institute of Health and Welfare, which aims to prevent the spread of coronavirus. In addition to the Finnish Institute for Health and Welfare, the Ministry of Social Affairs and Health, the Social Insurance Institute of Finland (Kela) and DigiFinland (SoteDigi) took part in the development. Solita Oy is responsible for the technical implementation of the program.

Koronavilkku allows you to track those who have experienced the virus faster and more comprehensively—also when they are not known to individuals who have suffered the infection. The application warns the user about the possible impact of coronavirus and provides instructions to it. The user can also anonymously notify others about the infection and thereby help stop the chains of infection. A user who has received a warning about possible exposure may ask health services to contact or contact the services themselves.

User privacy is securely protected in the application. Koronavilkku does not store name, date of birth or contact information. This cannot determine the identity of the user or the identity of the person he contacted. The application does not collect location data [27].

The use of the program is voluntary, and special attention is paid to the implementation of fundamental rights, data protection and information security. An act was adopted with provisions on the purpose of the application, the powers associated with it, and the processing of personal data. Koronavilkku was published to the App Store and Google Play on August 31, 2020. The application has made it possible.

- track a user who may have been exposed to coronavirus.
- support users in following the current instructions, contacting the health services and, if necessary, passing the test.

As long as the application determines potential exposure situations, the actual probability of exposure will be assessed together with healthcare professionals. The application will not transmit data on close contacts with authorities or personal data that can be used to connect subjects to the chain of infection without a separate interview. The impact notification provided by the program does not provide the user with benefits, such as assistance in paying for the disease due to an infectious disease. The user will not have access to the person who has the confirmed infection or to information about the time of possible exposure.

The app doesn't track the location or movements of individual users. Instead of geospatial data, it, relying on Bluetooth technology, will determine situations where two mobile devices were closer than 2 m apart for at least 15 min.

The application used in each state should be able to interact with back-office information systems and national health system solutions, which is why solutions developed in other countries are not directly suitable for Finland's needs. Designers of the Finnish program, as far as possible, rely on international experience. Any ready-made interface-capable codes and standards will be used where possible.

Privacy, data protection and information security are the most important priorities in the development of the application used in Finland, and special attention is paid to these properties throughout the life cycle of the program. The law will contain detailed provisions on how the application should work in terms of information security and data protection. For example, it will not collect any sensitive personal data, even on back-office systems.

The functioning of the program is based on random, periodically changing aliase tags, which will not allow to directly identify users of applications. The app also cannot be used to identify the person who reported the infection. The security of the program is assessed by the National Cyber Security Center. In addition, the source solution codes will be published openly, allowing anyone to rate their content.

The application is used on a voluntary basis. Users can uninstall the program at any time and withdraw their consent to the processing of data. The data will be automatically deleted from mobile devices and the back office system within 21 days.

The application can only send warnings about laboratory-confirmed coronavirus infection. Health services will give infected people a one-time key code that they can use to send a warning through their programs. All users of apps that may have been exposed will receive notifications on their phones.

4.16 SwissCovid (Switzerland)

The app was given to the Swiss public in its final form on June 25, and was downloaded on July 4 by 1.6 million. It is designed to complement traditional tracking carried out by cantons. However, the number of active users began to decline steadily: with 1,017,504 users on July 8, the program reached a minimum of 959,815 customers on July 11, according to the Federal Statistical Office.

The SwissCovid app is the first in Europe to use Google and Apple's application programming (API) interfaces. They allow devices that use Android and iOS operating systems to work together and give users direct control over their own data from their device. This approach uses Bluetooth Low Energy technology and encryption to establish contact between two smartphones that are at least 1.5 m apart for a period of at least 15 min, providing both confidentiality and low energy consumption [28].

As the SwissCovid app issues a warning:

- If you have installed the SwissCovid program and found a positive result for coronavirus, you will receive a Covid code (release code). Codes can be issued by cantonal organs and doctors conducting tests.
- allows you to activate the notification function in the application, thereby allowing you to warn other users of applications with whom you have come into contact in a close state during the infectious phase (starting two days before the onset of symptoms) - automatically and anonymously. The person to whom the warning is sent is not informed about the identity of the person who initiated the message. However, it is possible that someone could determine the identity of the contact, based on the date.

You will receive a notification under the following two conditions:

- Within 24 h you have been in contact with an infected person:
- at a distance of less than 1.5 m
- for at least 15 min

Within 24 hours you have been in contact with several infected individuals:

- at a distance of less than 1.5 m
- less than 15 min per person, but in total more than 15 min

If someone detects a positive coronavirus test, they will receive a code from the cantonal health service, which they enter into the program to activate the ID. Anyone who receives a warning from a program confirming contact with an infected person can decide for themselves how they will react. The application displays the phone number of the FOPH hotline, where they can receive additional information anonymously. The app also recommends that anyone experiencing symptoms consult the coronavirus check online or seek medical attention and voluntarily enter self-isolation.

Entering code into the program and calling the hotline are completely voluntary. Personal data is not recorded when you install or use the program. An appointment with another user of the program calls the exchange of encrypted code. It is stored locally on both devices and automatically deleted after 21 days. This applies to both the data in the local mobile phone storage and the codes of the infected user on the state server. The warning is generated automatically and anonymously, and it is not possible to draw conclusions about information about a person or place of infection. When the SwissCovid app is uninstalled, the corresponding data from the mobile phone is automatically deleted [29].

SwissCovid devices constantly transmit an ephemeral identifier that changes every 10–12 min. They are called RPI, as for a moving intimacy identifier. Each ephemeral identifier comes from a daily key that is randomly selected. This daily key is called TEK as a temporary exposure key. Given the ephemeral ID, we cannot recover the daily key. However, given the TEC, we can restore all RPI days. If a user is diagnosed, his program receives the daily keys for the last few days from GAEN and uploads them to a public server. Those loaded folders are called diagnostic keys.

This server can be watched by everyone. Programs regularly download diagnostic keys and transmit them to GAEN. GAEN checks to see if these keys receive one of the collected ephemeral RPI identifiers to check if the diagnosed user has been in contact. GAEN reports on contacts with diagnosed users about the date, duration and distance to the program. The application decides to raise the message or not.

To estimate the distance of a contact, phones compare the power of a Bluetooth signal when sent and received. However, the transmitter signal power is encrypted (using the daily key) inside the Bluetooth message. Consequently, he is not seen by the accepting phone until he receives the sender's daily key (since he has been diagnosed and notified). So, only when comparing can the distance be estimated. Similarly, ephemeral identifiers change every 10–12 min. Typically, the recipient cannot bind the two IDs until they receive the day key that received them. Consequently, the duration of the contact (and for it to last at least 15 min) can only be determined during the comparison.

The server is actually the infrastructure of several servers on the network. The application regularly downloads reports from the server, as well as configuration parameters used to calculate risk notifications. Both types of downloads must be performed by millions of devices every day. To provide a reliable download service, SwissCovid uses the Content Delivery Network (CDN) provided by Amazon. When we try to download from anywhere in the world, it matches amazon's local server. The content is obtained from the servers of the Swiss federal offices and signed by them so that Amazon cannot interfere with the content. The server also includes another site where daily keys should be loaded after diagnostics. Since this operation is rare, there is no need for third-party service for this operation. In fact, the diagnosed user contacts health authorities who provide them with an access code called covidcode, out of 12 decimal digits. With this 12-digit figure, the application receives another code from another server (namely JWT code), which allows the application to load daily keys to the server. One problem is that the network sees all server requests made by the app and may be able to identify the phone. One way to protect your privacy is to make random "fake" requests. Requests are encrypted, so the network cannot understand if they are false or genuine.

In June 2020, researchers Serge Woudena and Martin Voagno published a critical analysis of the program, noting that it relies heavily on Google and Apple's iOS notification system, which is integrated into the respective Android and iOS operating systems. Since Google and Apple have not released the full source code of this system, this calls into question the truly open source program. The researchers note that the dp3t team, which includes the developers of the program, asked Google and Apple to release their code. Moreover, they criticize the official description of the program and its functionality, as well as the adequacy of the legal basis for its effective functioning [30].

Professor Serge Woudena and Martin Voagno also identify various vulnerabilities in the program. It is noted that the system will allow the third party to track the movements of the phone using the application using Bluetooth sensors scattered along its path, for example in a building. Another possible attack could be copying IDs from the phones of people who may be ill (for example, in a hospital), and reproducing these IDs to receive notifications about the impact of COVID-19 and illegally taking advantage of quarantine (thus entitling them to paid leave, an exam rescheduled or other benefits). The system will also allow a third party to use the phone using the app using Bluetooth sensors scattered along the way.

4.17   TraceTogether (Singapore)

For Singaporeans, the covid-19 pandemic was closely intertwined with technology: two technologies, specifically. The first is a QR code whose small black-and-white squares were ubivorous across the country as part of a SafeEntry contact tracking system deployed in April and May. According to SafeEntry, anyone entering a public place - restaurants, shops, shopping malls - must scan the code and register with a name, identification number or passport number and phone number. If someone gives a positive result to covid-19 infection, contact inspectors use it to track down those who are close enough to a potential infection. There is also a TraceTogether program, which was launched in March 2020. It uses Bluetooth to set the ping on close contacts; if two users are nearby, their devices trade anonymous and encrypted user IDs that can be decrypted by the Ministry of Health if one person has a positive test for Covid-19. For those who cannot or do not want to use the smartphone app, the government also offers TraceTogether tokens, small digital codes that serve the same purpose. Although TraceTogether is currently voluntary, the government has announced that it is going to merge the two systems, which will make it mandatory to either download the program or collect the token [29].

TraceTogether works by exchanging Bluetooth signals between nearby mobile phones that have an app installed, assessing the distance between users and how much time they spent on contact. This information will be stored locally on users' mobile phones. When tracking contacts, they can consent to sending data to the Ministry of Health. TraceTogether detection of other TraceTogether members works through smart use of the General Bluetooth Attribute Profile (GATT) and Low Bluetooth Energy (BLE). This is the essence of TraceTogether. The algorithm and protocol behind this magic have been packaged and sold as BlueTrace, and from writing on the official website, the authors intend to make it available to third parties. Unfortunately, I was not able to dive deeply into the components of GATT and BLE BlueTrace. However, it is necessary to highlight and applaud the new use of the protocol by the development team, since BLE peripherals can only connect to one central device at a time: the algorithm effectively scans all BLE peripherals in close proximity and loops each for connection. Interestingly, BlueTrace internally refers to "StreetPass" for data parcels. StreetPass is a Nintendo 3DS feature for passive Wi-Fi communication, functionally similar to BlueTrace.

When TraceTogether works on the phone, it creates a temporary identifier that is generated by encrypting the user ID with a private key stored by the Ministry of Health. The temporary ID is then exchanged with nearby phones and is updated regularly, making it difficult for someone to identify or bind temporary IDs to a user, said GovTech, the state agency behind the contact tracking program. He noted that a temporary ID can only be decrypted by the Ministry of Health. He added that the TraceTogether app shows connections between devices, not their location, and this data log is stored on the user's phone and transmitted to the ministry - with the user's consent - when needed to track contacts [30].

When the two systems were launched, there was not much room for the public to discuss detentions: they were considered necessary to combat the pandemic, and the Singapore government acted typically from top to bottom. However, he tried to calm fears by repeatedly reassuring Singaporeans that data collected using such technology would only be used to search for contacts during a pandemic.

However, in early January 2021, it turned out that the government's claim was false. The Interior Ministry confirmed that in fact the police can access data to investigate criminal cases; the day after that confession, the minister revealed that such data had in fact already been used in the murder investigation. It quickly became clear that, despite what ministers had previously said, Singapore's legislation meant law enforcement could use TraceTogether data all the time. These exposes have caused anger and criticism from the public, not necessarily because Singaporeers are particularly attentive to privacy—in fact, state oversight in the country is largely normalized—but because people felt they were being lured. Many people had reservations about TraceTogether when it was first launched, and began using it in large numbers only after the government indicated that it would soon become mandatory.

(According to the co-chair of the covid-19 working group, almost 80% of Singapore residents have adopted TraceTogether.) Since then, the government has announced that it will introduce new legislation that will limit law enforcement's use of contact tracking data to investigate seven specific categories of crimes, including terrorism, murder, kidnapping and the most serious drug trafficking cases. "We acknowledge our mistake without noting that TraceTogether's data is not exempt from the Criminal Procedure Code," the Office of Intellectual Nation and Digital Governance said in a statement. The new law, it states, "will state that personal data collected through digital contact tracking solutions …can only be used for specific contact tracking purposes, unless there is a clear and urgent need to use this data to investigate criminal cases of serious offenses." Not in the original spirit There are no deadlines for when the proposed law will be submitted to parliament for consideration, and the details are meager. "In Singapore, where laws give broad executive and legislative powers to state actors, I believe that any commitment to accountability and restraint is welcome," says digital rights activist Lee Ying- "But it remains to be seen whether the bill will make significant commitments to these proposed restrictions. For example, if state actors violate these norms, what investigative bodies will come into force, and to what consequences will state actors be detained? Some doubt how useful such data can be for police investigations, and are concerned that even the proposed restrictions still formally expand their use beyond contact tracking. It should be noted that concerns that governments may abuse contact tracking systems have been expressed around the world.

4.18   Comparison of World Applications in Terms of Technologies Used

After analyzing the information described in the previous section, summarize the data about the applications in Table 2.

Thus, analyzing the principles of the functioning of world applications, it should be noted that all applications use BLE technology to exchange anonymous user IDs. Part of the app uses GPS technology to build maps of the spread of the virus on the ground. However, at the same time, download statistics state that trust in applications using GPS is much greater than in the case of applications using Bluetooth.

5   Potential Attacks on Contact Monitoring Apps Using GPS, Bluetooth and Wi-Fi Technologies

The biggest differences in the security of contact tracking protocols are their approach to building them in terms of centralization. Serge Hydrogen noted in his work 12 potential attacks on decentralized systems and noted that, unfortunately, most attacks

Table 2 Features of contact tracking applications

Application	Country	GPS	Bluetooth	Centralization	Data required at the entrance
COVIDSafe	Australia	−	+	−	First name, last name, age, postal code, phone number
Stopp Corona	Austria	−	+	In the original version	−
NHC	United Kingdom	−	+	−	Zip code
HaMagen	Israel	+	+	+	Location buzz data from the last 14 days
Aarogya Setu	India	+	+	+	Phone Number
COVID Tracker Ireland	Ireland	−	+	−	Are you older than 16 years old, phone number (optional))
AsistenciaCovid19	Spain	+	+		Region, Phone Number
StayHomeSafe	Hong Kong	+	+	−	Phone Number
NZ COVID Tracer	New Zealand	−	+	−	Phone Number, Email, First Name, Last Name
TraceTogether	Singapore	−	+	−	Name, surname, date of birth, nationality, passport number, phone number
VirusRadar	Hungáry	Some devices	+	−	Phone Number
eRouška	Czech Republic	−	+	+	Phone Number
Apturi Covid	Latvia	−	+	−	−
Immuni	Italy	−	+	−	Region of residence, province
STAYAWAY COVID	Portugal	−	+	−	−
Koronavilkku	Finland	−	+	−	−
SwissCovid	Switzerland	−	+	−	−

on such systems are not detected and can be carried out on a large scale, which will lead not only to a decrease in public trust in the systems, but also to the disclosure of end-user data. However, proponents of decentralized approaches note that 8 out of 12 attacks are equally effective in relation to centralized systems. Consider the main effective and common attacks on both types of systems [31].

5.1   Power Reduction Attacks

Cryptanalystic attacks a device nearby, sending a large number of messages from the source. The device is forced to wake up to process each such message Such an attack is especially effective in a crowded area, where each message affects many mobile devices. Moreover, an attacker can arbitrarily change the strength of the message signal to make it difficult to determine that they come from the same source. The attack does not pose a danger in terms of the effectiveness of the system, but the end user can refuse to use the system due to a rapid loss of power on the device. Such an attack completely discredits the benefits of BLE and threatens the public's massive refusal to use contact tracking applications.

5.2   Relay Attack

The attacker uses two devices to implement a relay attack. One device is used to generate a message, the other to send it. The victim of the attack, who is near the device of the attacker who advertises the message, receives an intimacy ID with a fake location or time. If one of the intimacy identifiers obtained in this way is subsequently consistent with the diagnostic key, the victim is mistakenly believed to have been in contact with the diagnosed person due to this correspondence. If relay attacks are allowed, many victims mistakenly believe they have been in contact with a diagnosed person. In addition to the concerns of the victims and their loved ones, the victims will be diagnosed without good reason. This will not only waste precious diagnostic resources, but can also lead to a loss of trust in the system, to diagnosis, or to both.

5.3   Trolling Attacks

The essence of the attack is that a person who is sick or suspects of the disease wants as many people as possible to think that they are also at risk, or a certain person believed that he could become infected. To this end, the attacker attaches his phone to the car, drone, etc. Other devices record contact and after confirming the disease, the attacker will also be at risk. This leads to an increase in the number of diagnoses without valid reasons, as well as a decrease in trust in applications by the population.

5.4 Connection Attacks

The purpose of such an attack is to distinguish between two messages created by one device. It is assumed that the device uses regular Bluetooth signal strength and a constant frequency of messaging. The conclusion about the origin of messages from one device is based on a comparison of the time difference between messages and the frequency of messages. Carrying out this attack leads to a decrease in the privacy of individuals, which further leads to the abandonment of users from the system.

5.5 Attack Tracking

The attacker is interested in revealing the identity of the person or several people during the meeting, whose mobile device posted messages, including special intimacy identifiers. To this end, the attacker monitors the bluetooth signal strength and location of the attacker's mobile device when receiving each message and uses this information to correctly determine the location of the target person's mobile device over time. The attacker can then identify the targeted person by physically observing who is present in certain locations over time. The result of this attack violates the privacy of people who have the disease. This is a threat from a moral and ethical point of view and will lead to the refusal to publish daily clues in case of disease.

6 Summary

After analyzing the information regarding the implementation of world applications, as well as the principles of the functioning of typical protocols for contact tracking applications, it is worth noting several important conclusions.

Firstly, for the correct functioning of applications, all, without exception, applications use an Internet connection, in particular Wi-Fi. This is because it is necessary to transfer anonymous identifiers to the central server for further analysis and notification of potentially infected users in the case of a centralized system. In turn, in the case of a decentralized approach, an Internet connection is necessary to download the identifiers of sick people from the central server and locally analyze the list of their contacts in order to identify contacts with the infected person. Also, the StayHomeSafe application uses analysis of the device's connection to Wi-Fi networks to monitor the isolation of the disease.

Secondly, world practice proves that it is more effective to use Bluetooth technology to exchange anonymous identifiers between users. In particular, low-energy Bluetooth is used, which reduces power consumption and prevents the irrational use of the device's power supply, and as a result reduces the negative attitude of the population to the functioning of applications on their devices.

Thirdly, at the moment, using only GPS technology, no application works. GPS is used as an additional feature for applications using Bluetooth to build terrain maps indicating areas of the disease, as well as monitoring the isolation of the disease. The absence of applications using GPS technology only is explained by the fact that GPS technology does not allow correctly recording the location of the device in all cases. For example, accuracy in moving and underground environments is low, as well as limited vertical accuracy, when most residents or employees of one high-rise building are fixed within an error, which is especially important in urban environments and can lead to false suspicion of the disease.

References

1. Covid-19 apps pose serious human rights risks. (On-line). https://www.hrw.org/news/2020/05/13/covid-19-apps-pose-serious-human-rights-risks
2. GPS vs Bluetooth technology for contact tracing. (On-line). https://returnsafe.com/gps-vs-bluetooth-technology-for-contact-tracing/
3. Foy. Bluetooth signals from your smartphone could automate Covid-19 contact tracing while preserving privacy. (On-line). https://c19priority.ai/contact-tracing/bluetooth-signals-from-your-smartphone-could-automate-covid-19-contact-tracing-while-preserving-privacy/
4. Using contact tracing and GPS to fight spread of COVID-19. (On-line). https://www.gpsworld.com/using-contact-tracing-and-gps-to-fight-spread-of-covid-19/
5. Bluetooth contact tracing apps built with Google and Apple's APIs still collect Android users' location data. (On-line). https://www.mobihealthnews.com/news/bluetooth-contact-tracing-apps-built-google-apples-apis-still-collect-android-users-location
6. Coronavirus contact-tracing: World split between two types of app. (On-line). https://www.bbc.com/news/technology-52355028
7. Exposure notifications: using technology to help public health authorities fight COVID-19. (On-line). https://www.google.com/covid19/exposurenotifications/
8. PEPP-PT documentation. (On-line). https://github.com/pepp-pt/pepp-pt-documentation
9. DP-3T documentation. (On-line). https://github.com/DP-3T/documents
10. COVIDSafe. (On-line). https://www.covidsafe.gov.au/
11. StoppCorona. (On-line). https://epicenter.works/sites/default/files/report_stopp_corona_app_english_v1.0_0_0.pdf
12. The NHS app. (On-line). https://www.nhs.uk/using-the-nhs/nhs-services/the-nhs-app/
13. NHS App. https://www.nhs.uk/nhs-app/. Last accessed 07 Nov 2021
14. HaMagen. (On-line). https://govextra.gov.il/ministry-of-health/hamagen-app/download-en/
15. Aarogya Setu. (On-line). https://www.bbc.com/news/world-asia-india-52659520
16. StayHomeSafe. (On-line). https://we-gov.org/wego-smart-health-responder/stayhomesafe-wristbands-and-app/?ckattempt=1
17. NZ COVID tracer. (On-line). http://web.archive.org/web/20200520004054/
18. VirusRadar. (On-line). https://virusradar.hu/
19. eRouška. (On-line). https://erouska.cz/
20. COVID Tracker app. https://covidtracker.gov.ie

21. Apturi Covid. (On-line). https://covid19.gov.lv/ru/covid-19/drosibas-pasakumi/prilozhenie-apturi-covid
22. Apturi vīrusu ar telefonu!. (On-line). https://www.apturicovid.lv
23. Scarica Immuni. (On-line). https://www.immuni.italia.it
24. Immuni-app. (On-line). https://github.com/immuni-app
25. STAYAWAY COVID. (On-line). https://stayawaycovid.pt
26. STAYAWAY COVID-SNS. (On-line). https://www.sns.gov.pt/apps/stayaway-covid/
27. Coronavirus: SwissCovid app and contact tracing. (On-line). https://www.bag.admin.ch/bag/en/home/krankheiten/ausbrueche-epidemien-pandemien/aktuelle-ausbrueche-epidemien/novel-cov/swisscovid-app-und-contact-tracing.html
28. The dark side of SwissCovid. (On-line). https://lasec.epfl.ch/people/vaudenay/swisscovid.html
29. TraceTogether. (On-line). https://www.bbc.com/news/world-asia-55541001
30. Broken promises: how Singapore lost trust on contact tracing privacy. (On-line). https://www.technologyreview.com/2021/01/11/1016004/singapore-tracetogether-contact-tracing-police/
31. Vaudenay S, Centralized or decentralized?the contact tracing dilemma. (On-line). https://eprint.iacr.org/2020/531.pdf

Analysis and Research of Threat, Attacker and Security Models of Data Depersonalization in Decentralized Networks

Lyudmila Kovalchuk(ID), **Roman Oliynykov**(ID), **Yurii Bespalov**(ID), and **Mariia Rodinko**(ID)

Abstract Blockchain technology is one of the most attractive models of a decentralized environment that can be used to store, process and forward information, while ensuring the maximum level of its depersonalization. All models of the offender and all attacks aimed at cryptographic transformations remain threats to any blockchain. Since hash functions and digital signature have been used in cryptology for more than a decade, many of these threat models are well studied and researched, with the necessary and strictly substantiated necessary and sufficient conditions for the success of various attacks, developed methods for assessing resilience and building pre-stable cryptographic mechanisms. In this section, we focused on relatively new attacks that occurred only after the appearance of the blockchain and how to protect against them. These attacks are aimed specifically at the way blockchain operates and are not attacks on the cryptographic mechanisms that serve it. Some of the blockchain attacks are common, meaning their main idea can be applied to any blockchain, regardless of the consensus protocol it uses. Other attacks are special attacks, meaning they depend heavily on the type of consensus protocol. We looked at two main models of common blockchain attacks because, unlike attacks on different cryptographic primitives, they are much smaller. In addition to reviewing and comparatively analyzing these attacks, the main ways to counter these attacks are also given, which significantly use the peculiarities of the functioning and maintenance of the blockchain.

Keywords Blockchain technology · Consensus protocol · Decentralized network

L. Kovalchuk
National Technical University of Ukraine "Igor Sikorsky Kyiv Polytechnic Institute", 37, Prosp. Peremohy, Kyiv 03056, Ukraine

R. Oliynykov (✉) · M. Rodinko
V. N. Karazin Kharkiv National University, 4 Svobody Sq., Kharkiv 61022, Ukraine

Y. Bespalov
M.M. Bogolyubov Institute for Theoretical Physics of the National Academy of Sciences of Ukraine, 14-b Metrolohichna str., Kyiv 03143, Ukraine

© The Author(s), under exclusive license to Springer Nature Switzerland AG 2022
R. Oliynykov et al. (eds.), *Information Security Technologies in the Decentralized Distributed Networks*, Lecture Notes on Data Engineering and Communications Technologies 115, https://doi.org/10.1007/978-3-030-95161-0_3

1 Introduction

As shown in the first section, blockchain is one of the most attractive models of a decentralized environment that can be used to store, process and forward information, while ensuring the maximum level of its depersonalization. The findings also indicate the extraordinary role of cryptology in ensuring the "operation and liveness" of this environment. Indeed, blockchain itself, as a technology, is based on cryptographic mechanisms such as digital signature and hash function. These cryptographic transformations can be considered basic because they ensure the correct functioning of any blockchain, regardless of its specificity, additional properties and purpose.

Consequently, all models of the offender and all attacks aimed at these cryptographic transformations remain threats to any blockchain. Since hash functions and digital signature have been used in cryptology for more than a decade, many of these threat models are well studied and researched, with the necessary and strictly substantiated necessary and sufficient conditions for the success of various attacks, developed methods for assessing resilience and building pre-stable cryptographic mechanisms.

In further research, we will return to the specifics of these cryptographic transformations that arise when they are used in a decentralized environment, in particular in blockchain. And in this section we want to focus on the relatively new attacks that arose only after the appearance of the blockchain and the ways to protect against them. These attacks are aimed specifically at the way blockchain operates and are not attacks on the cryptographic mechanisms that serve it. Some of the blockchain attacks are common, meaning their main idea can be applied to any blockchain, regardless of the consensus protocol it uses. Other attacks are special attacks, meaning they depend heavily on the type of consensus protocol.

Hereinafter we keep some "habitual" terminology that belongs to the cryptocurrency blockchain, but we will provide these terms with new, broader meaning. For example, under the "transaction" we will understand not only (and not so much) the "real" cryptocurrency transaction, but any transformation of information that is performed in the blockchain. A similar remark applies to some other terms, the essence of which will usually be understandable from context.

2 Two Models of Attacks on Blockchain

At the moment, the literature describes two types of attacks on Bitcoin, and both of them are attacks on blockchain.

We will call them "Attack with visible Fork" (Attack I) and "Attack with invisible Fork" (Attack II). Below are their description and comparative analysis.

2.1   Attack I

The essence of this attack is that the enemy tries to create as long a fork as possible, trying to lay out its blocks in such a way as to maintain the existence of two branches of the same (or almost the same) length as far as possible. At the same time, it does not hide the very fact of building a fork, and an outside observer can easily see that a longer branch consists of one block or another. That is, some transactions then disappear, and then appear. Honest miners who are also involved in the construction of blockchain must adhere to the mining protocol, so they have to continue one or the other chain, thus unwittingly participating in the maintenance of this fork.

The work [1], which is dedicated to building blockchain resistance assessments for this attack, considers only the case of the PoS protocol. In this work, with certain assumptions, it is shown that if the share of the attacker $\frac{m}{m+n}$ is strictly less than the share of the honest $\frac{n}{m+n}$, then the probability of fork exponentially tends to zero if the length of this fork goes to infinity. In other words, for a fairly large l probability the fork with the length l has the form $e^{-f(l)}$ where $f(l)$ some positive function. At the same time, the result obtained is exclusively asymptotic, so it is impossible to say what length is "large enough" in order to be able to use the estimates obtained. Accordingly, the obtained results do not answer the question: when asked m, n and (small) $\varepsilon > 0$ what should be the length l of the fork, so that its probability is less than this specified $\varepsilon > 0$.

Another work [2] considers the same task, i.e. the same attack model, but for the PoW protocol. In this work for bitcoin and GHOST models the following results are obtained:

1. obtained analytical upper estimates of the probability of fork of different lengths l when set m, n and (small) $\varepsilon > 0$;
2. according to these estimates, the obtained numerical results for different particles of $\frac{m}{m+n}$ attacker.

Although the obtained estimates are not expressed directly through exponential functions from the length of the input, but nevertheless they are suitable for obtaining numerical results and plotting relevant graphs. Moreover, the graphs show the exponential probability of fork length increase (with fixed m and n).

2.2   Attack II

This attack is analyzed in at least five publications, as well as in countless popular articles on the Internet, and what is very important—in the course of Princeton University "Bitcoin and Cryptocurrency". In English sources, it is called "Double Spend Attack". The idea of the attack is the following. An attacker in a block with a number, for example, 5, performs a transaction by transferring money to a service provider or goods for a certain purchase. The supplier receives this money and, accordingly, supplies the buyer with the goods. After receiving the goods, the attacker quickly begins to mine another block with the same number 5, that is, the block that follows unit number 4, but in which either there is no such transaction, or he translates this money to another address. And to ensure that honest miners accept this particular alternative chain, it is on an alternative block with the number 5 trying to "hook" as many blocks as possible. If he succeeds in making an alternative chain longer, then it is he who, according to the mining protocol, will be considered correct. Obviously, the greater the share of the attacker (no matter the computing power in the case of PoW or the share of steak in the case of PoS), the more chances he has to perform this attack. In particular, if the share of the attacker is more than half, then the probability of success of this attack is 1.

To protect against this attack, Nakamoto in his very first job [3] offered not to provide the goods as soon as the transaction occurred, but to wait a while, more precisely, a few blocks after this transaction, and only then, if the transaction did not disappear from the blockchain, provide the goods. In this case, the attacker cannot build the fork immediately after payment, because then the supplier will see that the transaction then disappears, then appears in the blockchain, and refuses the transaction. Therefore, the enemy first waits until the required number of confirmation units grows on the block with the transaction. At this time, it can subtly generate fork, which begins to block with a transaction, that is, in our designations it can generate an alternative fifth block and subsequent blocks behind it, but in no case teach this alternative chain so that the supplier does not suspect anything wrong. This is the first stage of the attack. But when the confirmation units have already been formed and the goods have been received, the attacker is trying to "catch up" with the existing chain, and this is the second stage of the attack. Supposable, while 6 confirmation units are generated, the enemy was able to generate 4 blocks of an alternative chain. And now it lags at least 2 blocks behind. If someday in the future he will be able to generate so many blocks to "catch up" with the existing chain, which, in turn, will also grow all this time, then the attack will be successful. In particular, if he managed to generate 7 or more blocks already in the first stage of the attack, while he waited out the confirmation units, then the attack had already succeeded, nothing needed to catch up. After receiving the goods, he simply lays out his, longer chain, in which the money remains with him the same.

Note that in Attack I there are no such two stages, there is simply an "obvious" fork all the time. At the same time, unlike Attack I, the benefit of the attacker in Attack I is obvious: he bought both the goods and did not spend money. But in the case of Attack And the maximum that he can do is compromise the cryptocurrency, which will not bring him any benefit; and if he also owns this currency, he will even bring losses.

Now we will analyze the results of various authors, which examine Attack 2 and assess the resistance of the blockchain to this attack. Note that in all of the following works, the task is set as follows: with the specified m and n, as well as small $\varepsilon > 0$, what should be the number of confirmation units l after the transaction, so that the probability of a successful attack is less than the same set ε?

As already mentioned, the first results of this task were obtained in the work of Nakamoto. However, they were obtained in assumptions that do not quite correspond to the real model. The first assumption that is also present in almost all other works is the assumption that the time of generation of the block and the time of its appearance on the network coincide. That is, the delay in distributing the block is zero. But from this assumption it follows that the probability of an "unintentional" fork is zero, and reality shows that such fork occur about 6 times a month. The second assumption is even more unsuccessful. It is as follows: if the probability of an event is p, the number of tests in which exactly n events equals exactly $\frac{n}{p}$. In fact, this means that the random variable was replaced by a mathematical expectation, which, to put it mildly, is not entirely correct.

In these simplified assumptions, analytical expressions are obtained for the probability of attack success, by which numerical values of this probability can be obtained. At the same time, the resulting expressions clearly do not show that the probability decreases exponentially with an increase in the number of confirmation units, but the constructed graphs look like exponential functions that are being extracted.

Rosenfeld's work [4] offered other, and as it turned out, more accurate analytical expressions for these probabilities, while a slightly different model was chosen to obtain them than Nakamoto's. However, this work did not provide any justification for this chosen model. Simply, the authors suggested that the appearance of "honest" / "dishonest" blocks on the network is described as a negative binomial distribution, but this assumption was not substantiated here. At the same time, in Rosenfeld's work, the results were also obtained in the assumption that the time of distribution of the block in the network is zero. As for Nakamoto's second admission, it is unclear to what extent the authors noticed his mistake; however, they themselves did without this assumption. Therefore, the numerical results in this article differ from the results of Nakamoto, well, that is, for the same probability of an attack in Rosenfeld's work, more confirmation blocks are required, which is natural.

Pinzon's work [5] first drew attention to the incorrectness of Nakamoto's second admission. More precisely, it said that the success of the attacker in the first stage of the attack will significantly depend on the time it took to generate confirmation units. Rosenfeld's results also had remarks that they were not entirely accurate, and some (incomprehensible) function was proposed to calculate the probability of success in the first stage. But the authors did not cede numerical results obtained by their formulas.

At the end of the work, it was said that the results of these three works [3–5] are approximately the same, since the formulas for the probability of the number of blocks in the first stage of the attack are approximately equal. However, calculations show that the numerical results of Nakamoto and Rosenfeld are significantly different. It should also be noted that the first assumption about the instantaneous distribution of the block in the network in Pinzon's work is also present.

Remarkable from a mathematical point of view, Grunspan's work [6] impresses with the mathematical rigor of its presentations and justifications. In this paper, the authors prove what Rosenfeld suggested without proof—that the process of generating "honest" / "dishonest" blocks on the network is described by negative binomial distribution. However, the authors could not, and did not even try to get rid of the same assumption about the instantaneous spread of the block on the network. By the way, in this work, using special functions, for the first time it has been proven that the probability of fork becomes exponentially with the height of its length.

For the first time, the question of what to consider the time of synchronization, sounded in the work of Zohar. However, further high-profile statements that "now we will fix it all", the authors did not go. This work proves some statements about the ratio between the rate of generation of the block and the growth rate of the main chain, but some of the statements contain errors and even serious errors that destroy the entire course of proof, while others simply accumulate a bunch of incomprehensible formulas, without any explanation. We would like to emphasize that there is no result in the work with analytical estimates of the probability of an attack.

Next, we managed to get analytical expressions for the probability of an attack, taking into account the time of synchronization of the network[1]. These expressions are much more cumbersome, but numerical values can be (and have been) derived along them. According to these numerical results, it can be assumed that the synchronization time has a significant impact on blockchain stability only with a fairly large share of the attacker (close to 40%).

And finally, the last of the works[2] in this direction considers the same statement of the problem, but not for PoW, but for PoS.

In this section, we will look at the parameters of the consensus protocol named Proof-of-Accuracy (hereinafter referred to as PoA)—Proof of Accuracy.

3 Proof-of-Accuracy Consensus Protocol Parameters

3.1 Problem Statement

For the first time, the main ideas and provisions of this protocol were described in the paper [7]. It will provide and strictly mathematically substantiated the choice of protocol parameters in which it is possible to ensure its resistance to various attacks on blockchain. First of all, this refers to the most common attack, which in cryptocurrency blockchains is called a double spend attack [3–6]. Since we consider the wider use of blockchain, the analogue of such an attack will be called an attack of a substination of the block, which accurately reflects its essence.

Note that almost all basic consensus protocols are subject to the attack of the block subsisting. However, strictly substantiated analytical results to calculate the

---

[1] Lyudmila Kovalchuk, Dmytro Kaidalov, Andrii Nastenko, Mariia Rodinko, Oleksiy Shevtsov, and Roman Oliynykov. 2020. Decreasing security threshold against double spend attack in networks with slow synchronization. Computer Communications 154 (2020), 75–81.

[2] Karpinski M, Kovalchuk L, Kochan R, Oliynykov R, Rodinko M, Wieclaw L. Blockchain Technologies: Probability of Double-Spend Attack on a Proof-of-Stake Consensus. Sensors. 2021; 21(19):6408. https://doi.org/10.3390/s21196408.

probability of this attack to date are obtained only for the most "ancient" consensus protocol—proof-of-work protocol, or abbreviated PoW proposed in the pioneering work of Nakamoto [3]. We will briefly analyze the works in which these results were obtained.

In his only work, Nakamoto was the first to score a double spend attack, but his calculations were incorrect. They were obtained in assumptions that do not quite correspond to the real model.

The first assumption that is in place in all subsequent works on this topic is the assumption that the moment of generation of the block and the moment of its appearance on the network coincide. That is, the time of distribution of the block to all nodes of the network is zero. But if this corresponded to reality, the probability of an "unintentional" fork (branch) would be zero, and in fact such fork occur about 6 times a month. However, it should be noted that if the generation time of the block is much longer than the time of distribution of the block in the network, then this assumption will not greatly distort the results.

The second assumption is even more incorrect. It is as follows: when calculating the probability of an attack, Nakamoto assumes that if the probability of some event is equal to p, then the number of trials until the moment of appearance is exactly equal to $\frac{n}{p}$. From a mathematical point of view, this means that a random variable with a negative binomial distribution was replaced by a mathematical expectation. Given the fact that the variance of such a random value, in general, is significantly different from zero, this assumption makes tangible errors in the result.

In these simplified assumptions, analytical expressions were obtained for the probability of success of a double spend attack, at which numerical values can be obtained at different parameters of the network. Experimental results showed that with an increase in the number of confirmation units, the probability of an attack becomes exponential, but there is no analytical evidence of such an exponential dependence.

In subsequent works [4–6] many authors tried to get rid of the second assumption, but to get mathematically justified results for the first time only in the work [6]. However, the authors could not bypass the first assumption about the instantaneous spread of the block on the network. The incorrectness of this assumption was first noticed in the work [5], and in it the author (without evidence and justifications) brought some general formulas that supposedly take into account the time of distribution of the block. However, due to the lack of evidence, it is not possible to check their correctness, and their appearance does not allow using them to obtain numerical results.

In addition to the attack of the the block substitution, the literature analyzes another attack—Splitting Attack [1, 8]. It is more powerful, but less common. The attack is that the attacker, having created a fork, tries to divide the network, supporting the fork as long as possible by adding new blocks to the shorter blockchain branch. A description of such an attack appeared much later [1, 8]. The purpose of this attack is to create conditions on the network under which consensus is impossible for a long time, which will lead to a loss of trust in the cryptocurrency and, accordingly, a drop in its value. At the same time, if the attacker previously took out a loan in this currency, then it will be much easier for him to give this loan. Unlike the block sub-subsete attack, when an attacker secretly builds an alternative branch of the blockchain, in an attack on the division of the network, he openly participates in the creation of the fork and its support. Under favorable conditions for him, honest miners acting in accordance with the consensus protocol will also be forced to unwittingly support the fork, continuing one branch or another. Since not only the resources of the attacker are involved in the support of the fork, but also, in part, honest miners, it is easier to succeed in it than in attacking the subseminal of the block.

For the first time, asyptotic estimates of the probability of this attack for the PoW and PoS (Proof-of-Stake) protocols were obtained in [1, 8]. Building upper (not asymphetical) scores of success probability for a splitting attack is significantly more difficult than attacking block subsemination. For the first time, the top success rate for both attacks obtained for the PoW protocol without admission of zero synchronization time was presented in [2]. It should be noted that in this work the model with discrete time is considered, the same as in [1, 8], which does not quite correspond to the real situation. Together with analytical estimates, numerous results obtained for these estimates at different parameters of the network were given. Analytical and numerical results have shown that the situation in which an attacker is synchronized significantly better than honest miners is extremely dangerous. In such a situation, the probability of an attack of block subsemination can be equal to units even in the case when the enemy's computing power is significantly less than that of honest miners. That is, the basic principle of the so-called "50% attack" is violated, stated in previous works - to guarantee the success of an enemy attack, you need to have 50% computing power. This is the danger of so-called "partial centralization" in mining, that is, situations when miners are united into large mining pools, especially if the participants of the same pool are geographically nearby or have some opportunities for instant forwarding of the block.

Based on the above, we will take into account the time of network synchronization when justifying the choice of parameters of the PoA protocol. Moreover, we will consider the most common model with continuous time: we assume that the synchronization time for honest miners and the attacker is different. We will make several assumptions in favor of the attacker, in particular, the assumption about his "centralization", which gives him the opportunity to send blocks almost instantly, and the assumption that he can in some way affect the time of delivery of blocks to honest miners.

It should be noted that the PoA protocol leaves less opportunity for the enemy to "partial centralize" than the PoW protocol. Therefore, one of the advantages of this protocol is that the enemy cannot significantly reduce its synchronization time, which reduces its ability to attack in the presence of small computing power.

The PoA protocol also has a number of other advantages, in particular, it is less excited about the oligopoly of large mining pools.

3.2 Markings, Definitions, and Assumptions

In this paragraph we will introduce the basic designations and summarize the main assumptions of our model. Many honest miners we will designate HMs (Honest Miners), and many attackers – MMs (Malisious Miners). We will enter the following values:

T_H—a random value equal to the time HMs spend creating a single block;

T'_H—a random value equal to the time that HMs spend on creating one block and distributing it to all nodes of the network;

T_M—a random value equal to the time MMs spend creating a single block;

T'_M—a random value equal to the time that HMs spend on creating one block and distributing it across all nodes of the network.

As shown in [6], random variables and exponential distributions:

$$F_{T_H}(t) = P(T_H < t) = 1 - e^{\alpha_H t},$$
$$F_{T_M}(t) = P(T_M < t) = 1 - e^{\alpha_M t}, \tag{1}$$

for some $\alpha_H > 0$, $\alpha_M > 0$. The physical meaning of these parameters is that it is the average generation intensity of the block for honest miners and for the attacker, respectively.

Let's denote $\alpha = \alpha_H + \alpha_M$ the overall intensity of unit generation in the network.

The model of network operation that we consider in this work will be based on two natural assumptions.

Assumption 1 We will assume that the delivery time of the block created by honest miners to all other miners (at least honestly) does not exceed D_H. Moreover, we will make assumptions in favor of the attacker and assume that he may influence the delivery of the block to other miners, for example, to delay it at any time that does not exceed D_H. Similarly, D_M—it is either the time of delivery of a block created by an unfair miner, other dishonest miners (if we consider it the same), or the maximum time of distribution of the block among dishonest miners.

Assumption 2 We make another assumption in favor of the attacker and believe that $0 \leq D_M \leq D_H$.

In our designations, equality is performed:

$$T'_H = D_H + T_H, T'_M = D_M + T_M. \tag{2}$$

Let's denote p_H likelihood that HMs will create the next block before it is created MMs, i $p_M = 1 - p_H$—probability of the opposite event. According to [6],

$$p_H = \frac{\alpha_H}{\alpha_H + \alpha_M}, p_M = \frac{\alpha_M}{\alpha_H + \alpha_M}. \tag{3}$$

The values p_H (p_M) will be called part of the common hashrate, which belongs to HMs (MMs). In the works [3–6] the ratio of values (3) plays a crucial role in determining how vulnerable the protocol is to an attack of block subseminal. However, in our case, taking into account the nonzero synchronization time, the degree of vulnerability to the attack will be determined by other values.

Definition 1 Use a 2-D definition With network settings specified α and $\Delta D = D_H - D_M$ will be called security threshold for the consensus protocol PoA the smallest of these values p_0, that provided $p_M \geq p_0$ the probability of success of the block subsemination attack is 1, regardless of the number of confirmation blocks.

Note that in a model with zero synchronization time $p_0 = \frac{1}{2}$, where did the name "50% attack" come from.

Let's denote p'_H the likelihood that HMs will create and distribute its unit across all nodes of the network (at least to all honest nodes) before MMs do; let's denote p'_M probability of an alternate event, $p'_M = 1 - p'_H$.

Then

$$p'_H = P(T'_H < T'_M), \, p'_M = P(T'_M < T'_H), \tag{4}$$

and $p'_H + p'_M = 1$.

Values (4) are more important when analyzing attack resistance than (3) because they take into account the synchronization time and more realistically reflect the state of the network. The following shows that the probability of a DSA attack depends on the magnitudes (4). In particular, if the value of the D_H large enough compared to D_M, then, for a successful attack, the attacker, in general, does not have to have an advantage in the hashrate.

3.3 Defining the Security Boundary for the PoA Protocol

In this paragraph, we will give accurate analytical expressions that bind, on the one hand, network parameters such as synchronization time, the intensity of the output of blocks and the share of the attacker's hashrate, and on the other hand, the security border for this protocol. Using these expressions, in the next paragraph we justify the choice of network parameters for the PoA consensus protocol.

We will use the basic result, which shows that to guarantee the success of an enemy attack, it is not necessary to own at least half of all the computing power of the network. That is, in a more adequate model, which is discussed in this article and takes into account the time of synchronization of the network, the so-called "50% attack" [9] can turn into a "45% attack" or even a "30% attack" if the attacker is very well synchronized compared to honest miners.

As shown earlier for the PoW consensus protocol [2, 9], in a model that takes into account the time of synchronization, the determining role will be played by values

Table 1 Security border value at different intensity values α generation of blocks (in blocks per second) and synchronization time D_H (in seconds)

α	D_H				
	$D_H = 2$	$D_H = 5$	$D_H = 10$	$D_H = 20$	$D_H = 60$
$\frac{1}{600}$	0.49916736	0.49792102	0.49585079	0.49173685	0.47564318
$\frac{1}{60}$	0.49173685	0.47961146	0.46014582	0.42408032	0.31492306
$\frac{1}{6}$	0.42408032	0.33756934	0.24626202	0.15678438	0.06282608

p'_H and p'_M, defined in (4). It is these quantities that determine the vulnerability to the block subsemination attack, and they also allow you to set a security limit for the PoW consensus protocol.

For the PoA protocol, these results remain fair and their evidence will be similar. Since these results will be used further, we will present them here without evidence.

Theorem 1 In our definitions:

$$p'_M = 1 - e^{-\alpha_M \Delta D} \cdot \frac{\alpha_H}{\alpha_M + \alpha_H} = 1 - e^{-\alpha_M \Delta D} \cdot p_H;$$

$$p'_H = e^{-\alpha_M \Delta D} \cdot \frac{\alpha_H}{\alpha_M + \alpha_H} = e^{-\alpha_M \Delta D} \cdot p_H.$$

Theorem 2 Let α—the intensity of unit generation, $\Delta D = D_H - D_M$. Then in our PoA protocol model, the security limit can be found as a solution to the equation

$$2(1 - p_0) = e^{\alpha p_0 \Delta D}. \tag{5}$$

In the further presentation, we will make another assumption in favor of the attacker. We will assume that it is well synchronized, that is $D_M = 0$. Note that the analytical expressions obtained in this assumption for the probability of an attack can be used when, but in this case they will be the upper estimates of the probability of an attack, and not its exact values.

Table 1 presents numerical results of security border values for different network parameters.

3.4 Selecting the Parameters of the PoA Consensus Protocol and Justifying the Choice

Additionally, we will introduce the following designations:
   n—the number of network nodes (or, the same, the number of IP addresses);
   m—number of "useful" network nodes, i.e. those that contain parts of the secret;
   k—number of secret pieces to collect to restore it;
   d—time required to visit the same IP address.

First, determine how the intensity of unit generation depends on the n, m, k and d.

Theorem 3 In our designations, the intensity of block generation is determined by the following equality:

$$\alpha = \frac{n}{mkd}.$$

Proof Define a random variable ξ on the probabilist space, which is a multitude of IP addresses, as follows:

$$\xi = \begin{cases} 1, & \text{if IP-adress contains information;} \\ 0, & \text{else.} \end{cases} \tag{6}$$

By definition, a random variable (6) has a Bernoulli distribution with a parameter $p = \frac{m}{n}$.

Define a random variable ζ as the number of attempts in the Bernoulli scheme to succeed, that is, as the number of IP addresses that the miner must visit before it collects the required number of pieces of secret. Then a random variable $\zeta - k$ equals the number of failures before success, so it has a negative binomial distribution (10) with parameters (k, p) and mathematical expectations $E(\zeta - k) = \frac{k(1-p)}{p}$. Then

$$E(\zeta) = \frac{k(1-p)}{p} + k = \frac{k}{p} = \frac{km}{n}.$$

Accordingly, the average time t (in seconds), required to create a single block equal to $t = \frac{kmd}{n}$, and the intensity of unit generation $\alpha = \frac{1}{t} = \frac{n}{kmd}$. Theorem proved.

Example 1 In a network consisting of $n = 1000$ nodes, at $m = 950$, $k = 500$ and $d = 0.3$ s, we get.

$$\alpha = \frac{1000}{500 \cdot 950 \cdot 0.3} \approx 0.007 \text{ blocks in 1 s,}$$

that is, it takes approximately 142 s, or 2 min 22 s, to generate one block.

Now we will define the general requirements for the work of the consensus protocol, which will impose appropriate restrictions on its parameters. Then we will outline the area of their significance more accurately. After that, we will analyze what will be the security limit of the protocol, that is, what share of the resource of the attacker is critical for the network.

To begin with, we want to reduce to a certain level "extra costs"—the so-called orphan blocks, which arise when two or more blocks are created at about the same time. In this case, branching occurs—fork, which leads to the temporary separation of the blockchain, since one part of miners will see one block earlier and the other part—the other. Therefore, different miners will build blocks on different branches of the blockchain. For example, bitcoin's largest fork in its entire existence is six blocks. After the network is synchronized again, one of the branches is lost, which leads to the loss of work done by miners.

Therefore, network parameters should guarantee a "small" probability of branching. To understand what probability is acceptable, you can contact the BTC network, which has been functioning successfully for more than 10 years.

According to statistical studies, "extra" blocks occur in this network about 6 times a month. That is, the probability of such a block is

$$\frac{6}{6 \cdot 24 \cdot 30} \approx 1.4 \times 10^{-3}.$$

We will take this value as a base value, but round it in a smaller direction, that is, we will demand that the probability of fork does not exceed (for BTC it would correspond to five "extra" blocks per month).

We will make an admission in favor of the attacker and assume that.
$\Delta_M = 0$, that means $\Delta = \Delta_H$.

Theorem 4 Let arbitrary $\varepsilon > 0$. In order for the probability of P_f fork was no more than ε, necessary and enough for the inequality to be carried out:

$$\alpha \cdot \Delta \leq x_f,$$

where x_f is the solution to the equation:

$$1 - e^{-x_f}\left(1 + x_f\right) = \varepsilon.$$

Proof Denote a random value that equals the time before the next block is created.

It should be noted that if any B_1 created, the time before the next block is created B_2 does not depend on how long it took to create the previous block.

Therefore, a random variable T has an exponential distribution with a parameter α:

$$P(T \leq t) = 1 - e^{-\alpha t}.$$

An unintentional fork can occur then and only if during a time that does not exceed the time Δ network synchronization, two or more miners have created blocks.

Since the generation time of blocks has an exponential distribution, the random value η_t, equal to the number of blocks created during the t, will have a Poisson distribution with the parameter $\alpha \cdot t$:

$$P(\eta_t = k) = e^{-\alpha t}\frac{(\alpha t)^k}{k!}, k = 0, 1, 2, \ldots$$

Therefore, the probability of fork is equal to

$$P_f = P(\eta_\Delta \geq 2) = 1 - P(\eta_\Delta = 0) - P(\eta_\Delta = 1)$$
$$= 1 - e^{-\alpha\Delta} - \alpha\Delta e^{-\alpha\Delta} = 1 - e^{-\alpha\Delta}(1 + \alpha\Delta).$$

To complete the evidence, we need to show that the function $P_f(x) = 1 - e^{-x}(1 + x)$ is growing. Indeed, its derivative

$$P'_f(x) = e^{-x}(1 + x) - e^{-x} = xe^{-x}$$

integral when $x \geq 0$ and positive at $x > 0$.

Therefore, if for some x_f equality is performed

$$P_f(x_f) = \varepsilon,$$

then when $x \in (0, x_f)$ inequality will be met $P_f(x_f) < \varepsilon$, and the theorem proved.

Let's calculate x_f on condition $P_f(x_f) = 10^{-3}$ and we get inequality

$$\alpha\Delta \leq 0.05, \tag{7}$$

which we will use to determine the parameters n, m, k network.

From the theorem 3 and inequalities (7) we get

$$\frac{m}{nkd} \leq 0.05. \tag{8}$$

Note that the number of network users is usually $n \geq 10^4$. We will take this value as the base value and show how other parameters are defined.

Since we, generally speaking, cannot influence the parameters of d and Δ, and the parameter n should be able to grow without any restrictions, the restrictions can only be imposed on the parameters of the m and k. It is also worth noting that the parameter should not be significantly less than n, that the probability of success in the relevant Bernoulli scheme is close to 1. Experimental results showed that it is best to choose $m \approx 0.95n$. From here, taking into account (8), we get:

$$\frac{0.95 \cdot n \cdot \Delta}{k \cdot n \cdot d} \leq 0.05, \text{ or } k \geq \frac{0.95 \cdot n \cdot \Delta}{0.05 \cdot n \cdot d} = \frac{19 \cdot \Delta}{d}$$

However, by definition, $k \leq m$, therefore, we have double inequality:

$$\frac{19\Delta}{d} < k < 0.95n. \tag{9}$$

Let's calculate the lower bound of the values that the parameter can take k at different, most common values d and Δ. The values of the lower limit are given in Table 2.

Table 2 Lower limit of parameter value k

d	Δ				
	5c	10c	20c	30c	60c
0.3c	316.7	633	1267	1900	3800
0.5c	190	380	760	1140	2280
1c	95	190	380	570	1140
2c	47.5	95	190	285	570
3c	32	63	127	190	380
5c	19	38	76	114	228

Analyze how the safety limit changes when changing the parameter k. According to the definition, the security limit p_0 is the smallest proportion of attackers who will allow him to attack the blockchain with the probability of success 1. As proved in the theorem 2, the security limit is defined as the solution of the.

Show that the safety limit decreases as the work grows $\alpha \cdot \Delta$. To do this, prove the following theorem.

Theorem 5 Let the implicit function be given by the formula

$$F(x, y) = 2(1 - x) - e^{xy} = 0 \tag{10}$$

Then function exists $x = x(y)$, $y \geq 0$, which takes the value in the interval $(0, 1)$, such that equality is performed for it:

$$F(x(y), y) = 0.$$

In this case, the function of $x = x(y)$ is descending.

Proof To prove, it is enough to show that the function $y = y(x)$ is continuous and descending. Then it will be a bicep, so it will have an inverse function, which will also be continuous and descending.

From (10) get:

$$y(x) = \frac{\ln 2(1 - x)}{x}.$$

We will show that the derivative of this function is negative, that is, that

$$y'(x) = \frac{\frac{x}{x-1} - \ln 2(1 - x)}{x^2} < 0. \tag{11}$$

To do this, it is enough to prove that the thinker in (11) less than zero: $\frac{x}{x-1} - \ln 2(1 - x) < 0$,

$$\text{or that } x - (x - 1) \ln 2(1 - x) > 0, \tag{12}$$

because $x - 1 < 0$.
Let's denote

$$f(x) = x - (x - 1) \ln 2(1 - x).$$

Then

$$f'(x) = 1 - \ln 2(1 - x) - 1 = -\ln 2(1 - x), \tag{13}$$

whence $f'(x) = 0 \Leftrightarrow x = \frac{1}{2}$, and with (13) it can be seen that $x = \frac{1}{2}$ is a minimum point.

Let's calculate the value of the function $f(x)$ at the point of minimum:

$$f = \frac{1}{2} + \frac{1}{2} \ln\left(2 \cdot \frac{1}{2}\right) = \frac{1}{2} > 0.$$

Besides

$$\lim_{x \to 0+0} f(x) = f(0) = \ln 2 > \frac{1}{2};$$

$$\lim_{x \to 1+0} f(x) = 1 > \frac{1}{2},$$

that is, at the point of $x = \frac{1}{2}$ function $f(x)$ takes the smallest value, and it is positive. Therefore $f(x) \geq \frac{1}{2} > 0$, and (12) is performed, and therefore (11) is performed, i.e. $y'(x) < 0$. Consequently, the function of $y(x)$ is descending. It is also continuous as a composition of continuous functions. Therefore, the reverse function $x = x(y)$ exists and is descending. Theorem proved.

Corollary 1. Function $p_0 = p_0(\alpha \cdot \Delta)$ is descending, i.e. security boundary p_0 is in ascending product $\alpha \cdot \Delta$.

We will analyze what security limit corresponds to the values we have chosen $\alpha \cdot \Delta = 0.05$. To do this, we need to calculate the value that is the solution to the.

The solution to this equation is approximately equal to $p_0 \approx 0.488$, while the maximum possible value of the security boundary (ideal for honest miners) is $p_0 = 0.5$. That is, under our restrictions $\alpha\Delta < 0.05$ the security limit value will be less than ideal by just 1.2%, which can be considered a slight security downsizing. In other words, the network will be stable until the share of the attacker does not exceed 48.8%.

We would like to emphasize that the theoretically achievable security limit - 50%—is possible only if there is zero synchronization time for honest miners, which is unattainable in practice.

Table 3 Intensity Values α at $n = 10^4$, $m = 9500$, $k = 5000$ and variable d

d	$5 \cdot 10^{-3}$	10^{-2}	$2 \cdot 10^{-2}$	$3 \cdot 10^{-2}$	$5 \cdot 10^{-2}$	0.1	0.2
α	$3.8 \cdot 10^{-2}$	$19 \cdot 10^{-3}$	$9.5 \cdot 10^{-3}$	$6.3 \cdot 10^{-3}$	$3.8 \cdot 10^{-3}$	$1.9 \cdot 10^{-3}$	$0.95 \cdot 10^{-3}$

Inequality (9) gives us a fairly wide range for the parameter k. For practical reasons that have been tested by experiments, it is convenient to choose the value of k from the ratio

$$k \approx \frac{1}{2} n.$$

This value satisfies inequalities (9) provided that $n \geq 10000$ and $\Delta < 263d$, that is almost always done in practice.

We calculate the intensity of the output of blocks at some values of parameters that satisfy our limitations. Let $n = 10^4$, $m = 0.95n = 9500$, $k = 0.5n = 5000$, and the option d Let it accept variable values. The results of calculations are shown in Table 3.

All the parameter values in this table, except the last, are bolder than the corresponding parameter value for BTC (which is equal to $1.6 \cdot 10^{-3}$). Note that the parameter $k = 0.5n = 5000$ provides a "small" probability of fork only if $\Delta \leq 263d$. For example, if $d = 0.1$, then such a selection of the parameter k is only possible when $\Delta \leq 26.3$. If this inequality does not work, then you need to list the parameter k, which, accordingly, will reduce the intensity of the exit of the blocks. But this is an objective problem that is inherent in any blockchain, regardless of the consensus protocol.

One of the advantages of our protocol is that by changing the value of the k, we can adapt the intensity of the output of blocks depending on the average network synchronization time. For example, if it is guaranteed "small", then accordingly it is possible to significantly increase the intensity of unit generation, and vice versa. Similar adaptation is possible in relation to the parameter d.

4 Summary

1. This section discusses the new blockchain consensus protocol. It is adaptive, that is, it allows you to select its parameters depending on the existing network parameters that we cannot influence. The choice of protocol parameters is based on the following considerations: the probability of fork is no greater than for the Bitcoin network; the intensity of blocks output may be greater than in the Bitcoin network; network stability to be almost the same as for the "ideal" network.

2. If this consensus protocol is used in some local network, for which the synchronization time is guaranteed to be small, then the intensity of unit generation can be increased many times.

3. For an "non-ideal" network, it is impossible to uncontrollably increase the intensity of unit generation. But this limitation is not a feature of this particular protocol, but is a common flaw in any consensus protocol for blockchain.
4. To be able to significantly increase the intensity of unit generation, you need to move from blockchain to another structure. Publications in recent years suggest using structures such as DAG instead of blockchain - directed acyclic graph (directed acyclic graph). But in almost all known works in this direction, the guarantee of consensus achievement and justification of the stability of the proposed structure to various attacks, as well as the formulation and justification of analogues of other necessary properties inherent in blockchain, are either not strictly mathematically justified, or contain errors in evidence, or have only asymphotic assessments.

From our point of view, the most promising structure to replace blockchain is DAG. But before it becomes possible to use it, it is necessary to solve the non-trivial mathematical problem of transferring the corresponding mathematics results from blockchain to a much more complex structure. First of all, it concerns the results of sustainability assessment.

References

1. Kiayias A, Russell A, David B, Oliynykov R, Ouroboros: a provably secure proof-of-stake blockchain protocol. https://eprint.iacr.org/2016/889.pdf
2. Kovalchuk L, Kaidalov D, Shevtsov O, Nastenko A, Rodinko M, Oliynykov R (2017) Analysis of spliting attacks on bitcoin and GHOST consensus protocols. In: Proceedings of the 9th IEEE international conference on intelligent data acquisition and advanced computing systems. Romania, Bucharest, 2017
3. Nakamoto S (2009) Bitcoin: a peer-to-peer electronic cash system/satoshi nakamoto, 9c
4. Rosenfeld M (2014) Analysis of hashrate-based double-spending/Meni Rosenfeld., 13p
5. Pinzon C (2016) Double-Spend attack models with time advantage for bitcoin. In: Pinzon C, Rocha C (eds) Electronic Notes in Theoretical Computer Science, vol 329, pp 79–103; Pinson P, Inclusive block chain protocols. In: Pinson C, Lewenberg Y, Sompolinsky Y (eds.), 20c
6. Grunspan C, Double spend races. In: Grunspan C, Perez-Marco R (eds), 35p
7. Kudin AM (2017) Blockchain and cryptocurrencies based on proof-of-accuracy//mathematical and computer modelling. Ser Tech Sci (15):104–108. (In Russian)
8. The bitcoin backbone protocol: analysis and applications/Juan Garay//Aggelos Kiayias//Nikos Leonardos (2015). 310p
9. Kudin A, Kovalchuk L, Kovalenko B (2019) Theoretical foundations and application of blockchain: implementation of new protocols of consensus and crowdsourcing computing. Mathematical and computer modelling. Ser Techn Sci 19:62–68. https://doi.org/10.32626/2308-5916. 2019-19. (In Ukrainian)

Cryptographic Transformations in a Decentralized Blockchain Environment

Alexandr Kuznetsov⬥, Dmytro Ivanenko⬥, Nikolay Poluyanenko⬥, and Tetiana Kuznetsova⬥

Abstract Distributed decentralized systems built using Blockchain technology are becoming increasingly popular and widespread. This is due to their reliability and security. But their transparency and openness, the lack of centralized administrative levers that can interfere and impose their decision, are the most important and attractive. Blockchain is able to provide completely anonymous, decentralized and free (uncontrolled) storage of any digital assets: from electronic money, such as cryptocurrencies, to secure property inventories and election registers. Therefore, further intensive implementation of this technology in various applications should be expected, as Blockchain implements a reliable and secure way to store distributed data in the decentralized environment. This is achieved through the use of various cryptographic transformations. This section analyzes the promising methods and mechanisms of cryptographic transformation that can be used in a decentralized Blockchain environment to provide various security services, including confidentiality, integrity, accessibility, indisputability, both users and elements/nodes of a decentralized system, etc. The section considers and investigates algorithms and protocols of homomorphic encryption, ring signatures, protocols with zero disclosure, principles of construction of anonymous secure networks, etc.

Keywords Blockchain technology · Monitoring systems · Contact tracking apps · Decentralized systems

1 Introduction

Blockchain—is just an approach to data storage, and the security of technologies built on it, in most cases, depends only on the software implementation of the latter. Therefore, it is wrong to believe that blockchain has no vulnerabilities.

A. Kuznetsov (✉) · D. Ivanenko · N. Poluyanenko · T. Kuznetsova
V. N. Karazin Kharkiv National University, Svobody sq., 4, Kharkiv 61022, Ukraine
e-mail: kuznetsov@karazin.ua

A. Kuznetsov · N. Poluyanenko
JSC "Institute of Information Technologies", Bakulin St., 12, Kharkiv 61166, Ukraine

© The Author(s), under exclusive license to Springer Nature Switzerland AG 2022
R. Oliynykov et al. (eds.), *Information Security Technologies in the Decentralized Distributed Networks*, Lecture Notes on Data Engineering and Communications Technologies 115, https://doi.org/10.1007/978-3-030-95161-0_4

Blockchain implies decentralization, and decentralization in turn, many nodes and clients in the network who constantly interact with each other. Most often, anyone can join such a network (having previously run the appropriate software on their site). Technologies built on blockchain have high security requirements. Such technologies should have as few software vulnerabilities as possible that can be exploited by an attacker, which could potentially be any new network member. It is worth noting that the problem of high network requirements is partially solved by the formation of a trusted circle of nodes that will support blockchain, thereby preventing attackers from using vulnerabilities even if they are present.

The vast majority of blockchain-built technologies are cryptocurrencies. One of the most famous vulnerabilities of cryptocurrencies is a 51% attack. Its essence lies in the fact that attackers gain control of the network by owning computing power, which is 51% of the power of the entire network. If this vulnerability is successfully exploited, attackers can actually dictate data to be added to blockchain in the future. With this vulnerability, a so-called double-spending attack can be carried out. It is worth noting that today quite a lot of projects have closed at the very beginning precisely because the attackers quite easily used this vulnerability (at the start of the project there are few nodes in the network and the total power is not too large). In blockchain solutions, the user is responsible for the availability of his data, and the attribute of access to the system is usually a key pair. Another vulnerability is theft or loss of a key pair. In fact, the fact of owning assets in blockchain is confirmed by the user's personal key. When stealing a key, the attacker will be able to freely use the user's assets. Another drawback is that the user, having on his hands all the evidence of loss or theft of keys, will not be able to recover them (this happens when the password of some service is lost or compromised). Therefore, users of decentralized systems built on blockchain should understand the responsibility for storing their data and take into account not only the security features of the technology, but also the vulnerability of their local machines or data storages.

Disadvantages associated with the loss of personal keys are solved by using third-party services that act as an intermediary between the user and the end system. Thus, they take responsibility for the preservation of personal data of users. On the one hand, it makes life easier for end users, and on the other hand, such services have repeatedly been attacked with the following theft of key data (which, accordingly, is a great harm to customers). Therefore, to make a transaction in the system, the best solution would be to use multi-signing or smart contracts.

In addition to vulnerabilities related to the aspect of decentralization and key pairs, cryptocurrencies have software vulnerabilities or features that, for example, allow a third-party observer to analyze the user's actions and find out confidential information about it. For example, in Bitcoin, all transactions and addresses are open to all network users and third-party observers. With proper effort, you can match the addresses from which transactions and events were made in the real world (for example, buying a car), and try to find out the identity of the client of the system or information that will eventually lead to the client. In addition to simple analysis, other algorithms that allow you to deonymize the user apply to Bitcoin.

However, not all cryptocurrencies work since Bitcoin. After all, cryptocurrency is just a software implementation of the protocol that works on blockchain. In some protocols, attention is paid to the openness and transparency of operations, while in others, attention is paid to the security of users of the system (which often complicates their lives for the benefit of preserving their personal data). An example of such a cryptocurrency is the Monero project. Monero decentralized system is a system in which the emphasis is primarily on ensuring the anonymity of the user. In the Monero system, information such as: the amount of transactions, the sender's identification data and the recipient's identification data are hidden in the chain of blocks, so the storage actions and costs of coins cannot be tracked. Monero also includes the wallet module. The wallet stores keys and performs complex cryptographic operations on asset management. To access the wallet, it is not necessary to control the private keys, it is enough to save the seed phrase that was used in the generation of the wallet. Seed is a secret number that the wallet uses to generate keys and access coins, although for convenience and human perception this number turns into a series of 12–25 words. Seed phrase (or basic secret) should be available only to the wallet user, since knowledge of the wallet's seed phrase determines the possession of the assets that the wallet stores.

Monero provides enhanced functionality, anonymity and confidentiality through the use of several unique cryptographic technologies that protect users and their activities from public access, such as:—RingCT (ring confidential transactions) — hides the amount of transactions;—Ring signatures—allows you to protect the user from disclosure;—Stealth address—ensures that the recipient's address will not be recorded in the block chain;—Kovri is an implementation of I2P written in C++, which allows you to break the connection between transactions and physical location, hiding network signs of Monero node activity.

Blockchain also uses the following encryption systems, encryption methods, and digital signatures:

- Homomorphic encryption;
- Milty captions;
- Ring and group labels;
- P2P anonymous network using encryption;
- Symmetric encryption (e.g. 256-bit AES algorithm in CBC mode with PKCS#5);
- Public-key crypto transformation (e.g. 2048 El Gamal bit scheme; ECDSA together with the elliptical curve secp256k1; 2048 Diffie-Hellman bit algorithm; 1024 bit DSA algorithm, etc.);
- Message authentication codes (e.g. HMAC 256-bit algorithm) to enhance the crypto-resistantness of other cryptoalhorites;
- Cryptographic hashing (e.g. SHA 256) and more.

This section is devoted to the analysis of modern methods and mechanisms for achieving security services—confidentiality, integrity, accessibility, irrelevance, both users of the decentralized system and elements/nodes of the decentralized system.

2 Homomorphic Encryption

Homomorphic encryption is a form of encryption that allows you to perform certain mathematical actions with encrypted text and get an encrypted result that corresponds to the result of operations performed with open text. For example, one person could make up two encrypted numbers without knowing the decrypted numbers, and then the other person could decrypt the encrypted amount—get the decrypted amount without having the numbers decrypted. Homomorphic encryption allows you to provide various services that provide open user data for each service.

There are cryptosystems partially homomorphic and completely homomorphic. Partially homomorphic cryptosystem allows to produce only one of the operations—either assembly or multiplication. A fully homomorphic cryptosystem supports both operations, that is, the properties of homomorphism in relation to multiplication and assembly are performed in the system.

2.1 General Model of Homomorphic Encryption

Homomorphic encryption is a form of encryption that allows for a certain algebraic operation on open text by performing an algebraic operation on encrypted text.

Let k—encryption key, t—open text (message), $E(k, t)$—performs encryption function.

The E function is called homomorphic for operation "$*$" (add or multiply) over open text (messages) t_1 and t_2, if there is an effective algorithm M (requires a polynomial number of resources and works for a polynomial time), which, having received the input of any pair of encrypted texts of the form $E(k, t_1)$ and $E(k, t_2)$, produces encrypted text (cipher text, cryptogram) $c = M(E(k, t_1), E(k, t_2))$ such that when decrypting c, open text will be obtained $t_1 * t_2$ [3].

In practice, the following separate case of homomorphic encryption is more often considered.

Let encryption for this function E above open texts t_1 and t_2 over encrypted text from encrypted text $c = (E(k, t_1) * E(k, t_2)$ decryption extracts open text $t_1 * t_2$. It is necessary that with the specified c, $E(k, t_1)$, $E(k, t_2)$, but with an unknown key k, it would be impossible to effectively verify that encrypted text c is derived from $E(k, t_1)$ and $E(k, t_2)$.

Any standard encryption system can be described using three operations: keyGen operation, encryption operation (encypt), and decrypt operation [4].

To describe the homomorphic encryption system, in addition to the three operations listed above, it is necessary to describe the calculation operation (eval). The use of homomorphic encryption involves the use of a sequence of four operations: key generation, encryption, calculation, decryption:

- key generation is the generation of a public key by the client (for decryption of encrypted public text) pk and a secret key sk (for public text encryption);

- encryption is plain text encryption PT using a secret key sk– calculation of Encrypted Text (cipher text) $CT = Esk(PT)$;
- sending encrypted text by the client an open key CT and an open key pk to the server;
- calculation is the server's receipt of a function F and use F and pk to perform calculations on encrypted text CT; sending the result to the client by the server;
- decryption is the decryption by the client of the value received from the server using the sk.

Let E—encryption function; D—decryption function; t_1 and t_2—open texts; symbols « $\otimes$ » and « $\oplus$ » denote multiplication and addition operations on encrypted texts, corresponding to multiplication and assembly operations over open texts.

The encryption system is homomorphic in relation to the multiplication operation (has multiple properties of homomorphism) if $D(E(t_1)) \otimes D(E(t_2)) = t_1 * t_2$.

The encryption system is homomorphic in relation to the addition operation (has additive homomorphic properties) if $D(E(t_1)) \oplus E(t_2)) = t_1 + t_2$.

The encryption system is homomorphic in relation to multiplication and addition operations, that is, completely homomorphic (has both multiple and additive homomorphic properties), if $D(E(m_1) \otimes E(m_2)) = m_1 * m_2$,

If a cryptosystem with such properties can encrypt two bits, then, since addition and multiplication operations form a full Turing-complete basis over bits, it becomes possible to calculate any Boolean function, and therefore any other calculated function.

2.2 Areas of Application

2.2.1 Cloud Computing

Homomorphic encryption opens up new opportunities for maintaining the integrity, availability and confidentiality of data when processing them in cloud systems. In cloud computing, where performance is the main priority, different algorithms should be used, each of which best copes with the task. For example, for operations of multiplication of encrypted data, it is advisable to use the RSA algorithm or the El Gamal algorithm, and for assembly—the Peye algorithm. When applying a fully homomorphic encryption system, it is necessary to limit the number of operations that can be performed on the data, since some error accumulates as a result of the calculations produced r. If the error value r will exceed the secret parameter value p, there may be a situation in which it will not be possible to correctly decrypt the data.

2.2.2 Electron Voting

Electronic voting is another promising area of application of homomorphic encryption. The system will be able to encrypt the votes of voters and make calculations on encrypted data, while maintaining the anonymity of voters. For example, in the Benalo electronic voting scheme, the voting process includes the following stages of [1]:

- each participant of the scheme divides his vote (secret) into components (partial secrets) according to the appropriate scheme of secret division with the property of homomorphism in assembly and sends partial secrets to elected representatives;
- representatives make up the received votes; scheme for adding, therefore, the amounts of votes are partial secrets of the relevant election result;
- the main trustee calculates the final result of the vote using a set of partial amounts of votes, which was transferred to him by elected representatives.

Let's consider an example of how this approach can be implemented.

Let there be a set of n candidates from which the ballot (list) is formed. The initiator of the voting has a cryptosystem, homomorphic about the addition operation, distributes the public key of the homomorphic encryption system among the participants of the secret ballot pk and newsletter as a vector $(p_1, p_2, \ldots, p_n)$, where p_i—last name i candidate. Each of the voters is a vector of advantages $(v_1, v_2, \ldots, v_n)$ where $b \in \{0, 1\}$. Then use the public key to pk each of the voters elementally encrypts the vector and sends the initiator of the vote. To sum up the results of voting, he performs calculations on the corresponding elements of the obtained vectors of advantages and produces decryption using a secret key sk. Since the cryptosystem is homomorphic about the addition operation, the indexes of the largest elements of the resulting vector will be the indexes of the winning candidates.

2.2.3 Secure Information Search

Homomorphic encryption can enable users to extract information from search engines while maintaining confidentiality: services will be able to receive and process requests, as well as issue processing results without analyzing or recording their actual content. For example, you can present a method for deleting records from a database by their indexes as follows.

Let $v_1, v_2, \ldots, v_n$—index of records to be extracted; $v_i \in \{0, 1\}, c_1, c_2, \ldots, c_{2n}$—all indexed records from database.

Then, in order to select the desired record, you need to calculate the following function F:

$$F(v_1, v_2, \ldots, v_n, c_1, c_2, \ldots, c_{2n})$$
$$= C_1 * ((v_1 \oplus 1) \otimes (v_2 \oplus 1) \otimes \ldots \otimes (v_n \oplus 1))$$
$$+ C_2 * ((v_1 \oplus 1) \otimes (v_2 \oplus 1) \otimes \ldots \otimes (v_{n-1} \oplus 1) \otimes v_n)$$

$$+ C_3 * ((v_1 \oplus 1) \otimes (v_2 \oplus 1) \otimes \ldots . \otimes v_{n-1} \oplus (v_n \otimes 1)$$
$$+ \ldots +$$
$$+ C_{2n} (C_2 * (v_1 \otimes v_2 \otimes \ldots . \otimes v_n)$$

If all C_i encrypted using a homomorphic cryptosystem, F can be homomorphic over encrypted texts. To do this, it is enough for the client to encrypt the index bitwise $v_1, v_2, \ldots, v_n$ record and send it to the server. Result of homomorphic calculation of a function F the encrypted value of the client's c_i record will be searched over the encrypted texts. Obviously, the cryptosystem should have both multiplier and additive homomorphic properties.

2.2.4 Protection of Wireless Decentralized Communication Networks

Wireless decentralized self-organized networks (MANET) are networks consisting of mobile devices. Each such device can independently move in any direction and, as a result, often disconnect and connect to neighboring devices. One of the main problems in building MANET is to ensure the security of the transmitted data. To solve this problem, homomorphic encryption that is embedded in routing protocols for increased security may be used. In this case, operations on encrypted texts can be safely performed by intermediate nodes. In particular, to find the optimal path between the two nodes, it is necessary to carry out linear operations on encrypted data without decryption. The presence of homomorphic encryption does not allow the attacker to find a connection between messages that are included in the node and messages that leave the node. Therefore, it is not possible to track the message transfer path using traffic analysis [2].

2.2.5 Outsourcing Services for Smart Cards

Currently, there is a tendency to develop universal maps with its own operating system, which can perform various functions and interact with several service providers. It is suggested that some applications may run outside the map on homomorphic encrypted data. Especially resource-efficient applications, for example, applications of service providers, as well as biometric checks (voice recognition, fingerprints or handwriting), which, as a rule, require a significant amount of data and a large number of relatively simple operations, can use external storage devices and external processors, more powerful than the processor built into the card.

2.2.6 Feedback System

Homomorphic encryption can be used, for example, in so-called secure feedback homomorphic systems. Secure feedback system) when it is necessary to preserve the anonymity of the user and hide the intermediate results of calculations. The systems help to carry out an anonymous collection of feedback (comments) of students or teachers about their work. The resulting reviews are encrypted and stored for further calculations. Feedback systems can be used to raise awareness of the state of affairs and to improve performance. It was established that reliable feedback of any system or process can be provided only in cases of maintaining the anonymity of the user, the invariability of the collected data, ensuring the security of internal operations for data analysis.

2.2.7 Obfuscation to Protect Software Products

The main purpose of obfuscation is to make it difficult to understand the functioning of the program. Since all traditional computer architectures use binary strings, using fully homomorphic encryption over bits, you can calculate any function. So, you can homomorphically encrypt the entire program so that it retains its functionality.

2.3 Partial Homomorphic Encryption

Partially homomorphic cryptosystems are such cryptosystems that are homomorphic in relation to only one operation—either adding operations or multiplication operations. In the following examples, the expression $E(t)$ indicates the use of encryption E to encrypt open text (message) t.

2.3.1 RSA Cryptosystem

The RSA cryptosystem is a public key cryptographic scheme, homomorphic by multiplication. Let n—RSA module, t—open text, k—public key (to encrypt open text). Encryption looks like $E(t) = t^k \bmod n$. Let's show homomorphism by multiplication:

$$E(t_1) * E(t_2) = t_1^k * t_2^k \bmod n = (t_1 t_2)^k \bmod n = E(t_1 t_2).$$

2.3.2 El Gamal CryptoSystem

The El Gamal cryptosystem is an alternative to the RSA cryptosystem and, at an equal value of the key, provides the same crypto-resistance. El Gamal improved Diffie's Hellman algorithm and obtained algorithms for encryption and authentication. The cryptosystem is a probabilist cryptosystem. Its homomorphic encryption function for the operation of multiplication of open texts: a multiplication cryptogram can be calculated as a product of a (paired) multiplier cryptogram. Let E—encryption function; t_1 and t_2—open texts. If $E(y, g, \{r_1, r_2\}) = (y_{t_1}^{r_1} t_1, g^{r_1})$ and $E(y, g, \{r_1, r_2\}) = (y_{t_2}^{r_2} t_2, g^{r_2})$ can be obtained in the form of $(y^{r_1} y^{r_2} t_1 t_2 g^{r_1} g^{r_2})$.

2.3.3 Goldwasser-Mikali Cryptosystem

In the Goldwasser cryptosystem, Mikali, if the public key is a module m, then encryption function E Bit b є $E(b) = x^b r_2 \bmod m$ for a random element $r \in \{0, \ldots, m-1\}$. Then this cryptosystem is homomorphic for multiplication operations: $E(b_1) * E(b_2) = x^{b_1} r_1^2 x^{b_1} r_2^2 = x^{b_1 + b_2} (r_1 * r_2)^2 = E(b_1 \oplus b_2)$, where the character « $\oplus$ » marked by modulo 2 addition operation.

2.3.4 Peye CryptoSystem

In the Paye cryptosystem, if the public key is a module m and g—random number, the encryption function E message (open text)) t presented as a $E(t) = g^t r^m \bmod m$ for random variables $r \in \{0, \ldots, m-1\}$. Then the homomorphism by adding is as follows:

$$E(t_1)E(t_2) = (g_1^t r_1^m)(g_2^t r_2^m) \bmod m = g^{t_1 + t_2}(r_1 + r_2)^m \bmod m = E(t_1 + t_2)$$

2.3.5 Benalo Crypto System

In the Benalo cryptosystem, if the public key is a module m, then the open text encryption function t presented as $E(t) = g^t r^c \bmod m$ for random $r \in \{0, \ldots, m-1\}$. Then the homomorphism by adding is as follows:

$$E(t_1)E(t_2) = (g_1^t r_1^c)(g_2^t r_2^c) \bmod m = g^{t_1 + t_2}(r_1 + r_2)^c \bmod m = E(t_1 + t_2)$$

2.4 Fully Homomorphic Encryption

Partially homomorphic cryptosystems allow for the production of homomorphic calculations for only one operation (or assembly or multiplication) of open texts. The cryptosystem that supports both addition and multiplication (thus preserving the structure of rings of open texts) is known as fully homomorphic encryption and is more powerful. Using such a system, any scheme can be homomorphic, allowing you to effectively create programs that can be run on input encryption to make encryption output. Since such a program will never decipher its inputs, it can be performed by an unreliable party, without showing its introduction and internal state. In the existence of an effective and completely homomorphic cryptographic system, there would be large practical implementations in outsourcing closed computing, for example, in the context of cloud computing [3]. Homomorphic encryption would allow to combine different services that provide data for each service. For example, merging into one whole services of different companies could consistently calculate the tax, apply the current exchange rate, send an application for an agreement that gives actual data for each of these services [4]. Homomorphic property of different cryptographic systems can be used to create secure voting systems [5], hash functions resistant to collisions, closed information of search engines and make widespread use of public cloud computing possible, ensuring the confidentiality of the processed data.

2.5   *The Problem of Homomorphic Encryption*

One of the significant problems known completely homomorphic cryptosystems is their extremely low productivity. Currently, there are two main ways to increase productivity: the use of "limited homomorphism" (англ. Leveled fully homomorphic encryption) [6] and the use of the so-called "cipher text packaging method" [7]. The first involves a cryptosystem that can perform addition and multiplication operations, but in limited quantities. The essence of the other is that several open texts are written into one cipher text at once, and at the same time in the process of single operation of such batch cipher text, all incoming ciphertexts are processed simultaneously.

3   RingCT

RingCT—is a cryptographic technology that hides the amount of money received in any transaction. In most cryptocurrencies, transaction amounts are sent in the form of open text, which is visible to any observer.

RingCT uses several mechanisms at once: multilayer related spontaneous group signatures (Multilayered Linkable Spontaneous Group Signature, hereinafter referred to as MLSAG), a commitment scheme (Pedersen Commitments) and range proofs.

RingCT introduces two types of anonymous transactions: simple and full. The first (simple) wallet generates when the transaction uses more than one login, the second (full)—in the opposite situation. They differ in validation of transaction amounts and are signed by an MLSAG signature (let's talk more about this below). Moreover, transactions of type full can be generated with any number of inputs, there is no fundamental difference. The book "Zero to Monero" on this occasion states that the decision to limit full transactions with one entrance was made on a quick hand and may change in the future.

3.1 MLSAG-Signature

Remember that represents the signing of transaction inputs. Each transaction spends some money. The generation of funds occurs by creating transaction exits, and the output that the transaction spends becomes an input.

The entrance refers to several outputs, but spends only one, thus creating a "smoke curtain" to complicate the analysis of the translation history. If a transaction has more than one login, then such a structure can be represented as a matrix, where rows are inputs and columns are outputs. To prove to the network that the transaction spends exactly its outputs (knows their secret keys), the inputs are signed with a ring signature. This signature guarantees that the signer knew the secret keys for all elements of any column.

Confidential transactions are no longer used classically for circular signatures, replaced by MLSAG—signatures that are adapted for multiple inputs. This is a version of similar single-layer ring signatures, LSAG.

Multilayered they are called because they sign several inputs at once, each of which is mixed with several others, that is, a matrix is signed, and not one row. This helps save on the signature size.

Let's look at how a ring signature is formed, using the example of a transaction that spends 2 real outputs and uses to knead $m-1$ random entry with blockchain. We denote the public exit keys that we use as

$$P_\pi^0 \text{ and } P_\pi^1,$$

a key images for them, respectively:

$$I^0 \text{ and } I^1$$

Thus, we get a matrix of size $2xm$. First we need to calculate the so-called challenges for each pair of outputs:

3.2 Calculation of Challenges for the Signature

We start the calculations with the outputs that we spend using their public keys:

$$P_\pi^0 \text{ and } P_\pi^1$$

and random numbers

$$\alpha^1 \text{ and } \alpha^0$$

As a result, we obtain the:

$$L_\pi^0, L_\pi^1, R_\pi^0 \text{ and } R_\pi^1$$

which we use to calculate the challenge

$$(C_{n+1})$$

the next pair of outputs. All subsequent values are calculated in a circle according to the formulas shown in Fig. 1. The latter calculates the challenge for a couple of real outputs.

In all pages, except for real outputs, randomly generated numbers are used.

For π column convert α into s (Fig. 2):

The signature is presented as a tuge of all these values (Fig. 3).

MLSAG

mix

output

$$L_n^0 = a_0 G$$
$$R_n^0 = a_0 H(P_n^0)$$

$$L_{n+1}^0 = s_{n+1}^0 G + c_{n+1} P_{n+1}^0$$
$$R_{n+1}^0 = s_{n+1}^0 H(P_{n+1}^0) + c_{n+1} I_0$$

$$L_n^1 = a_1 G$$
$$R_n^1 = a_1 H(P_n^1)$$

$$L_{n+1}^1 = s_{n+1}^1 G + c_{n+1} P_{n+1}^1$$
$$R_{n+1}^1 = s_{n+1}^1 H(P_{n+1}^1) + c_{n+1} I_1$$

$$C_{n+1} = H(m, L_n^0, R_n^0, L_n^1, R_n^1)$$

$$c_n = H(m, L_{n-1}^0, R_{n-1}^0, L_{n-1}^1, R_{n-1}^1)$$

Fig. 1 Forming challenges, P—public output keys, a, s random keys, I—key images

Fig. 2 Recovery process

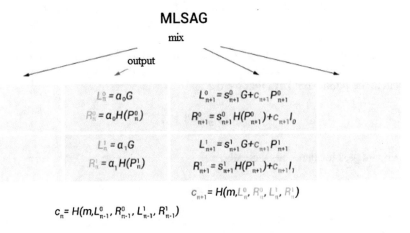

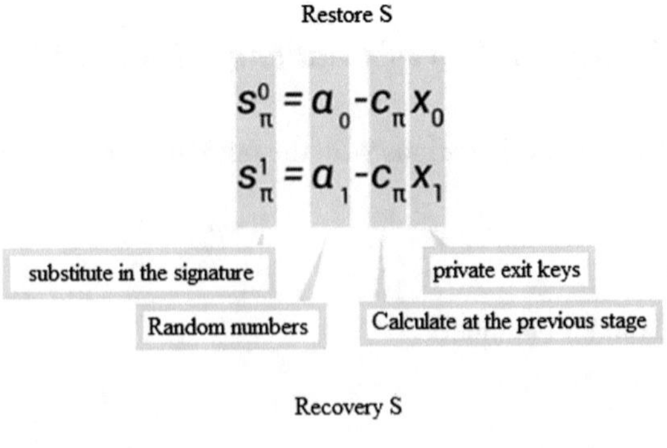

Restore S

$$s_\pi^0 = a_0 - c_\pi X_0$$
$$s_\pi^1 = a_1 - c_\pi X_1$$

substitute in the signature   private exit keys

Random numbers   Calculate at the previous stage

Recovery S

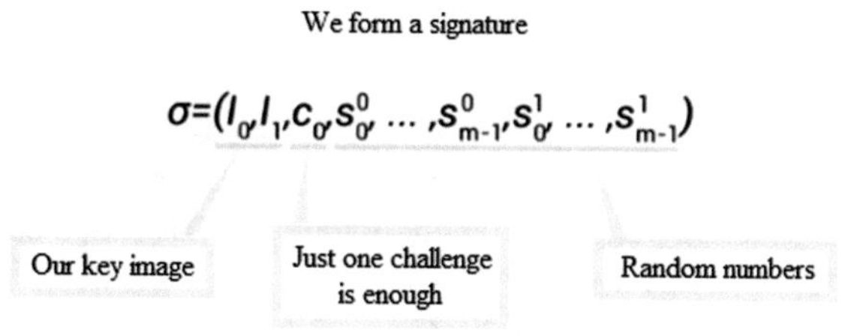

Fig. 3 MlSAG signature
formation

We form a signature

$$\sigma = (I_{o'} I_{1'} c_{o'} s_{o'}^0 \ldots, s_{m-1'}^0 s_{o'}^1 \ldots, s_{m-1}^1)$$

Our key image   Just one challenge   Random numbers
                     is enough

Further, this data is written to the transaction. MLSAG contains only one challenge c_0, which saves on the signature size (which already requires a lot of resources). Next, any inspector using the data C_0^i and S_0^i, restores value $c_0, \ldots, c_m$ and checks that $c_0 = c_m$.

Thus, our ring was locked and the signature was checked. For RingCT transactions of type full, another line is added to the matrix with shuffled outputs.

3.3 Pedersen Commitments

Commitment schemes (more often use the English term—commitments) are used so that one party can prove that it knows a certain secret (number), practically without revealing it. For example, you throw away a number on the bones, think of it and pass it to the checker. Thus, at the time of disclosure of the secret number, the inspector independently considers it, thereby making sure that you did not deceive him.

Monero is used to hide transfer amounts and use the most common option— Pedersen. By the way, an interesting fact is that at first the developers offered to hide the amounts with ordinary kneading, that is, add exits to arbitrary amounts to introduce uncertainty, but then switched to commitments (not a fact that saved on the size of the transaction).

In general, it looks like this:

$$C = xG + aH$$

where C—the value of the most important, a—hidden amount, H—Fixed point on elliptical curve (optional generator), but x—some arbitrary mask that hides the randomly generated factor. The mask is needed here so that the third party cannot simply overset to pick up the value of the goal.

When generating a new exit, the wallet calculates for it, and when spending takes either calculated when generating a value, or recalculates it again—depending on the type of transaction.

3.4 RingCT Simple

In the case of Simple RingCT transactions in order to ensure that the transaction has created outputs equal to the amount of inputs (did not spend money from the air) it is necessary that the amount of the first and second transactions be the same, that is:

$$\sum C_{input} - \sum C_{output} - \sum C_{fee} = 0$$

The commission is considered a little different—without a mask:

$$C_{fee} = aH$$

where a—commission amount, it is publicly available.

This approach allows us to prove to the party that checks that we use the same amounts without disclosing them.

To make everything clearer, let's look at an example. Supposing that a transaction consumes two outputs (i.e. they become inputs) on 10 and 5 XMR and generates three outputs worth 12 XMR: 3, 4 and 5 XMR. At the same time pays a commission of 3 XMR. Thus, the amount of money spent plus the amount generated and the commission is equal to 15 XMR. Let's try to calculate the sums and see the difference in their amounts:

x_1, x_2—inputs masks
y_1, y_2, y_3—outputs masks (Fig. 4).

In order for the equation to come down—the sums of the masks of inputs and outputs, we need the same. To do this, the wallet generates randomly x_1, y_1, y_2 and y_3, but x_2 is calculated as:

$$x_2 = y_1 + y_2 + y_3 + x_1$$

Using these masks, you can prove to any inspector that the funds are generated no more than spent, without disclosing the amount.

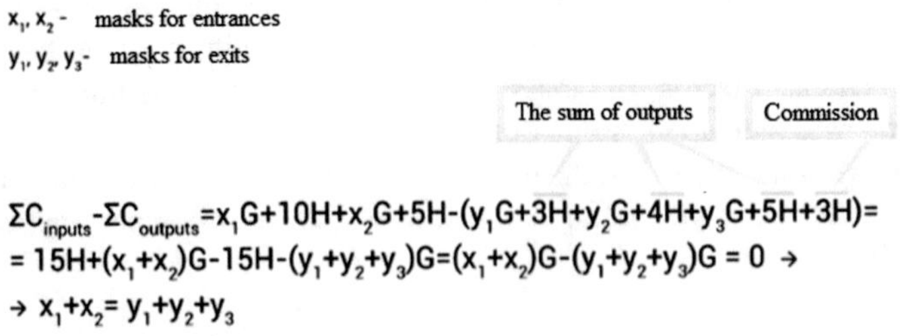

x_1, x_2 - masks for entrances
y_1, y_2, y_3- masks for exits

| The sum of outputs | Commission |

$\Sigma C_{inputs} - \Sigma C_{outputs} = x_1 G + 10H + x_2 G + 5H - (y_1 G + 3H + y_2 G + 4H + y_3 G + 5H + 3H) =$
$= 15H + (x_1 + x_2)G - 15H - (y_1 + y_2 + y_3)G = (x_1 + x_2)G - (y_1 + y_2 + y_3)G = 0$ →
→ $x_1 + x_2 = y_1 + y_2 + y_3$

Fig. 4 Deduction of difference in amounts

3.5 RingCT Full

In full RingCT transactions, checking the amount of transfers takes place some more intricately. In these transactions, the wallet does not list the inputs, but uses the wallets calculated during their generation. At the same time, one should think that the difference between the amount is not zero, but instead:

$$\sum C_{input} - \sum C_{output} = zG$$

Here z—difference mask inputs and outputs. If we consider $A = zG$ as a public key, then z—it's a private key. Thus, we know the public and his corresponding private keys. With this data, you can use it in mlsag's ring signature next to the public exit keys, which is kneaded (Fig. 5).

Thus, a valid ring signature will guarantee knowledge of all the private keys of one of the columns, and the private key in the last row can only be known if the transaction does not generate more than it spends. In the event that $zQ = 0$, then a column with real outputs will be revealed.

And how does the recipient of the funds find out how much money was sent to him? The sender of the transaction and the recipient exchange keys according to the Diffie-Hellman protocol, using the transaction key and the recipient's view key and calculate the general secret. The sender writes data to special transaction fields about the exit amounts encrypted by this public key.

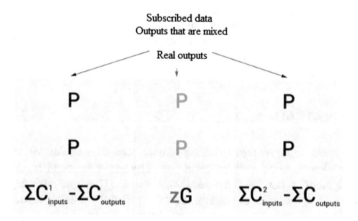

Fig. 5 MLSAG output replacement model

3.6　Range Proofs

What happens if you use a negative number as an amount in your account? This can lead to the generation of additional coins. This result is unacceptable, so the guarantee we use is not negative. In other words, you need to prove that the amount is in the interval $[0, 2^2 - 1]$.

To do this, the amount of each exit is divided into executive discharges and is considered to be for each category separately.

Supposing that the amounts we have are small and fit in 4 bits (in practice it is 64 bits) and an output of 5 XMR is created. We consider the total amount of money for each discharge and the total amount:

$$5 \text{ XMR} = 0101$$

$$C_3 = x_3 G \qquad\qquad C_0 = x_0 G + 2^0 H$$
$$C_2 = x_2 G + 2^2 H \qquad C_1 = x_1 G$$

$$C = \sum_i C_i = (x_0 + x_1 + x_2 + x_3)G + 5H$$

Next, each is mixed with surrogates—and paired with Borromeo's ring signature (another ring signature) proposed by Greg Maxwell in 2015 (Fig. 6).

All together, this is called range proof and allows you to ensure that amounts are used in the interval $[0, 2^n - 1]$.

4　Ring Signature

A ring signature—is one of the mechanisms for implementing an electronic signature, in which it is known that the message was signed by one of the members of the list of potential signatories, but does not disclose who exactly. The signatory independently forms a list of any number of persons (including himself). The signature does not

Fig. 6 Borromeo circular signature model

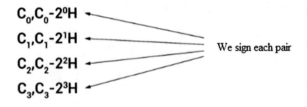

$$C_0, C_0 - 2^0 H$$
$$C_1, C_1 - 2^1 H$$
$$C_2, C_2 - 2^2 H$$
$$C_3, C_3 - 2^3 H$$

We sign each pair

require the permission, assistance or assistance of the listed persons, only the public keys of all members of the list and their own private key are used.

The mathematical ring signature algorithm was developed by Ronald Rivest, Adi Shamir, and Yael Tauman and presented at the 2001 Asiacrypt International Conference. In the simplest example, a ring signature is formed based on the keys of the sender and recipient of the message. The signature then means that the recipient can be sure that the message was created by the sender. But for a stranger, such a signature loses its persuasiveness and unambiguity—there will be no certainty who exactly formed and signed the message, because it could be the recipient himself.

Later, other modifications of the ring signature appeared:

(1) Threshold ring signatures. Unlike the standard "t-out-of-n" threshold signature, where t from n users must cooperate to decrypt a message, this ring signature option requires t users to cooperate in the signing process. To do this, t participants $(i_1, i_2, \ldots, i_t)$ must calculate the signature σ for the message m, giving t private and n public keys to the input $(m, S_{i_1}, S_{i_2}, \ldots, S_{i_t}, P_1, \ldots, P_n)$ [9].
(2) Linked ring signatures. The connectivity property allows you to determine if any two ring signatures were created by the same person (the same private key was used), but without specifying who exactly. One of the possible applications may be an autonomous system of electronic money [10].
(3) Traceable ring signature. A protocol that allows the implementation of secret electronic voting systems, which allow only one signature to be placed anonymously, but disclose a participant who has voted twice, may be disclosed when re-used by the public key of the person who signed.

4.1   The Essence of the Ring Signature Algorithm

The ring signature for a message will be generated based on a list of r public keys (marked in the diagram as $pk_1, \ldots, pk_r$), among which the signer's key has a serial number s. Public keys allow you to encrypt arbitrary information (information block x_1, encrypted with a key pk_1, marked on the diagram as $Enc_{pk_1}(x_1)$. "Information blocks" are not part or result of the processing of a signed message and have no independent meaning, they are generated random data that become components of the signature.

There is a combination function from any number of arguments, according to the value of which and the values of all arguments, except one, you can uniquely restore one missing argument. An example of such a function is sequential summation. If you know the total amount and all components except one, then the missing term can be calculated.

The authors of the algorithm of the combination function proposed the following sequence of actions (see Fig. 7): take some starting value (the scheme is denoted by v, which is formed randomly), on which the first argument is executed bitwise OR (denoted by the symbol ⊕). Then a certain reversible transformation is applied to the result (denoted in the diagram as E_δ), mutually uniquely related to the hash sum

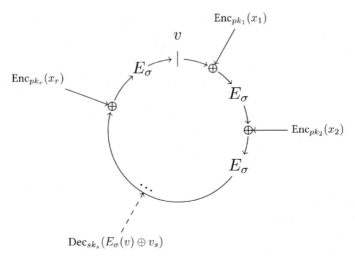

Fig. 7 Model of the ring signature

of the signed message. The obtained result participates in the operation of bitwise exclusive "or" with the second argument, the transformation is applied again E_δ etc. Appropriate public key encrypted information blocks are used as arguments $x_1, \ldots, x_r$.

The selected random value is both the starting and target (final) value of the combination function: the result of all transformations must "pass through the ring" and become equal to the initial value. Information blocks $x_1, \ldots, x_r$. for each of the keys except the block x_s, corresponding to the signer key itself are set as random values. The signer encrypts the information blocks with the appropriate public keys. The signer now has the target value of the combination function and all but one argument, which corresponds to its own key. Thanks to the properties of the combination function, the user can find out the missing argument and, using his own private key sk_s, "decipher" this argument (Dec_{sk_s}), having received the missing information block x_s.

Components of the finished ring signature:

1. the message to be signed;
2. list of used public keys;
3. values of all information blocks $x_1, \ldots, x_r$;
4. the value of the combination function v.

To verify the signature you need [1]:

- use public keys to encrypt information blocks;
- using the hash sum of the signed message to calculate the value of the combination function, using as arguments v (sets the starting value) and encrypted information blocks (the results of the previous step);
- make sure that the value obtained matches v.

5 Zero-Disclosure Protocols

Zero-knowledge proof in cryptography is an interactive cryptographic protocol that allows one of the interacting parties ("the verifier") to verify the validity of any statement without having no other information from the other party ("the prover"— proves). Moreover, the latter condition is necessary, as it is usually to prove that a party has certain information in most cases trivial, if it has the right to simply disclose information. The whole difficulty is to prove that one of the parties has information, rather than disclosing its content. The whole difficulty is to prove that one of the parties has information, rather than disclosing its content. The protocol should take into account that the proof can convince the verifier only if the statement is really proved. Otherwise, it will be impossible to do so, or extremely unlikely due to computational complexity [11].

The proof with zero disclosure should have three properties:

- completeness: if the statement is true, the one who honestly proves (one who fully follows the protocol) will always convince an honest examiner;
  $$\forall x \in L, \Pr[(P, V)[x] = accept] \geq 1 - negl(k)$$
- soundness protects against making a false statement: if the statement is false, the probability of deception in any case should be very low;
  $$\forall \notin L, \forall \hat{P}, \Pr[(\hat{P}, V)[x] = accept] \leq negl(k)$$
- zero disclosure.

Zero disclosure: if the statement is true, no examiner who plays by the rules can know anything but the truth.

The first two properties are fulfilled for the general interactive proofing system. The third property leads to zero disclosure.

A proof with zero disclosure is not a proof in a mathematical sense, because there is a small probability (soundness error) that a dishonest proofr will be able to convince the examiner of the truth of a false statement. In other words, proof is more likely than deterministic. However, there are techniques available to reduce the possibility of error to negligibly small.

The formal definition of zero disclosure requires the use of some computational model, the most common being the Turing machine. Let P, V and S be Turing automata. An interactive proof-of-speech system with (P, V) for the language L is zero-disclosed if any probabilistic polynomial time (PPT) V tester has a simulator with an expected PPT S, such that

$$\forall x \in L, z \in \{0, 1\}*, View_{\hat{V}}[P(x) \leftrightarrow V(x, z)] = S(x, z)$$

where View:

1. The distribution of probabilities $P, V(x)$ over that is, over all possible negotiations that P and V may have in relation to x
2. $View_v[P(x) \leftrightarrow V(x)] \leq$ a message from P, accidental V>.

The closer P is modeled in such a way that has unlimited computing power (in practice, P is usually a probabilistic Turing machine). Intuitively, the definition states that an interactive proofing system (P, V) has zero disclosure if there is an effective simulator V for any tester S that can reproduce a conversation between P and V at any input. Thus, a conversation with P cannot teach V the calculation of something that he could not calculate before.

Auxiliary line z plays the role of "prior knowledge". The definition implies that V cannot use any prior knowledge of z to obtain information from a conversation with P, because we require that if it also has this knowledge, then it can reproduce the negotiations between P and V as before.

5.1 Different Types of Zero Disclosure

Execution of the evidence protocol with zero disclosure leads to the conclusion of the Accept/Reject result and also generates a transcript of the evidence. Different variants of zero disclosure can be defined by formalizing the concept itself and comparing the dissemination of information of different models with the protocol in the following ways [12]:

- Ideal zero disclosure protocol—if the random variables in the transcript of the evidence of this model are evenly distributed and do not depend on the total input data [13].
- Statistically zero disclosure [12] means that the distribution is not necessarily the same, but it is at least statistically close, and the statistical difference is a small function [13].
- Computationally zero disclosure is called such a model, if there is currently no such effective algorithm that could distinguish the distribution of quantities from the propagation of information in an ideal protocol [13].

5.2 Zk-SNARK

The Zk-SNARK algorithm is used in the Zcash cryptocurrency network to hide data—from which address the payment came, to which address the payment was sent and what amount. It also allows you to prove that the transaction has actually taken place and the recipient's account has the correct amount.

Zk-SNARKs consists of three algorithms G, P and V which are as follows:

The key generator takes the lambda secret parameter and program C, and generates two public keys: the proof key pk and the verification key. These keys are public settings that only need to be created once for this C program.

The algorithm takes as input values: key pk, public value x and secret value w. The algorithm generates proof $pr_f = P(pk, x, w)$ p that proves knowledge of the secret value of w and that the secret value satisfies program C.

The verification algorithm V calculates $V(vk, x, pk_f)$ the result which will be true if the proof is true, and false otherwise. Thus, this function returns true if the tester knows that the secret value w satisfies $C(x, w) == true$.

Note the secret lambda parameter used in the generator. This parameter sometimes complicates the use of zk-SNARKs in real applications. The reason for this is that anyone who knows this parameter can generate fake evidence. In particular, for any program C and public value x, a person who knows lambda, but does not know the secret w, can create proof $fake\_prf$ to which the algorithm $V(vk, x, fake\_prf)$ will return true.

The main disadvantage of this family of zk protocols is the need to generate initial (trusted) system parameters—this process is also called a ceremony. After all, secret parameters are used for generation, which must be destroyed—they are called toxic. The main problem is that if the toxic parameters are preserved, the person who owns them will be able to prove false facts (in the case of Zcash—to generate cryptocurrency from the air).

5.3   Zk-STARK

ZK-STARK (Zero-Knowledge Scalable Transparent Argument of Knowledge— "short transparent argument of knowledge with zero disclosure")—a cryptographic protocol that uses public presumptively verifying evidence with zero disclosure. This technology allows users to share verified information without disclosing it or perform calculations with a third party without disclosing calculations. ZK-STARK is a transparent protocol, i.e. it does not require prior configuration and disclosure to third parties, such protocols are also called Arthur-Merlin protocols.

ZK-STARK is based on the ZK-SNARK algorithm, but ZK-STARK technology greatly speeds up the information exchange process and eliminates the need for pre-trust configuration, which in previous protocols endangered the confidentiality of the entire system. The new technology is currently being developed by leading experts from StarkWare, co-founded by Eli Ben-Sasson [11].

ZK-STARK technology can be used for:

- electronic voting;
- identity verification (verification is a secret);
- performing calculations and checking the results of calculations, for example for blockchain transactions.

Differences between ZK-STARK and ZK-SNARK:

- ZK-SNARK uses elliptical cryptography, which makes this technology vulnerable to attack by quantum computers. Whereas ZK-STARK uses post-quantum cryptography [14];
- when implementing ZK-SNARK it is necessary to carry out a preliminary trust setting with a third party, in ZK-STARK there is no such need due to the use of

probable verifiable evidence. This eliminates the vulnerability that arises due to the need to trust the data to a third party [9];

- unlike ZK-SNARK, systems using ZK-STARK are easily scaled in terms of computing speed and memory size for tasks with a large number of calculations [14];
- the size of the proof for ZK-SNARK was 288 bytes, for ZK-STARK it increased to several hundred kilobytes.

6 I2P—Anonymous Network

I2P (Invisible Internet Project)—is open source software designed to organize a highly persistent anonymous, overlay, encrypted network and used for web surfing, anonymous hosting (creation of anonymous sites, forums and chats, file sharing servers, etc.), instant messaging systems, blogging, as well as for file sharing (including P2P—Torrent, eDonkey, Kad, Gnutella, etc.), e-mail, VoIP and network interaction in the cryptocurrency system (Kovri).

I2P—it is an anonymous, self-organized distributed Network database that uses a modified DHT Kademlia, but differs in that it retains network node address hashing, encrypted AES IP addresses, and public encryption keys, and network database connections are also encrypted. The network provides applications with a simple transport mechanism for anonymous and secure messaging. Although the I2P network is focused purely on determining the path of packet transmission, thanks to the Streaming lib library, their delivery in the original sequence without errors, losses and duplication is also implemented, which makes it possible to use IP-telephony, Internet radio, IP-TV in the I2P network. video conferencing and other streaming protocols and services.

Within the I2P network there is its own directory of sites [11], electronic libraries, as well as torrent trackers. In addition, there are instructions for accessing the I2P network directly from the Internet [15], which are designed specifically for users who for various reasons can not install I2P on a computer.

The I2P network is similar in structure to the traditional Internet and differs only in the impossibility of censorship through the use of encryption and anonymization mechanisms. Therefore, it is not possible for third parties to find out what the user is viewing, which sites he visits, what information he downloads, what his range of interests, acquaintances, etc.

The I2P network has no central servers and no conventional DNS servers, and the network is completely independent of external DNS, which makes it impossible to destroy, block and filter the network, which will exist and function as long as there are at least two computers on the planet. online. Also, the lack of DNS servers and the use of DHT Kademlia—a mechanism for naming in the I2P network—allows any user of the I2P network to create their own site, project, torrent tracker, etc., without having to register anywhere, pay for any to whom a domain name or wait for someone's permission. Everyone is free to create any site for free and freely, and it is almost impossible to know the location of the server and the person.

Each new entrant that receives inbound connections from other routers increases the reliability, anonymity, and speed of the entire network.

To get into the I2P network, you only need to install a router program on your computer, which will decrypt/encrypt all traffic and redirect it to the I2P network. To work with I2P sites, you need to configure the browser in advance. When accessing a site or other resource on the regular (external) Internet, a router program automatically, like Tor, creates a "tunnel" to one of the external gateways and allows you to visit and use external Internet resources by hiding your IP address. Also, internal sites in the I2P network are accessible from the external Internet through special gateways.

At first glance, it may seem that the I2P router, due to the constant need to encrypt outgoing and decrypt incoming packets and the use of a large number of encryption algorithms with long keys, should negatively affect the CPU load and computer memory, in fact the load does not affect even on low-power office computers and is calculated in percentage units.

The network was originally designed with the assumption that all intermediate nodes are compromised or malicious (captured by an attacker and collect information that passes through them), so a number of active measures were introduced to counter.

All network traffic is encrypted from sender to recipient. In total, four equal layers of encryption (end-to-end encryption), garlic (Garlic Routing), tunnel, and transport layer encryption are used when sending a message, and a small random layer is automatically added to each network packet before encryption. the number of random bytes to further anonymize the transmitted information and complicate attempts to analyze the content and block transmitted network packets. IP addresses in the I2P network are not used anywhere and never, so it is not possible to determine the true address of a node in the network. Each network application on the computer builds for itself separate encrypted, anonymous tunnels. Tunnels are mostly one-way (outbound traffic goes through some tunnels, and inbound—through others)—the direction, length, as well as what application or service created these tunnels, it is almost impossible to find out. All transmitted network packets have the ability to diverge in several different tunnels, which makes it pointless to try to listen and analyze with a sniffer the flow of data passing.

There is also a periodic change (approximately every 10 min) of already created tunnels to new ones, with new digital signatures and encryption keys, and digital signatures and encryption keys for each tunnel.

For these reasons, you don't need to worry about encrypting your traffic. Or, if there is a lack of trust in encrypting programs that have closed source code (such as Skype). Also, for example, there are IP telephony programs (such as Ekiga) that do not know how to encrypt your traffic and transmit it in the open. In any case, the I2P network will perform four-level encryption of all packets and ensure the transmission/reception of all data.

In the I2P network, all packets are encrypted on the sender's side and decrypted only on the recipient's side, and unlike Tor, none of the intermediate exchange participants is able to intercept the decrypted data and none of the participants knows who the sender is and who the recipient is. who transmits packets may be the sender, and may be the same intermediate node, and the next node to which this packet should be sent, may be the recipient, and may also be the same intermediate node, to know the endpoints of the sender and recipient intermediate node can not, just as it cannot find out what happened to the packet just passed to the next node—it processed it, or passed it somewhere else, it is impossible to find out.

7 Summary

The openness of transactions in decentralized systems, on the one hand, is a plus for the community, and on the other—there are risks of disclosing the address of the subject, to identify the identity of the subject. The analysis of the existing decentralized systems allows to draw a conclusion that on the basis of a chain of blocks it is possible to identify belonging of some addresses to concrete subjects. Therefore, to perform a transaction in the system, it is optimal to use multi-signature or smart contracts, homomorphic encryption, cryptoprotocols to hide the personal information of the subject unnecessarily.

The use of a zero-disclosure evidence protocol can be used in confidential calculation protocols that will allow some participants to claim that the other party is behaving fairly.

All I2P network traffic passes through special tunnels, the direction of which changes periodically. Typically, multiple one-way encrypted tunnels are created and packets are routed through these tunnels in random order. Encryption occurs on the sender's side and decrypts on the recipient's side. It is not possible for an attacker to intercept all traffic unless the attacker controls all I2P nodes. The intercepted encrypted packet will not be useful to the attacker, because it is not clear the sequence of data received and to whom the packet belongs.

A circular signature scheme can ensure the anonymity of the signatory as well as its independence from other participants, while ensuring the integrity and authenticity of the imprint with which the message is signed.

The homomorphic property of cryptosystems can be used to create secure voting systems, hash functions, collision-resistant, closed search engine information, and will make public cloud computing widespread, ensuring the confidentiality of the data processed.

References

1. Benaloh JDC (1987) Verifiable secret-ballot elections
2. rfc2501. https://datatracker.ietf.org/doc/html/rfc2501. Last accessed 07 Nov 2021
3. Micciancio D (2010) A first glimpse of cryptography's holy Grail (англ.). Association for computing machinery (1 Mar 2010)
4. Stuntz C (2010) What is homomorphic encryption, and why should i care? (англ.) (18 Mar 2010). Архивировано 24 мая 2013 года. Ron Rivest. Lecture Notes 15: voting, homomorphic encryption (англ.)
5. Rivest R (2002) Lecture Notes 15: Voting, homomorphic encryption (англ.) (29 Oct 2002)
6. Brakerski Z, Gentry C, Vaikuntanathan V (2011) Fully homomorphic encryption without bootstrapping
7. Homomorphic encryption schemes and applications for a secure digital world. J Mobile, Embed Distrib Syst (2013)
8. Rivest RL, Shamir A, Tauman Y (2001) How to leak a secret. In: Boyd C (ed) Advances in cryptology—ASIACRYPT 2001. Springer, Berlin, Heidelberg, pp 552–565
9. Liu JK., Wong DS (2005) Linkable ring signatures: security models and new schemes. In: Computational science and its applications—ICCSA 2005, vol 2. Springer, Berlin, New York, pp 614–623. (Lecture Notes in Computer Science[en]. vol 3481)
10. Fujisaki E, Suzuki K (2007) Traceable ring signature. In: Public key cryptography—PKC 2007. Springer, Berlin, Heidelberg, New York, pp 181–200 (Lecture Notes in Computer Science[en]. vol 4450)
11. Santis AD, Micali S, Persiano G (1988) Non-interactive zero-knowledge proof systems (англ.). In: Advances in cryptology—CRYPTO '87: a conference on the theory and applications of cryptographic techniques. Santa Barbara, California, USA, 16–20 Aug 1987, Proceedings, Pomerance C (ed). Springer, Berlin Heidelberg,pp 52–72 (Lecture Notes in Computer Science; vol 293)
12. Sahai A, Vadhan S (2003) A complete problem for statistical zero knowledge (англ.). J ACM 50(2):196–249, ISSN 0004-5411; 1557-735X; 1535-9921. https://doi.org/10.1145/636 865.636868
13. Modern cryptography: theory and practice: Mao, Wenbo: 0076092025399: Amazon.com: Books. https://www.amazon.com/Modern-Cryptography-Practice-Wenbo-Mao/dp/013066 9431. Last accessed 07 Nov 2021
14. Blum M, Feldman P, Micali S (1988) ACM special interest group for automata and computability theory Non-interactive zero-knowledge and its applications (англ.). STOC'88: Proceedings of the twentieth annual ACM symposium on Theory of computing. ACM, NYC, pp 103–112

Statistical and Signature Analysis Methods of Intrusion Detection

Tamara Radivilova⬭, Lyudmyla Kirichenko⬭, Abed Saif Alghawli⬭,
Dmytro Ageyev⬭, Oksana Mulesa⬭, Oleksii Baranovskyi⬭,
Andrii Ilkov⬭, Vladyslav Kulbachnyi⬭, and Oleg Bondarenko⬭

Abstract Existing models and methods of intrusion detection are mostly aimed at detecting intensive attacks, do not take into account the security of computer system resources and the properties of information flows. This limits the ability to detect anomalies in computer systems and information flows in a timely manner. The latest monitoring and intrusion detection solutions must take into account self-similar and statistical traffic characteristics, deep packet analysis, and the time it takes to process the information. An analysis of properties traffic and data collected at nodes and in the network was performed. Based on the analysis traffic parameters that will be used as indicators for intrusion detection were selected. A method of intrusion detection based on packet statistical analysis is described and simulated. A comparative analysis of binary classification of fractal time series by machine learning methods is performed. We consider classification by the example of different types of attack detection in traffic implementations. Random forest with regression trees

T. Radivilova (✉) · L. Kirichenko · D. Ageyev · V. Kulbachnyi · O. Bondarenko
Kharkiv National University of Radio Electronics, 14 Nauky Ave, Kharkiv, Ukraine

L. Kirichenko
e-mail: lyudmyla.kirichenko@nure.ua

D. Ageyev
e-mail: dmytro.aheiev@nure.ua

O. Bondarenko
e-mail: oleh.bondarenko@nure.ua

A. S. Alghawli
Prince Sattam Bin Abdulaziz University, Aflaj, Kingdom of Saudi Arabia

O. Mulesa
Uzhhorod National University, Narodna Sq., 3, Uzhhorod, Ukraine
e-mail: oksana.mulesa@uzhnu.edu.ua

O. Baranovskyi
Blekinge Institute of Technology, Karlskrona, Blekinge, Sweden

A. Ilkov
Ivan Kozhedub Kharkiv National Air Force University, Sumska str, Kharkiv, Ukraine

R. Oliynykov et al. (eds.), *Information Security Technologies in the Decentralized Distributed Networks*, Lecture Notes on Data Engineering and Communications Technologies 115, https://doi.org/10.1007/978-3-030-95161-0_5

and multilayer perceptron with periodic normalization were chosen as classification methods. The experimental results showed the effectiveness of the proposed methods in detecting attacks and identifying their type. All methods showed high attack detection accuracy values and low false positive values.

Keywords Intrusion detection · Self-similar traffic · Attacks · Machine learning · Statistical analysis · Security · Classification

1   Introduction

The world is gradually approaching the stage of direct dependence on information technology and online access to the Web. E-mail, VoIP- SIP-calls, communication and continuous online presence, making payments or transfers, online purchases—have long been available from phones and tablets. It precisely nowadays the threat of stopping a lot of services is a topical issue for intruders. Therefore, the use of cyber attack protections is becoming as relevant to network protection as a firewall, intrusion detection/prevention system or unified threat management [1–4].

During a distributed denial of service (DDoS) attack, infected hosts anywhere in the world overwhelm the victim's hardware or software resources (server, network device, network), causing a denial of service for legitimate clients. Thus disrupting the operation of online services, information portals, and electronic payments. Detecting, preventing or significantly impeding cyber attacks is one of the central areas of cybersecurity in information networks. One security solution is anomaly-based intrusion detection systems (IDS) [5–7]. For the effective functioning of information networks in modern conditions and means of their protection, as well as for reliable detection and assessment of cyber attacks it is necessary to develop new approaches and methods and their implementation. The purpose of this paper is to analyze statistical intrusion detection methods and machine learning based methods for different traffic implementations and attack types.

2   Intrusion Detection Systems Approaches

Considering the intrusion detection methods, IDSs can be divided into two main types of systems: signature-based and anomaly-based [1, 3, 8].

Anomaly analysis method. Anomaly-based IDSs create a profile by tracking normal activity in the system [1]. This profile is used to find deviations of the observed activities from the normal behavior of the system and report them as anomalies. In this type of system, it is very important to correctly identify the normal profile of the system offline or online [2, 9, 10]. Otherwise, any activity that differs from the created profile may be identified as an intrusion, and thus the system may suffer from a high rate of false positives caused by labeling any non-intrusive abnormal behavior as harmful. Unlike signature-based systems, anomaly-based systems are capable of detecting new types of attacks.

An intrusion detection system that uses anomaly detection techniques to detect malicious activity is called an anomaly detection system. Anomaly detection systems are built on the assumption that an intrusion is a subset of anomalies [2, 11, 12]. Ideally, every anomaly detected is actually an intrusion, which means there are no false positives in the system. However, a major disadvantage of anomaly detection systems is the high rate of false positive errors that occur when observing any legitimate abnormal activity on a monitored system or network. In real-world applications, the burden of false positive errors can lead to an IDS failure [6]. On the other hand, during the detection of abnormal activity, the system administrator must spend a lot of time identifying the root cause of the generated alarms [13–15]. As the number of alarms increases, the system administrator may not be able to keep track of all initiated alerts or simply disable the alarm generator. Consequently, a number of valid alarms may be lost among the large number of false alarms in the system traffic.

The overall process of the anomaly detection system can be divided into two phases: training and testing. In the training phase, the normal profile of the system is built by the IDS [16–19]. This profile constitutes the normal behavior of the system in terms of certain characteristics and metrics. In the testing phase, the IDS monitors the current actions of the system or network and tries to find anomalies by comparing the trained profile with new data [20–25]. Anomaly detection systems can use a variety of available data feature methods. These include methods based on statistics and machine learning.

The method is based on statistical data analysis. In statistical-based methods, the system maintains two profiles for normal and current system or network behavior. In each of these profiles, the behavior of the system is modeled by calculating some statistical measures such as mean, standard deviation, data distribution, etc. [15, 25–27]. When a new set of actions is triggered in the system or network, the IDS tries to find the deviations between the two profiles according to the mentioned measures. Using some statistical tests, the system calculates an anomaly score for the current profile, which indicates the degree of deviation from the norm. If the abnormality score is above a certain threshold, the system gives an alarm as a sign of abnormality [13, 28].

Methods based on statistics have several advantages. First, they do not require any prior knowledge of the system or network, since a normal profile can only be created by observing the actions on the system over a period of time. In addition, because these approaches use statistical models, they are particularly effective in detecting attacks that occur over long periods of time [25–27]. In addition, statistically-based approaches have two major disadvantages. In these methods, an attacker can train the system to assume that compulsive behavior is normal. Moreover, the basic principle of these approaches is to model data points as stochastic distributions, assuming a quasi-stationary process [26].

Machine learning methods. In machine learning approaches, the system tries to learn data patterns and build a model to classify the patterns in question [26–29].

Using this model, machine learning-based anomaly detection systems can improve their performance according to previous results. Machine learning-based techniques [30–32] can use supervised or unsupervised learning. In supervised learning, a set of labeled data is used by the system during the training phase. Each label is a class of actions, such as malicious or normal. Once a model is created based on the characteristics of the data and their corresponding labels, the system is able to make predictions about the classification of new data available in the testing phase.

In addition, the available dataset is not affected by learning without teacher. In this method, the system tries to analyze the data structure and identify similar data points, which makes it quite effective in solving clustering problems [28, 30].

Thus, all the above traffic management and IDS methods can be combined into the concept of universalization and centralization of traffic analysis and network state when using IDS by developing DPI/DCI (Deep packet inspection/Deep content inspection) tools. This concept can be designated as DPI as a service—under this name it was cited in [29]. The essence of the concept is that if the network uses a large number of different tools that implement some kind of traffic analysis (firewalls, IDS systems, traffic optimizers, etc.), it makes sense to put all the analysis in a separate device. This device will perform full parsing of network data and send the results of the analysis to all devices depending on their needs, and they, in turn, will implement only the response to incoming data [15].

An analysis of traffic management and IDS methods has been carried out, which showed that they can be combined in the concept of universalization and centralization of traffic analysis and network state when using IDS by developing DPI/DCI tools. In the future, the DPI concept should be used. Since traffic has self-similar properties, they must also be taken into account when analyzing traffic to detect intrusions.

3   Fractal Random Processes and Models

If the process $a^{-H}X(at)$ has the same finite-dimensional distribution laws with $X(t)$ then a random process $X(t)$ is self-similar. The parameter $H, 0 < H < 1$, is called the Hurst exponent. It is the self-similarity degree and the measure of the long-term dependence of the process. The moments of the self-similar process satisfy the scaling relation $E[|X(t)|^q] \propto t^{qH}$.

Multifractal random processes are inhomogeneous fractal processes and have more flexible scaling relation: $E[|X(t)|^q] \propto t^{\tau(q)+1}$, where $\tau(q)$ is a nonlinear function of scaling exponent [19].

One of the most used characteristics of multifractal processes and time series is the generalized Hurst exponent $h(q)$, which is associated with the function $\tau(q)$ by the ratio [23]:

$$h(q) = \frac{\tau(q) + 1}{q}.$$

The value $h(q)$ at $q = 2$ corresponds to the value of Hurst exponent H. The self-similar process are monofractal, their scaling exponent $\tau(q)$ is linear.

The popular models of the multifractal processes are the stochastic conservative binomial multiplicative cascades [24]. Such multifractal models are constructed using an iterative algorithm, where the values of the cascade realization are the values of some specially selected random variable. The conservatism of the cascade consists in the fact that for any number of iterations, the sum of the cascade values remains the same.

Mandelbrot B. proposed a multifractal model of financial time series which is based on fractional Brownian motion in multifractal time by operation of subordination. The subordination is a random substitution of time and it can be represented in the form $Z(t) = Y(T(t))$, where $T(t)$ is a nonnegative nondecreasing random process called subordinator, $Y(t)$ is a random process, independent of $T(t)$.

In [25] it is proved that if $X(t)$ is the process of subordination

$$X(t) = B_H(\theta(t)). \tag{1}$$

where $B_H(t)$ is fractional Brownian motion with Hurst exponent H and $\theta(t)$ is conservative binomial multiplicative cascades, then $X(t)$ is the multifractal process. The scaling function $X(t)$ is defined by

$$\tau_X(q) = \tau_\theta(Hq), \tag{2}$$

where $\tau_\theta(Hq)$ is the scaling function of the multiplicative cascade $\theta(t)$.

4 Selection of the Optimal Set of Features for Detecting Intrusions

Today, a heuristic definition (selection) of the set of measurement parameters of the protected system is used, the use of which should give the most effective and accurate intrusion detection. The complexity of choosing the set can be explained by the fact that its constituent subsets depend on the types of intrusions detected. Therefore, the same set of this installation will not be adequate to all types of intrusions [18, 33].

Any system consisting of familiar hardware and software can be viewed as a unique set with its own characteristics. This is an explanation for the possibility of skipping specific intrusions for a system protected by IDSs that use the same set of evaluation parameters. The most preferable solution is to determine the necessary evaluation parameters on the fly. The difficulty with effective dynamic formation of evaluation parameters is that the size of the search area depends exponentially on the power of the initial set. If there is an initial list of parameters N, relevant for intrusion identification, then the number of subsets of this list is 2 N. Therefore, it is impossible to use brute force algorithms to find the optimal set.

Data is collected at nodes and on the network.

Since existing attacks cause deviations in resource utilization, the following parameters are collected at the nodes and network and combined into a vector:

- CPU utilization (percent);
- disk IO operations (operations per second);
- memory consumption (in percent);
- network activity (Mbit per second).

In [14, 18, 24] was proposed to collect the following network parameters:

- TCP packet sequence, flags, sender and receiver port numbers and sender IP addresses;
- ICMP packet sequence, sender IP addresses, ICMP_ID;
- Source IP addresses, packet length and UDP source and destination ports.

In [33] there is a description of the traffic dataset containing attacks and has an extensive list of parameters, some of which will be used as attributes in the future. Traffic and attack parameters are taken from data packets and statistical traffic parameters.

The attributes can be selected by the network administrator. The sender port number does not usually make sense to consider, because it is automatically generated.

When developing the concept and methods of security and intrusion detection, the database for network activity emulation was used, because it is convenient to work with [34]. Accordingly, the parameters are taken from there. In a real system, however, as follows from the review, it is desirable to analyze in detail the packages of application protocols.

The following parameters were selected:

Duration—duration of the connection;
protocol_type—protocol type (TCP, UDP, etc.);
service—network service of the recipient (HTTP, TELNET, etc.);
flag—the connection state;
src_bytes—the number of bytes sent from the source to the destination.
dst_bytes—number of bytes sent from the receiver to the source;
land—1—if the connection is on an identical port; 0—in other cases;
wrong_fragment—number of incorrect packets;
urgent—number of packets with URG flag;
srv_count—number of connections to this host for the last 2 s;
srv_count—number of connections to this service in the last 2 s;
serror_rate—percentage of connections with syn errors;
diff_srv_rate—percentage of connections to different services;
srv_diff_host_rate—percentage of connections to different hosts;

dst_host_srv_count—number of connections to the local host established by the remote party and using the same service.

An analysis of the many existing parameters of the system and network functioning has been performed, showing that it is desirable to reduce the number of input variables both to reduce the computational cost of the model and to improve the performance of the model. Feature selection methods based on statistics involve estimating the relationship between each input variable and the target variable using statistics and selecting the input variables that have the closest relationship to the target variable. These methods can be fast and efficient, although the choice of statistical measures depends on the type of data of both the input and output variables. Thus, the selection of filtering-based features selected a set of features to develop methods for monitoring and detecting intrusions.

5 A Study of Statistical and Machine Learning Methods of Intrusion Detection

5.1 A Study of Statistical Methods for Analyzing Protocols

Stateful flow-based network packet analysis. The concept of stateful protocol analysis is simple: add stateful characteristics to normal protocol analysis. When protocol analysis is performed, TCP and UDP payloads containing protocols such as DNS, FTP, HTTP and SMTP are checked. IDS sensors that perform protocol analysis understand how each protocol should work based on the RFCs and implementations of those protocols. In this way, an IDS sensor can detect many suspicious values in protocol application payloads. Protocol analysis signatures can also be designed to detect attempts by attackers to hide their actions.

With a statistical approach, not only input and output packets are monitored, but also the state of individual connections, which is stored in dynamic tables. Thanks to this, the analysis of the next packet can take into account not only the specified rules and policies with respect to packet addresses and contents, but also the state of the connection to which the packet belongs and the previous packets belonging to it, as well as other data-related, connections.

While protocol analysis itself is a very powerful method, it is limited to looking at a single request or response. Typically, many attacks cannot be detected by looking at a single request—an attack may involve a series of requests. The best way to detect such attacks is to add state characteristics to the protocol analysis.

When stateful protocol analysis is performed, all events in a connection or session are monitored and analyzed. The IDS sensor can "remember" meaningful events and data during a session. This allows the sensor to find correlations between different events during a session, identifying attacks with multiple components that otherwise cannot be detected. Without the ability to store state, we can only check each packet, request, or response independently of the rest of the session.

An example of stateful protocol analysis. The concept of stateful protocol analysis is best explained by looking at a few examples. There are several ways to think about "state" in communications. One of the simplest forms of state is the ability to associate a request with a corresponding response. Because the IDS sensor that performs stateful protocol analysis can consistently track each request and response in a session, it can easily match the response to the request it has generated. In many cases, this is extremely valuable because many responses contain some sort of status indicator that tells us what the result of the request was.

An example of stateful protocol analysis.

The concept of stateful protocol analysis is best explained by looking at a few examples. There are several ways to think about "state" in communications. One of the simplest forms of state is the ability to associate a request with a corresponding response. Because the IDS sensor that performs stateful protocol analysis can consistently track each request and response in a session, it can easily match the response to the request that generated it. In many cases, this is extremely valuable because many responses contain some sort of status indicator that tells us what the result of the request was.

1.   For example: when executing an FTP command, the server response starts with a three-digit code. A status code beginning with "2" indicates that the command was successful, and "5" indicates a failure.

FTP sensors can use this information in several ways.

The most obvious way is to determine if certain attacks have succeeded or failed. If an attacker attempts to retrieve a sensitive file, and you can see that the response status code starts with "2", then the attempt was successful.

Because you can identify each failed request and keep track of how many requests failed in a session, you can find a large number of failures. This is very useful in detecting brute force attacks, such as password attacks.

A more difficult use is to identify cases where a request has no response. Some attacks involve sending the same command to a server hundred or thousands of times at speed. In these cases, the IDS sensor may see many requests from a client before seeing one in response from the server.

Thus, the simple ability to remember the last command and associate it with the response allows multiple types of attacks to be identified. However, there is no limit to remembering the previous command; it is possible to remember previous commands and responses. It is even more possible to simply retrieve principle data from certain commands and responses and save that data for future use.

2.   Example of saving data from an FTP sensor session to be used in the last part of the session. The FTP user logs in using four steps:

The user sends USER kkent to the FTP server, where kkent is the user name.
The FTP server responds with "331 Send password".
The user sends a PASS foobar to the FTP server.
The FTP server confirms that the password is correct and corresponds to "230 User logged in".

Table 1 Attack detection accuracy for some types of attacks

Attack type	Precision	Recall	F1
DDoS	0.97	0.95	0.96
UDP-flood	0.97	0.9	0.94
MAC flooding	0.99	0.92	0.96
ARP-spoofing	0.96	0.93	0.95
HTTP flood	0.98	0.88	0.93

The FTP sensor detects the USER command and stores "kkent" as the previous username for this FTP session. It then waits for the PASS command, which has a response with a status code that starts with 2, indicating that the authentication was successful. The FTP sensor also monitors for another USER command indicating that the user is trying to authenticate again. Once successful authentication has been detected, the IDS sensor knows that the session being tracked is the kkent username session.

This information is useful in two main ways. First, if an attack occurs during a session, the IDS can report that the user name kkent was used for that session. This information can be very useful when investigating the incident. Second, if the name root or administrator was used. By analyzing all authentication-related requests and responses, the IDS sensor can detect attempts to use such accounts, and record whether all attempts were successful or not. And this statistical method can be applied in many different ways to different types of attacks.

Thus, the analysis shows that a stateful protocol is, in many cases, the only way to detect various complex attacks.

Table 1 shows the values of the attack detection property of the protocol statistical analysis method for different types of attacks. The proposed method has false positives. This can occur due to dropped connections, high traffic speed, due to the restriction of packet header information (encrypted data), etc. The results obtained coincide with those obtained by other scientists [4, 13, 16].

Thus, the intrusion detection method based on statistical protocol analysis showed high values of attack detection accuracy (about 94%) and low values of false positive index (about 10%). Consequently, the statistical packet analysis method can be used to detect different types of attacks.

This approach has several advantages. First, it provides administrators with a multidimensional view of network traffic based on parameters such as packet classification according to the set of attributes transmitted by the packets. Second, it detects anomalies that cause spikes in network traffic as well as changes that increase network traffic. Significant deviations from the underlying distribution can only be caused by packets that make up an unusual portion of the traffic. If an anomaly occurs, no matter how slowly the traffic value increases, it can be detected as soon as the value increases to a certain level. Third, this method gives information about the type of anomaly detected, but requires a constant amount of memory.

5.2 A Study of Behavior Analysis Methods Using Machine Learning

During the experiment, several types of features derived from time series values were used for classification:

- time series value;
- statistical characteristics of the series;
- fractal characteristics of the series;
- recurrence characteristics.

5.2.1 Decision Tree Method

Can be applied to classification problems that arise in a wide variety of fields, and is considered one of the most effective.

The decision tree method for a classification or prediction problem consists in carrying out the process of distributing the original data into groups until homogeneous (or nearly homogeneous) subsets of them are obtained. The set of rules yielding this separation then allows the prediction obtained by evaluating some of the input features for the new data to be made.

The tree learning (or formation) algorithm operates on the principle of recursive sectioning. Sectioning of the dataset (i.e., partitioning into unremarkable subsets) is performed based on the use of the most appropriate attribute for this purpose. A suitable decision node is created in the tree, and the process continues recursively until the stopping criterion is met.

There are different numerical methods for constructing decision trees. One of the better known is an algorithm called C5.0. In fact, the C5.0 algorithm is the standard procedure for building decision trees. This application is implemented on a commercial basis, but the version embedded in the Python package (and some other packages) is available for free.

The algorithm implements the principle of recursive partitioning. The algorithm starts with an empty tree and a full dataset. In the nodes, starting from the root node, a feature is selected, the value of which is used to split all the data into two classes. After the first iteration of the algorithm, a tree with one node appears, splitting the dataset into two subsets. This process can then be performed repeatedly, with respect to each of the subsets, to create subtrees.

To separate the data, can be used the conditions of the form: $x < a, x > a$, where x is factor value, a is some fixed number. These are "axis-parallel splits", i.e. "parallel axis splits". Essentially, each time a condition is checked, the data samples are sorted so that each data item is defined as corresponding to only one branch. Decision criteria partition the initial data set into unremarkable subsets. The process continues until the stopping criterion is met.

In order to build decision trees, it is necessary to determine the attributes by which the partitioning will be performed. In the case of classification of data samples, these features can be the same sample values. From the set of attributes for partitioning, you need to select those that will allow you to get as many homogeneous (pure) groups as possible. Algorithm C5.0 uses entropy as a protective homogeneity (purity) of a group, which is a measure of data disorderliness.

Using entropy as a measure of purity of groups resulting from partitioning, the algorithm can choose the feature, the partitioning by which will give the cleanest group (i.e. the group with the lowest entropy). These calculations are called information gain. This attribute is determined by brute force method. For each feature F, the value of informational gain is calculated as a difference between entropies of the group (segments) before and after the division. The greater the value of informational growth for a selected feature, the better this feature is suitable for making a partition, since such a partition provides for the most homogeneous group. If the value of informational growth for a selected attribute is close to zero, it means that the partitioning by this attribute is unpromising, because it will not lead to a decrease in entropy. Besides, the very probable value of the value of information gain is equal to the value of entropy to the partition. This means that entropy after partitioning will be equal to zero, i.e. the groups obtained as a result of partitioning will be completely homogeneous (pure).

5.2.2 Tree Ensembles

Solver tree models are very sensitive to noise in the input data, the whole model can change dramatically if the training sample changes slightly. In other words, by making even minor changes in the training data, we will always get a different model. But the original and modified models will function approximately the same and with comparable accuracy: insignificant changes in training data will not lead to changes in the basic regularities. In this case it is advisable to use ensembles of models. An ensemble of models as a whole can be regarded as a complex, constitutive model consisting of individual basic models. The constituent models may be of the same type or of different types.

One of the first and best known types of ensembles is the method of begging and is based on the statistical bootstrap method. Bootstrap is a computer method for studying the distribution of statistics of probability distributions, based on multiple generation of samples on the basis of one.

Bootstrapping is a classification technique where all elementary classifiers are trained and operate independently of each other. The idea is that the classifiers do not correct each other's errors, but compensate during voting.

At the heart of how begging works is a classification technique called perturbation and combination. By perturbation we mean making some random changes in the training data and constructing several alternative models on the changed data, followed by combining the result. Several subsamples are extracted from a single training set by sampling, each of which is used to train one of the ensemble models. If the ensemble is based on different types of models, each type will have its own training algorithm.

Usually the following combining methods are used: voting (the class that was issued by a simple majority of the ensemble models is chosen) or averaging, which can be defined as a simple average of the outputs of all models (if weighted averaging is performed, the outputs of the models are multiplied by the corresponding weights). The efficiency of begging is achieved due to the fact that the basic algorithms trained on different subsamples are quite different, and their errors are mutually compensated during voting, and also due to the fact that the outlier objects may not fall into some training subsamples.

Random forest is also a method of begging, but unlike its basic version, random forest adds several features:

1. Uses within itself an ensemble of only regression or classification decision trees.
2. In addition to randomly selecting training objects, the sampling algorithm also randomly selects traits.
3. For each selection, the decision tree is built until the training examples are fully exhausted and are not subjected to the branch cutoff procedure.

5.2.3 Neural Network Method

We can say that any neural network (NM) acts as follows: iteration by iteration deforms a vector of input data so that as a result of the deformation the data provided at the input gets to the areas where we expect to see it at the output. If we consider conventional NMs, each individual example is perceived without taking into account the influence of past information on the current result.

Recurrent Neural Networks (RNN) have been developed to solve this problem. These are networks that contain feedbacks and allow to take into account the previous iterations.

A recurrent network can be thought of as several copies of the same network, each copy carrying information about a further copy. RNNs resemble a chain, and their architecture is well suited for dealing with data such as sequences, lists, and time series.

Recently, a recurrent NM architecture called long short-term memory neural networks (LSTM) has become popular. The idea is that the proposed method that certain elements in the context in previous iterations have more influence on the result, and certain elements have less. In LSTM NM it is proposed to extend the classical scheme of recurrent NM with such notion as a gate—an element within NM, which is "memory gate" and "forget gate", and defining with what probability it is necessary to forget or remember this example for next iterations.

A full-coherent multilayer cross-layer neural network was used to perform the classification. The neural network had ten hidden layers, five of which were full-link layers with a sigmoidal neuron activation function. To prevent the effect of overtraining, the network included regularization layers, one layer after each full-link layer. The batch normalization method has been used as a method of regularization. This method allows to increase the performance and stabilize the neural network. Normalization of the input NM layer is usually performed by scaling the data fed into the activation functions. Data normalization can also be performed in the hidden layers of neural networks. The PU method has the following advantages: it reduces the magnitude by which the values of nodes in hidden layers are shifted (the so-called covariance shift); faster convergence of models is achieved despite performing additional calculations; batch normalization allows each network layer to learn more independently from other layers; it becomes possible to use a higher learning rate, because batch normalization ensures that outputs of neural network nodes without too big or small values; the regularization mechanism is fulfilled.

Adam (Adaptive Moment Estimation) stochastic optimization method was chosen as the training method. The Adam optimization algorithm is an extension of the stochastic gradient descent method with iterative updating of network weights based on training data. The advantages of this method are: it is easy to implement; computationally efficient; it does not require large amounts of memory; and it is well suited for tasks that are large in terms of data and/or parameters. It combines the idea of motion accumulation with the idea of weaker updating weights for typical features. It also shows the frequency of gradient variability. The authors of the algorithm proposed to estimate also the average uncentered variance for this purpose.

The developed neural network had a total of 250 training neurons. The network input was a vector of attributes: statistical, fractal and recurrence characteristics, while the output was an indicator of the DDoS attack presence in the traffic time series.

Thus, machine learning algorithms that can be applied to classification problems arising in a wide variety of domains were improved. The following most effective algorithms were selected: decision tree method, tree ensembles and neural network method.

5.3 Simulation Results of Machine Learning Methods

The objects were dataset traffic implementations [33, 34] containing a DDoS attack and traffic implementations without an attack. The classification method (classes: there is an attack or no attack) was chosen a meta-algorithm based on random forest decision trees using regression decision trees [10, 18]. The output of the model was the ability to match traffic to a given class.

The decision tree models were built using the Python language with libraries that implement machine learning techniques. Models for each class were trained on 500 time-series training examples and tested on 50 test examples.

The features according to which the time series were classified were the estimates of statistical, fractal, and recurrence characteristics calculated for each time series. The following characteristics were used: statistical, such as mean and RMS value of the time series; fractal, such as range of the generalized Hurst index, Hurst index value, RMS value and others; periodic, such as average length of diagonal and vertical lines, measure of determinism, laminarity, etc. [23]. Classification was performed for time series of different lengths (value 1000 5000), but the results do not differ significantly, so data are presented for a series size of 1000 values. Python with appropriate machine learning method libraries was used to classify and apply the random forest method.

Classification was performed using different methods: a multilayer perceptron-type neural network, a random forest method with different parameters [19, 22, 23]. After a series of experiments, a neural network structure containing 10 hidden layers with a total number of 250 neurons was chosen, and 100 regression trees were taken to create a random forest.

To classify time series by random forest method using regression decision trees, estimates of statistical, fractal, and recurrent characteristics calculated for each time series were used. The following characteristics were used: statistical, such as the mean and RMS value of the time series; fractal, such as the range of the generalized Hearst index, Hearst index value, RMS value, and others; periodic, such as average diagonal and vertical line length, a measure of determinism, a measure of laminarity, etc. [32].

Two samples were chosen for classification: training and test. They consisted of traffic realizations containing and not containing attacks. The DDoS attack implementations were chosen at random times and their durations ranged from 1 to 1/2 the length of the traffic.

Statistical, fractal, and recurrence characteristics calculated from the time series were features and fed to the input of the classifier. The input values were 1 or 0: the presence or absence of DDoS attacks in the traffic time series.

To compare the classification results, we focused on the following metrics: True Positive (TP), False Negative (FN), and the proportion of correctly identified objects in both classes (accuracy). True Positive value corresponds to the probability of correct attack detection, False negative value corresponds to the false detection of normal traffic as attacked, F-measure value means cumulative test of attack detection.

Table 2 shows the average probabilities of attack detection. True Positive value corresponds to the probability of correct attack detection, False negative value corresponds to false detection of normal traffic as attacked, F-measure value means cumulative attack detection test. Table 3 shows as expected, that the probability of attack detection depends significantly on the type of attack.

It should be noted that the probability of detecting an attack depends significantly on the value of the Hurst parameter. The highest probability is that the value of the Hurst parameter for normal traffic is the most different from the Hurst value for an

Table 2 Values of true-positive, false-negative and F-measure for machine learning methods

Probability	Methods			
	Random forest	Random forest + regression trees	Random forest + self-similar	Neural network
True positive	0.83	0.87	0.92	0.76
False negative	0.1	0.12	0.29	0.09
F measure	0.77	0.79	0.8	0.85

Table 3 Values of true-positive, false-negative and F-measure for machine learning methods for different types of attack

Probability	attacks	Methods			
		Random forest	Random forest + regression trees	Random forest + self-similar	Neural network
True positive	DDoS	0.83	0.87	0.92	0.76
	UDP-flood	0.89	0.92	0.9	0.88
	MAC flooding	0.88	0.94	0.95	0.92
	ARP-spoofing	0.9	0.95	0.96	0.94
	HTTP flood	0.88	0.9	0.91	0.89
False negative	DDoS	0.3	0.12	0.29	0.09
	UDP-flood	0.19	0.2	0.16	0.18
	MAC flooding	0.12	0.1	0.2	0.08
	ARP-spoofing	0.21	0.2	0.17	0.19
	HTTP flood	0.22	0.21	0.18	0.17
F measure	DDoS	0.92	0.94	0.96	0.95
	UDP-flood	0.89	0.9	0.96	0.91
	MAC flooding	0.77	0.79	0.8	0.85
	ARP-spoofing	0.78	0.82	0.84	0.8
	HTTP flood	0.82	0.83	0.84	0.86

attack. Since attacks have high values H, the probability of detecting an attack is highest for traffic with small values of the parameter H.

A comparative analysis of binary classification of fractal time series by machine learning methods is performed. We consider classification by the example of DDoS attack detection in traffic implementations. Random forest with regression trees, multilayer perceptron with periodic normalization, Random forest using regression decision trees, were chosen as classification methods. The random forest method has proven to be better at detecting normal traffic than a neural network, but neural networks have a lot of potential for improvement.

References

1. Jeong HDJ, Ahn W, Kim H, Lee JSR (2017) Anomalous traffic detection and self-similarity analysis in the environment of ATMSim. Cryptography 1(3):1–24
2. Scarfone K, Mell P (2007) Guide to intrusion detection and prevention systems (IDPS). NIST Special publication 800–94
3. Common Vulnerability Scoring System v3.0: Examples, forum of incident response and security teams. https://www.first.org/cvss/examples

4. Schaelicke L, Wheeler KB, Freeland C (2005) SPANIDS: a scalable network intrusion detection load balancer. In: Computing Frontiers: proceedings of the second conference,. Ischia, Italy, 4–6 May 2005. https://doi.org/10.1145/1062261.1062314
5. Barracuda Load Balancer ADC. Secure application delivery & load balancing. Barracuda. https://www.barracuda.com/products/loadbalancer/features
6. Deka R, Bhattacharyya D (2016) Self-similarity based DDoS attack detection using Hurst parameter. Secur Commun Netw 9:4468–4481. https://doi.org/10.1002/sec.1639
7. Wu M, Moon Y (2019) Alert correlation for cyber-manufacturing intrusion detection. Procedia Manuf 34:820–831. https://doi.org/10.1016/j.promfg.2019.06.197
8. Daradkeh YI, Kirichenko L, Radivilova T (2018) Development of QoS Methods in the Information Networks with Fractal Traffic. Int J Electr Telecommun 64(1):27–32. https://doi.org/10.24425/118142
9. Weber M, Pistorius F, Sax E, Maas J, Zimmer B (2019) A hybrid anomaly detection system for electronic control units featuring replicator neural networks. In: Arai K, Kapoor S, Bhatia R (eds) Advances in information and communication networks, FICC 2018, Advances in intelligent systems and computing, vol. 887. Springer, Cham, pp 43 62. https://doi.org/10.1007/978-3-030-03405-4_4
10. Kirichenko L, Radivilova T, Ryzhanov V (2022) Applying visibility graphs to classify time series. In: Babichev S, Lytvynenko V (eds) Lecture notes in computational intelligence and decision making. ISDMCI 2021. Lecture notes on data engineering and communications technologies, vol 77. Springer, Cham, pp 397–409. https://doi.org/10.1007/978-3-030-82014-5_26
11. Kumar V, Sinha D (2021) A robust intelligent zero-day cyber-attack detection technique. Complex Intell Syst 7:2211–2234. https://doi.org/10.1007/s40747-021-00396-9
12. Ageyev D, Radivilova T, Mohammed O (2020) Traffic monitoring and abnormality detection methods analysis. In: 2020 IEEE international conference on problems of infocommunications. Science and Technology (PIC S&T), pp. 823–826. https://doi.org/10.1109/PICST51311.2020.9468103
13. Monshizadeh M, Khatri V, Atli BG, Kantola R, Yan Z (2019) Performance evaluation of a combined anomaly detection platform. IEEE Access 7:100964–100978. https://doi.org/10.1109/ACCESS.2019.2930832
14. Radivilova T, Lyudmyla K, Lemeshko O, Ageyev D, Tawalbeh M, Ilkov A (2020) Analysis of approaches of monitoring, intrusion detection and identification of network attacks. In: 2020 ieee international conference on problems of infocommunications. Science and technology (PIC S&T), pp 819–822. https://doi.org/10.1109/PICST51311.2020.9467973
15. Jyothsna V, Prasad KM (2019) Anomaly-based intrusion detection system. Computer and network security. IntechOpen. https://doi.org/10.5772/intechopen.82287
16. KhanM A, Karim MR, Kim Y (2019) A scalable and hybrid intrusion detection system based on the convolutional-LSTM network. Symmetry 11:581–585. https://doi.org/10.3390/sym11040583
17. Khraisat A, Gondal I, Vamplew P et al (2019) Survey of intrusion detection systems: techniques, datasets and challenges. Cybersecur 2:20. https://doi.org/10.1186/s42400-019-0038-7
18. Kirichenko L, Alghawli ASA, Radivilova T (2020) Generalized approach to analysis of multifractal properties from short time series. Int J Adv Comput Sci Appl (IJACSA) 11(5):183–198. https://doi.org/10.14569/IJACSA.2020.0110527
19. Kirichenko L, Bulakh V, Radivilova T (2020) Machine learning classification of multifractional Brownian motion realizations. In: Proceedings of the third international workshop on computer modeling and intelligent systems (CMIS-2020), vol 2608. Zaporizhzhia, Ukraine, April 27–May 1, pp 980–989
20. Elsayed MS, Le-Khac N, Dev S, Jurcut AD (2020) DDoSNet: A deep-learning model for detecting network attacks. In: 2020 IEEE 21st international symposium on "A world of wireless, mobile and multimedia networks" (WoWMoM), pp 391–396
21. Sharafaldin I, Habibi Lashkari A, Hakak S, Ghorbani AA (2019) Developing realistic distributed denial of service (DDoS) Attack dataset and taxonomy. In: 2019 international carnahan conference on security technology (ICCST), pp 1–8

22. Kirichenko L, Zinchenko P, Radivilova T (2021) Classification of time realizations using machine learning recognition of recurrence plots. In: Babichev S, Lytvynenko V, Wójcik W, Vyshemyrskaya S (eds) Lecture notes in computational intelligence and decision making. ISDMCI 2020. Advances in intelligent systems and computing, vol 1246. Springer, Cham, pp 687–696. https://doi.org/10.1007/978-3-030-54215-3_44
23. Kirichenko L, Radivilova T, Bulakh V (2019) Machine learning in classification time series with fractal properties. Data 4(1), 5:1–13. https://doi.org/10.3390/data4010005
24. Radivilova T, Kirichenko L, Ageyev D, Tawalbeh M, Bulakh V, Zinchenko P (2019) Intrusion detection based on machine learning using fractal properties of traffic realizations. In: 2019 IEEE international conference on advanced trends in information theory (ATIT). Kyiv, Ukraine, pp 218–221. https://doi.org/10.1109/ATIT49449.2019.9030452
25. Kelley T, Amon MJ, Bertenthal BI (2018) Statistical models for predicting threat detection from human behavior. Front Psychol 9:466. https://doi.org/10.3389/fpsyg.2018.00466
26. Li Y, Sperrin M, Ashcroft DM, van Staa TP (2020) Consistency of variety of machine learning and statistical models in predicting clinical risks of individual patients: longitudinal cohort study using cardiovascular disease as exemplar BMJ 371:m3919. https://doi.org/10.1136/bmj.m3919
27. Srinivasa Reddy L, Vemuru S (2020) A survey of different machine learning models for static and dynamic malware detection. Europ J Mol Clin Med 7(3):4299–4308
28. Magán-Carrión R, Camacho J, Maciá-Fernández G, Ruíz-Zafra Á (2020) Multivariate statistical network monitoring–sensor: an effective tool for real-time monitoring and anomaly detection in complex networks and systems. Int J Distrib Sens Netw 2020. https://doi.org/10.1177/1550147720921309
29. Zhu S, Li S, Wang Z, Chen X, Qian Z, Krishnamurthy SV, Chan KS, Swami A (2020) You do (not) belong here: detecting DPI evasion attacks with context learning. In: Proceedings of the 16th international conference on emerging networking experiments and technologies (CoNEXT'20). Association for computing machinery, New York, NY, USA, pp 183–197. https://doi.org/10.1145/3386367.3431311
30. Zhao J, Shetty S, Pan J et al (2019) Transfer learning for detecting unknown network attacks. EURASIP J. on Info Secur 1. https://doi.org/10.1186/s13635-019-0084-4
31. Radivilova T, Kirichenko L, Vitalii B (2019) Comparative analysis of machine learning classification of time series with fractal properties. In: 2019 IEEE 8th international conference on advanced optoelectronics and lasers (CAOL). Sozopol, Bulgaria, pp 557–560. https://doi.org/10.1109/CAOL46282.2019.9019416
32. Kirichenko L, Radivilova T, Bulakh V (2020) Binary classification of fractal time series by machine learning methods. In: Lytvynenko V, Babichev S, Wójcik W, Vynokurova O, Vyshemyrskaya S, Radetskaya S (eds) Lecture notes in computational intelligence and decision making. ISDMCI 2019. Advances in intelligent systems and computing, vol 1020. Springer, Cham pp701–711. https://doi.org/10.1007/978-3-030-26474-1_49
33. Sharafaldin I, Habibi Lashkari A, Ghorbani AA (2018) A detailed analysis of the CICIDS2017 Data Set. ICISSP
34. Intrusion detection evaluation dataset (CIC-IDS2017). https://www.unb.ca/cic/datasets/ids-2017.html

Criteria and Indicators of Efficiency of Cryptographic Protection Mechanisms

Alexandr Kuznetsov⬤, **Yurii Horbenko**⬤, **Kostiantyn Lysytskyi**⬤, and **Oleksii Shevtsov**⬤

Abstract Solutions to build monitoring systems to prevent the spread of infectious diseases already exist and are developing rapidly in many countries around the world. In this regard, in the means of tracking contacts as the main subsystem of population monitoring, there are practical and theoretical tasks for the mechanisms of cryptographic protection of personal data of users, providing: efficiency; objectivity of decision-making; depersonalization; generalized processing, reliability, availability of decentralized storage. This section substantiates the criteria and indicators of the effectiveness of cryptographic protection mechanisms. In particular, general theoretical information on criteria and indicators for assessing the security of symmetric cryptocurrencies, hash functions, generation of pseudo-random sequences, asymmetric cryptographic transformations, etc. is presented. This section discusses general approaches that can be used to determine the criteria and performance indicators of cryptographic protection mechanisms for users' personal data and interoperability protocols that provide various security services when tracking contacts with system users.

Keywords Efficiency criteria and indicators · Cryptographic transformations · Blockchain technology · Monitoring systems · Decentralized systems

1 Introduction

Solutions for building monitoring systems to prevent the spread of infectious diseases already exist and are developing rapidly in many countries around the world. In this regard, the means of tracking contacts as the main subsystem of global monitoring of

A. Kuznetsov (✉) · Y. Horbenko · K. Lysytskyi · O. Shevtsov
V. N. Karazin Kharkiv National University, Svobody sq., 4, Kharkiv 61022, Ukraine
e-mail: kuznetsov@karazin.ua

Y. Horbenko
e-mail: gorbenkou@iit.kharkov.ua

A. Kuznetsov · Y. Horbenko
JSC "Institute of Information Technologies", Bakulin St., 12, Kharkiv 61166, Ukraine

© The Author(s), under exclusive license to Springer Nature Switzerland AG 2022
R. Oliynykov et al. (eds.), *Information Security Technologies in the Decentralized Distributed Networks*, Lecture Notes on Data Engineering and Communications Technologies 115, https://doi.org/10.1007/978-3-030-95161-0_6

the population there are practical and theoretical problems regarding the mechanisms of cryptographic protection of personal data of users, in particular, for example, in interaction protocols that provide:

- efficiency;
- objectivity of decision-making;
- depersonalization;
- generalized processing;
- certainty;
- availability of decentralized storage.

This paper substantiates the criteria and indicators of efficiency of cryptographic protection mechanisms of personal data of users and protocols of interaction, in particular the substantiation of criteria and indicators of efficiency, objectivity of decision-making, depersonalization, generalized processing, reliability, availability of decentralized storage.

2 General Theoretical Information on the Criteria and Indicators for Assessing the Security of Symmetric Cryptocurrencies

The general provisions of the theory and problematic issues of cryptographic transformations help to formulate the *main indicators for assessing cryptographic stability*.

Solving the main problem of cryptanalysis, ie determining the key K_j, or a pair of keys $(K^\partial j, K^c j)$, which were used in symmetric cryptographic transformation, possibly provided the use of key indicators regarding cryptographic transformation [1].

In the future we will choose the following as the main indicators:

- the size of the set of keys N_k, that is, the number of keys that can be used in cryptographic transformation;
- entropy of the key source $H(K)$;
- secure time (mathematical expectation of the key determination time used) t_6;
- the power of the crypto-analytical system γ in the form of the number of group operations that can be performed by it;
- the probability of successfully solving the problem of cryptanalysis P_y;
- unity distance for symmetric crypto conversion (cipher) l_0.

Other indicators and criteria will be introduced and applied as needed.

The entropy of the key source will be defined as

$$H(K) = - \sum_{k=1}^{n_k} P(K_i) \log P(K_i) \tag{1}$$

where $P(K_i)$—probability of occurrence K_i key in the system.

The secure time t_6 will be defined as the mathematical expectation of the time of discovery of the cryptosystem, for example, finding a secret key, using a specific method of cryptanalysis:

$$t_6 = \frac{N_e}{\gamma K} P_y, \tag{2}$$

where:

N_e the number of group operations that the cryptanalyst must perform;
γ power of crypto analytical system (group operations per second);
K the number of seconds per year is approximately 3.15×10^7 (s/year).
P_y the probability with which the cryptanalysis problem must be successfully solved;
l_0 unity distance for symmetric crypto conversion (cipher).

Thus, the assessment of the complexity of cryptanalysis can be performed using a tuple of parameters

$$(N_k, H(K), t_6, \gamma, P_y, l_0) \tag{3}$$

One of the important tasks in the theory of stability is the classification of cryptographic systems according to the level of guarantees of stability. It can be solved on the basis of using the parameters of the tuple (3), but in our opinion, the leading indicator is the safe time t_6. The following is the classification of ciphers by this indicator.

Certainly stable or theoretically not decrypted are cryptosystems (ciphers), in which the requirements for the choice of parameters and mechanisms of key management, according to modern views, provides a condition:

$$t_6 \rightarrow \infty \tag{4}$$

Cryptosystems (ciphers) are computationally stable, for which the complexity of cryptanalysis with certain parameters is estimated as exponentially complex, and the value of secure time is much longer than the value of information, i.e.

$$t_6 \gg t_{yi} \tag{5}$$

Probably stable cryptosystems (ciphers), for which the complexity of cryptanalysis is assessed as exponentially or subexponentially complex, and proving the stability of which is reduced to solving mathematical problems, the complexity of which is not proven and the value of secure time is much longer than the value of information, i.e.

$$t_6 \gg t_{yi} \tag{6}$$

Computationally unstable cryptosystems (ciphers), in which the complexity of cryptanalysis is at the limit of the cryptanalyst's capabilities, or in which weaknesses are forced or incompetent due to design, and the value of a secure order of magnitude or less is the value of information, i.e.

$$t_6 \leq t_{yi} \tag{7}$$

In our opinion, the proposed classification allows to practically evaluate and compare crypto transformations, cryptosystems and cryptographic algorithms of different nature.

Conditions and examples of implementation of cryptosystems with an unconditional level of stability. Following a formalized approach, we consider the general conditions and requirements for building cryptosystems with unconditional, computational (guaranteed), probably stable and computationally unstable cryptosystem.

Determining the conditions for the implementation of cryptosystems of unconditional stability is difficult and ambiguous. But at the formal level they can be identified by making demands on practical implementation. In [2] at a sufficiently formal level, the necessary and sufficient conditions for ensuring unconditional stability are defined. Cryptographic transformations that provide unconditional stability can be used to protect particularly critical information in terms of ensuring a high level of confidentiality.

A necessary and sufficient condition for ensuring the unconditional stability of symmetric cryptocurrency is the condition that:

$$P(C_j/M_i) = P(C_j), \tag{8}$$

that is, the probability of occurrence C_j cryptograms at the output of the means of symmetric crypto conversion of the encoder should not depend on which M_i the message appears at the output of the message source. Otherwise, the probability of the cryptogram appearing should be the same for all keys and for all messages. This means that any message can be encrypted in any cryptogram with the same probability.

Thus, for an unconditionally stable system l_0 not less than the length of the message and not less than the length of the key since $d < 1$. Therefore, for a successful cryptanalysis, the cryptanalyst must receive at least l_K characters, i.e. the whole key.

For an unconditionally stable crypto conversion, the relation that also holds

$$l_K \log m_K \geq l_m \log m_M, \tag{9}$$

where we get that

$$l_M \le l_K. \tag{10}$$

Determination of computationally stable cryptocurrency (cryptosystem) and conditions of its implementation. According to (5), under computationally stable cryptosystems (ciphers) we will consider those for which the complexity of cryptanalysis is exponential and the value of secure time is much longer than the value of information, i.e.

$$t_6 \gg t_{yi}$$

In practice, computationally stable cryptosystems (ciphers) are called ciphers with guaranteed cryptographic stability. Examples of such ciphers are block and stream symmetric ciphers. We define the model and consider the conditions for the implementation of computationally stable cryptocurrency (cryptosystem) on the example of symmetric ciphers.

A feature of computationally stable cryptosystems, as opposed to unconditionally stable, instead of a key sequence K_i, used in unconditionally stable ciphers, gamma encryption is used Γ_i. Encryption is performed by composing message characters M_i and encryption gammas Γ_i for the module m and has the appearance:

$$C_i = (M_i + \Gamma_i) \bmod m \tag{11}$$

$$\Gamma_i = \Psi(K_j, \mathrm{Pr}), \tag{12}$$

Function (12) determines the gamma generation rule Γ_i, K_j—source (initial, working, current) key, and the gamma of encryption is generated in accordance with this key and the current parameters Pr cipher (block or current).

1. As an indicator of the complexity of the scale, you can use the concept of its structural secrecy, for example, in the form [1, 3]:

$$S_C = \frac{l_H}{l_n}, \tag{13}$$

where l_n—the number of gamma characters you need to know to reveal the remaining ones, i.e. $l_n - l_н$. According to (13) gamma Γ_i will be completely unpredictable if $l_H \to l_n$, which means $S_c \to 1$.

Indicators and a tuple of parameters can be used to assess the stability of a computationally stable cryptosystem $(N_k, H(K), t_6, \gamma, P_y, l_0)$, previously introduced.

2.1 General Criteria for Protection of Promising Block Symmetric Transformations with Respect to Cryptographic Stability

The results of research [1, 3] allow us to identify the following as the main criteria.

1. According to current views, cryptographic symmetric block transformations (hereinafter SBT) and block symmetric ciphers developed on their basis (hereinafter BSC) are the main cryptographic mechanism for ensuring confidentiality and integrity in information processing in modern information and telecommunication systems (ITS). In addition, BSC are used to ensure integrity, as well as a basic element in the construction of other cryptographic primitives, such as pseudo-random sequence generators (PRSGs), stream ciphers, and hashing functions. The level of stability and properties of BSS significantly determine the stability of cryptographic protection of information, the security of cryptographic protocols and the security of ITS in general.

2. As a rule, in ITS with cryptographic protection of information and information resources, the length of the message protected using BSC significantly exceeds the length of the encryption key, i.e. the entropy of the message source significantly exceeds the entropy of the key source. In this case, the criterion of unconditional stability is not met with respect to BSC, and in such conditions it is advisable to introduce a polynomial criterion, which implies the existence of restrictions on the computing resources of the attacker and the time during which the cipher remains stable.

    Such a polynomial criterion leads to a practical criterion of stability—the impossibility of an attack on the cipher in the modern computer base, including taking into account the progress of increasing the power of computers and the advent of quantum computers over the long term (for example, 10^{10} years).

3. The cryptographic stability of the cipher depends on the complexity of the attack on the symmetric block cipher. As indicators of the complexity of cryptanalysis, as a rule, can or should be used as follows.

    Temporary—mathematical expectation of time (safe time) tb required to implement an attack on available/promising computing facilities.

    Spatial complexity is the amount of memory required to perform cryptographic analysis.

    The minimum number of ciphertext/plaintext pairs or the number of ciphertexts required for a successful attack.

4. Preliminary analysis allows us to conclude that if at least one of these indicators, the implementation of the attack in practice is impossible with a significant margin of safety, the encryption algorithm can be considered stable.

5. As a rule, the initial assessment of resilience should be done in relation to force attacks: complete search of keys, attack on the dictionary, the creation of collisions, and so on. Provided that the required level of resistance to force

attacks is provided, an assessment of resistance to analytical attacks must be performed.

6.   The results of the analysis show that in relation to modern BSC, as criteria for assessing resistance to analytical attacks, it is recommended to use the following:

- the set of encrypted/plaintext required to perform a cryptanalytic attack must exceed the capacity of the set of valid encrypted/plaintext;
- the complexity of any analytical attack must be greater than or equal to the complexity of the force attack;
- to implement an analytical attack, the required number of group encryption operations must be not less than with a full search of keys;
- the amount of memory required to store intermediate results when carrying out an analytical attack must be not less than when implementing a full-cipher dictionary attack;
- given the possibility of improving crypto-analytical methods, it is necessary to use the criterion of "margin of resistance" to analytical attacks— according to which the complexity of the attack on the whole algorithm should be much higher than the complexity of force attacks. Typically, this criterion considers a version of the encryption algorithm with a reduced number of cycles, or when excluding or replacing operations with a less complex full-cycle version of the simplified cipher must remain resistant to analytical attacks.

7.   Most modern analytical attacks, primarily such as differential and linear cryptanalysis, are statistical. Therefore, a necessary (but not sufficient) condition for the stability of the cipher to analytical attacks is to provide "good" statistical properties of the original sequence (cipher texts).

8.   It must also be proved that in order to protect the SBT (BSC) from algebraic attacks, it is necessary that there is no way to practically construct a system of equations linking plaintext, cryptogram and encryption key, or that there is no way to solve such systems in polynomial time.

9.   When constructing means of cryptographic protection of information (CPI) it is necessary to take into account the possibility of organizing and carrying out attacks on the implementation (change in temperature of the electronic device, input voltage, the appearance of ionizing radiation, measurement of currents, PEMIN, execution time, etc.)

10. In general, the following criteria for the sustainability of modern SBT (BSC) can also be formulated.

- ensuring resistance to force attacks, for example according to the temporal or spatial criterion of the amount of memory required to store intermediate results;
- lack of ways to build or solve a system of equations that connects plaintext, cryptogram and encryption key;

- the impossibility of implementing known analytical attacks on the cipher, or their complexity must be higher than the complexity of the implementation of force attacks (for example: one of the following criteria: the power of the required set of open/encrypted messages; the required number of encryption operations; the amount of memory required to store intermediate results);
- the presence of a "margin of stability" of the cipher (additional encryption cycles), which ensures the safe use of the algorithm in the case of improving crypto-analytical attacks;
- the stability of a simplified version of the cipher, in which some operations are excluded or replaced by simpler ones;
- providing "good" statistical properties of the original cipher sequence (cryptograms or encryption scales), in which cryptograms and encryption scales do not differ in properties from the random sequence.

11. In addition to the required level of cryptographic stability, SBT (BSC) is required to ensure a high level of performance (complexity of encryption, decryption and key deployment).

2.2 Special Security Criteria for Promising Block Symmetric Transformation

The main criteria of protection given in paragraph 1 and results of researches [1, 3] allow to define such criteria of an estimation of protection of perspective symmetric transformation.

1. Protection of SBT from cryptanalytic attacks. The main methods of cryptanalysis are: differential cryptanalysis, extensions for differential cryptanalysis, search for the best differential characteristic, linear cryptanalysis; interpolation intrusion; intrusion with partial guessing of the key; intrusion using a bound key; intrusion-based intrusion handling; search for loopholes and potential attacks, etc.
2. Statistical security of the encryption algorithm. Statistical security means the statistical independence of encrypted message blocks and encryption scales from the public information blocks and keys used.
3. Features of a design and openness of structure of SBT. The cryptoalgorithm must have a clear, easily analyzed structure and be based on a reliable mathematical apparatus.
4. Modification stability, when all candidates for SBT are tested for resistance to various modifications: resistance to cryptanalytic attacks while reducing the number of cycles, reduced components used in SBT, etc.
5. Computational complexity (speed) of encryption/decryption of the block symmetric transformation algorithm. The complexity of software, hardware and software-hardware implementation should be assessed by the amount of

memory, both for software and hardware implementation. Including, in software implementation—the amount of required RAM, the size of the source code, the speed of the program on different platforms in the implementation of known programming languages. When hardware is estimated by the number of valves and speed in Mb/s.

6. Universality of cryptoalgorithm: possibility of work with various lengths of initial keys and information blocks; security of implementation on various platforms and applications; the possibility of using a cryptographic algorithm in the necessary reasonable modes of operation of the BSC.

7. Reliability of the mathematical base in relation to the lack of ability to carry out attacks "universal disclosure" due to imperfections or the intentional specific mathematical base. It is believed that such an attack of universal disclosure has a much smaller complexity than the complexity of the attack "brute force".

8. The absence of weak initial keys and the suspicion of the existence of keys in which the complexity of the cryptanalytic attack is less than the complexity of the attack "brute force".

9. Practical protection of the encryption algorithm from force attacks, which can be achieved through the use of symmetric block cryptocurrencies, with a block length of at least 128 bits and a key length of at least 128 bits (in some cases 256 or 512 bits). Allocate the necessary criteria for assessing the stability of BSC:

- ultra-high stability, when the lengths of the information block and the length of the output key are not less than 512 bits;
- high stability, when the length of the information block and the length of the key is not less than 256 bits;
- normal level of stability, when the length of the information block and the length of the key is not less than 128 bits;
- satisfactory level of stability, when the length of the information block is not less than 64 bits, and the key length is not less than 128 bits.

10. Implementation of the cryptoalgorithm: the cryptoalgorithm should be oriented for the possibility of implementation on 32-bit or 64-bit processors; provide for the possibility of parallel execution of several operations (if possible).

    It should be emphasized that these criteria are necessary but not sufficient.

3 Basic Security Criteria for Hashing Functions for Cryptographic Applications in Pseudo-Random Sequence Generation Algorithms

Hash functions are also widely used to generate pseudo-random sequences (PRS) [1, 3]. In fact, all such algorithms are described by a formula.

$h_i = H(h_{i-1}, i, Par)$. The output of the algorithm is one or more bits of the value h_i.

The main criteria for assessing the protection of PRS generators are:

- good statistical properties of the generated sequence, ie ensuring indistinguishability;
- the impossibility of predicting PRS on the intercepted fragment, ie ensuring the property of unpredictability;
- a given period of gamma;
- allowable complexity of production (speed).

The indistinguishability (good statistical properties) of PRS means the inability to distinguish PRS from a truly random sequence using a finite number of tests with a finite length of the message [1, 3]. To fulfill this requirement, the output of the hash function must look like a random string of bits and change unpredictably when at least one bit in the incoming message changes. The hash values themselves must be evenly distributed.

To ensure that the output of the PRS generator cannot be predicted, the hash function must be resistant to prototype recovery attacks. The essence of the prototype recovery attack is in finding the message that was used to calculate the specified hash value. Since the set of incoming messages has more power than the set of outgoing hash values, it is not possible for an attacker to establish which of the found prototypes was used to calculate the hash value. They can establish this only by the results of an attempt to predict the release of PRS. If the predicted output of the PRS generator differs from that obtained, then the second prototype was found.

A PRS generator acceptable for practical use must have a sufficiently long gamma period. Short-range gammas appear at the output of the PRS generator if i—that value of the state of the generator coincided with one of the previous ones. For the hash function, this means finding two incoming different messages with the same hash value, i.e. collision.

To ensure sufficient performance of the PRS generator, the hash function in its basis must have sufficient speed, and the time of initialization of the hash function is of particular importance, because the initialization of the hash function occurs at each calculation of the next fragment of the gamma.

Thus, in the case of use in PRS generators, the hash function must have the following properties:

- the complexity of finding the prototype must be exponentially complex;
- all hash values must be equally probable;
- the probability of collision should not exceed the specified value;
- the output of the hash function should not differ from the random sequence;
- the hashing function must have a minimum initialization time.

3.1 Basic Security Criteria for Hash Functions for Cryptographic Applications in Key Installation Protocols

In key installation protocols, hash functions are typically used to convert a shared secret generated during protocol execution into a shared key. In most cases [1, 3], a shared secret cannot be used as a shared key, usually because of its length or representation (such as the point of an elliptic curve) or because of its statistical properties.

At the same time, adding a fixed pattern to the desired length or trimming the excess bits of the secret is not acceptable, as it leaves the attacker with some information about the key. An attacker can use this information to attack BSC.

Thus, for the standard Diffie-Hellman protocol, the common secret is an element of the field, so there is a high probability that one or more high-order bits will be zero. This fact can be taken into account by an attacker when organizing a forceful attack on a symmetric cipher. In addition, the length of the shared secret in this protocol is usually not less than 1024 bits, so it cannot be used as an encryption key for symmetric systems in which the key lengths do not exceed 512 bits.

Thus secure cryptographic protocols must provide:

- random appearance of keys;
- exactly the probability of occurrence of keys;
- unpredictability of keys;
- allowable complexity (speed) of key generation;
- do not create leakage channels for key information.

Therefore, the following security criteria are applied to the hash functions used in key installation protocols:

- all hash values must be random and equally probable;
- there should be no correlations between the input and output bits of the hash function;
- the hash function should not create conditions for the appearance of side channels of information leakage.

Application of hash functions for control of integrity and authenticity of messages (in schemes of formation of MAC-codes of authentication of messages) and criteria of their protection.

Hashing features are used to control the integrity of messages that are transmitted over a network or stored outside a secure environment. Due to the fact that cryptographic hashing functions are usually keyless, they are used to build algorithms for calculating message authentication codes with the HMAC key or similar [1]. According to the HMAC scheme, the message authentication code is calculated as:

$$HMAC(K, m) = H((K \oplus opad)||H((K \oplus ipad)||m)).$$

The model of authentication systems includes three objects (subjects): sender, verifier and violator. The sender sends a message to the verifier via a communication channel controlled by the violator. The intruder may distort the messages transmitted over the communication channel. The intruder can be a source of the following threats:

- impersonalization—an attempt to initiate a relationship with the verifier, so as to pretend to be a real sender;
- substitution—an attempt to intercept and modify messages from the sender so that the verifier takes them for real;
- spoofing—a threat during which the violator observes r messages from the sender to the verifier, after which it tries to generate $r + 1$ a message that the verifier must recognize as authentic. This threat is called r-th spoofing.

According to [1, 3, 4], the resistance of HMAC to the listed threats is optimal, in the case when the MAC values are equally probable.

The main attacks on MAC codes are [4]:

- The ability to recover the key;
- Forgery based on internal conflict;
- Fake Exor.

Both cryptanalytic attacks on the MAC hash function and analysis of side channels can be used to recover the key.

The key can be restored either by finding a prototype for the hash function or by searching for internal collisions of the hash function.

Counterfeiting based on internal collisions can be implemented if the hashing function based on HMAC is not conflict resistant.

Forgery of Exor is possible if the hashing function does not end with a certain final transformation. Then for some unknown violator message M, hash value of which $h(M)$ known to the violator, they can calculate $h(M||M')$ as $h(h(M)||M')$. For hashing functions, this attack is known as "length extension attack". The HMAC scheme is not vulnerable to such an attack even if a vulnerable hashing function is used, but other MAC calculation schemes may be vulnerable to such an attack. For example, if $MAC(K, M) = h(K||M)$ then the violator can calculate $MAC(K||M||M')$ as $H(MAC(K||M)||M')$, thus carrying out the forgery of Exor.

Thus, for the hashing function used as message authentication codes, the following security criteria can be formulated:

- Reproducibility in space and time;
- Resistance to collisions;
- Resistance to finding a prototype;
- Resistance to message extension attack;
- No side channels for information leakage.

3.2 Basic Security Criteria for Hash Functions for Cryptographic Applications in Data Integrity Control Schemes Without the Use of Shared Secrecy

For this case, the threat model does not differ from the threat model for schemes for monitoring the integrity of messages using a shared secret [4].

The only difference is that the control hash value is transmitted over an authenticated channel (integrity control when transmitted in space) or stored in an authenticated storage (integrity control when transmitted over time).

The following security criteria must be met for the hash function used in message authentication codes:

- Reproducibility in space and time;
- Resistance to collisions;
- Resistance to finding a prototype;
- Resistance to message extension attack.

If the message is confidential, then there is an additional requirement for the absence of side channels of information leakage.

3.3 Basic Security Criteria for Hash Functions for Cryptographic Applications in Password Authentication Schemes

Hash functions are widely used in MAC password authentication schemes [1, 3, 4]. A password is a string of confidential characters that a user presents to the system as an access attribute. If the password is correct, the user receives certain rights in the system, defined by his role. To determine the correctness of the password, the system must compare the value obtained with the reference stored in the system itself. However, keeping the password open in the system is dangerous. An attacker could gain unauthorized access to the password store by causing the system to crash when the store is not protected, such as turning off the power and booting from removable media. Encrypting the repository is not the solution, as the encryption key must also be stored on the system, so it can also be compromised. The way out of this situation is to store not the passwords themselves, but their hash values. In this case, the attacker in the event of system compromise will receive a hash value of the password, but the shell when logging in will require the password, not its hash value. If the task of recovering a password from a hash value is difficult, then the system is securely protected from such attacks.

With network authentication, such a scheme is no longer secure. An attacker could intercept and save a hash value from a password, using it later to initiate another session on behalf of a legitimate user. To prevent such an attack, the system sends a random number R when initiating each session, and the user responds by sending a hash value from the password and that number $H(R||pass)$. In this case, the hash value in each session will be different and unpredictable. And considering that in system it is better to store not passwords, and their hash values, the user in response sends not $H(R||pass)$, but $H(R||H(pass))$. Because the value $H(pass)$ and R system are known, then comparing the obtained hash value with the calculated, you can unambiguously determine the correctness of the password.

The length of the password can be different, including one that is less than the length of the hash value. Therefore, the hash function must be able to operate in "stretch" mode when the length of the input message is less than the length of the hash value.

An extension of the described schemes is the implementation of unidirectional display of data in order to provide privacy services, which will protect personal data in directories in case of system compromise.

Thus, in password authentication systems, the following security criteria are put forward for the hash function:

- Unpredictability of hash values;
- The ability to work in the mode of "stretching";
- The hash function should not create conditions for the appearance of side channels of information leakage.

3.4 Generalization of the Main Criteria and Security Indicators for Hash Functions for Cryptographic Applications

Summarizing the analysis of existing requirements in view of the above in [1], we substantiate them and consider using a system of unconditional and conditional criteria. Unconditional criteria include those that are mandatory for cryptographic applications such as hashing, ie unconditional.

When substantiating and generalizing the security criteria, we will assume that the hash function allows to calculate the original hash value (hash code) with the length *hlen* at an arbitrary length lm of the incoming message M.

Studies have shown that modern and promising hash functions necessarily have security indicators:

- difficulty of finding a collision

$$C_{col} \geq 2^{hlen/2};$$ (14)

- the difficulty of restoring the prototype M

$$C_{col} \geq 2^{hlen};$$ (15)

- the difficulty of finding another prototype

$$C_{sec\_preim} \geq 2^{hlen}; \tag{16}$$

- the difficulty of finding a collision truncated to ht characters

$$C_{tr\_col} \geq 2^{(hlen-hr)/2}. \tag{17}$$

These requirements allow us to introduce unconditional criteria for evaluating the j-th hash function for cryptographic applications. In this case, the evaluation by the unconditional criterion will be performed using Boolean variables that take the value "1" (true), if the corresponding condition is met, and "0" otherwise. Thus, the stability of the j-th hash function can be described by a vector of Boolean variables $\left(w_1^{(j)}, w_2^{(j)}, w_3^{(j)}, w_4^{(j)}\right)$, which can be formally defined as:

$$w_1^{(j)} = \begin{cases} 1, C_{col} \geq 2^{hlen/2} \\ 0 \end{cases}$$

$$w_2^{(j)} = \begin{cases} 1, C_{preim} \geq 2^{hlen} \\ 0 \end{cases}$$

$$w_3^{(j)} = \begin{cases} 1, C_{sec\_preim} \geq 2^{hlen} \\ 0 \end{cases}$$

$$w_4^{(j)} = \begin{cases} 1, C_{tr\_col} \geq 2^{(hlen-ht)/2} \\ 0 \end{cases} \tag{18}$$

Vector $\left(w_1^{(j)}, w_2^{(j)}, w_3^{(j)}, w_4^{(j)}\right)$ we will call with some clarification the integral unconditional criterion.

Taking into account the above integral unconditional criterion, the correspondence function of the j-th cryptographic hash function is written in the form:

$$f_6^{(j)}\left(w_1^{(j)}, w_2^{(j)}, w_3^{(j)}, w_4^{(j)}\right) = w_1^{(j)} \wedge w_2^{(j)} \wedge w_3^{(j)} \wedge w_4^{(j)}, \tag{19}$$

where the symbol «∧» indicates the conjunction operation of Boolean variables.

Thus, the quality of the hashing function for cryptographic applications can be evaluated using an unconditional integral criterion—the function of compliance of cryptographic hashing requirements:

$$f_6^{(j)}\left(w_1^{(j)}, w_2^{(j)}, w_3^{(j)}, w_4^{(j)}\right) \in [0,1].$$

At

$$f_6^{(j)}\left(w_1^{(j)}, w_2^{(j)}, w_3^{(j)}, w_4^{(j)}\right) = 1 \tag{20}$$

cryptographic hashing is evaluated as meeting the unconditional criteria.

Regarding such an indicator as the complexity (speed) of hashing, it is no longer unambiguous. Therefore, it is necessary to define and justify new indicators and relevant conditional criteria.

Given [1, 3] that the calculation of the hash function can be performed in parallel, we consider the general issues of parallel calculations and the requirements for them.

Thus, according to Amdal's law [3, 4], the acceleration coefficient when performing N parallel calculations is determined by the formula:

$$K_N = \frac{1}{(1 - P) + \frac{P}{N}} \tag{21}$$

where P—the share of parallel calculations, $1 - P$—respectively, the share of sequential calculations.

The physical meaning of this acceleration factor K_N can be interpreted as the ratio of the normalized speed of the algorithm using parallel calculations when performing it on N computing cores to the speed of the same algorithm when running on a single core.

At the same time, it is almost impossible to determine the share of parallel computations in the algorithm based on its specifications. But it is possible to calculate the acceleration factor K_N when using the appropriate number of computing cores, if you use the results of measurements of the speed of the hash function on one and more computing cores

$$K_N = \frac{S_N}{S_1}, \tag{22}$$

where S_1—speed of the hash algorithm on one computing core, but S_N—speed of the hash algorithm on N computing cores.

Using (22), we can obtain Formula (23), which allows us to calculate the fraction of parallel computations, knowing the acceleration factor on N computing cores and the number of these cores. The formula has the form

$$P = \frac{(K_N - 1)N}{(N - 1)K_N}, \tag{23}$$

where, as before, K_N—acceleration factor on N computing cores, P—the share of parallel calculations.

Direct analysis (23) shows that increasing the number of computing cores makes sense to a certain value. Moreover, each new added computing core gives a smaller increase in speed than the previous one. The acceleration factor.

K_{max} at the same time limited from above by value $\frac{1}{1-P}$.

Really, if $N \to \infty$, then

$$K_{max} = \lim_{N \to \infty} \left(\frac{1}{1 - P + P/N} \right) = \frac{1}{1 - P}. \tag{24}$$

But in practice it has already been proved that the maximum value of the acceleration coefficient is unattainable, so from Formula (24) in the general case it is impossible to calculate the maximum number of cores that can be effectively used by the algorithm. The problem can be solved as follows.

The calculation of the maximum possible acceleration can be done if instead K_{max} use some $K'_{max} < K_{max}$. At the same time as far as K'_{max} is close to K_{max} can be determined using the value of the approximation factor r, which is close to one but less than it:

$$K'_{max} = r K_{max} \tag{25}$$

It is recommended to use values for practical calculations $r = 0.95$. The choice of such value r can be explained by the fact that the increase in the number of computing cores after reaching K'_{max} has almost no effect on the speed. Therefore, it is advisable to use this value to calculate other indicators [1, 3].

4 Criteria and Indicators for Evaluating the Properties and Quality of Asymmetric Cryptographic Transformations on the Example of Electronic Signature Transformations

At the international, regional and national levels, electronic signatures (hereinafter ESs) have already become very widespread in electronic document and electronic document management systems, e-mail, payment systems, e-government, e-reporting, e-commerce, etc. Therefore, both research and evaluation of ES at different stages of the ES life cycle—from development to modification and, as well as application at different stages of the life cycle of providing electronic trust services are important theoretical and practical tasks. At the first stage, the problems of determining and selecting criteria and indicators for evaluating the ES should be solved [1, 5].

In the future, the criterion will be understood as a feature on the basis of which the evaluation, definition or classification of something [1], ie, in essence, we will understand the measure of evaluation. Previous studies have concluded that the comparison of cryptographic curves can be done using two sets of criteria: unconditional and conditional [3, 5]. Taking into account [5], the evaluation of cryptoconversions of the ES type is recommended to be performed in 2 stages.

At the first stage, their compliance with unconditional criteria is checked, and at the second stage, appropriate assessments are obtained using conditional criteria. It is through the use of conditional criteria that it is possible to compare different cryptographic conversions of the ES type according to the integral conditional and general criteria. Moreover, in the future, under the compliance of one or another ES with unconditional criteria, we will understand the fact that expert assessments of unconditional criteria are positive, ie they are clearly met.

Unconditional criteria for evaluating cryptographic conversions of the ES type. Unconditional criteria will include those criteria, the fulfillment of which is mandatory for cryptographic conversions of the ES type, ie unconditional.

Analysis of the state of application, experience in developing and evaluating the properties of cryptoconversion such as ES, primarily in the group of EK points [1, 3, 5, 6], the results achieved in the practical solution of cryptanalysis and implementation of various attacks allow to choose as the main at least such unconditional evaluation criteria [1, 5]:

- reliability of the mathematical base used for ES in cryptocurrency conversion;
- practical protection of cryptographic transformations of the ES type from known attacks;
- real protection of ES from all known and potentially possible cryptanalytic attacks;
- statistical security of cryptographic transformation of ES type;
- theoretical security of cryptographic transformation of ES type in a group of points of an elliptic curve;
- lack of weak personal keys of cryptographic transformation of ES type;
- complexity of direct $I_{\text{пр}}$ and reverse $I_{\text{зв}}$ cryptographic transformations for ES is not higher than polynomial [5].

Let's consider these criteria in more detail both in terms of concepts and definitions, and in terms of application features.

1. Under the reliability of the mathematical base we understand the practical lack of the violator's ability to carry out relative ES attacks such as "universal disclosure" due to the imperfections of the mathematical apparatus used, such as groups of elliptic curves or weaknesses that can be laid due to specific properties of general parameters and keys. The criterion for assessing the reliability of the mathematical basis is the fact that the complexity of the attack "universal disclosure" $I_{\text{ур}}$ has an exponential character, and the criterion of unreliability—subexponential or polynomial complexity. Let us denote this criterion as W_{81}.

2. Under the practical security of cryptoconversions ES we will understand the security of ES from force and analytical attacks, which is achieved by choosing the size of the general parameters and keys, as well as the means of their generation. That is, the criterion of practical security of cryptoconversions such as ES there is a choice of such sizes of the general parameters and keys at which complexity of attack I_{ca} significantly (by the required number of orders of magnitude) exceeds the existing capacity of cryptanalytic systems of technologically advanced countries (third-level violator). Including taking into account the forecast of increasing the capacity of cryptanalytic systems due to the development of mathematical and software, as well as hardware and software and

hardware, including tools based on quantum computing. Let us denote this unconditional criterion as $W_{\delta 2}$.

3. Real protection of ES from all known and potentially possible cryptanalytic attacks, where protection means the fact that all known cryptanalytic attacks of the type "full disclosure" have exponential complexity I_{ec}, and under the criterion of insecurity—subexponential I_{ce} and below the nature of the complexity of the "full disclosure" attack. We will mark this unconditional criterion as $W_{\delta 3}$.

4. Statistical security of cryptographic conversion of the ES type, which means the statistical independence of the result of cryptographic conversion (output), such as ES (cryptogram), from the encrypted input block (signed) and the private key used. We will mark this unconditional criterion as $W_{\delta 4}$.

5. Theoretical security of a cryptographic conversion of the ES type, which uses general parameters with appropriate properties and lengths, for which there are no (unknown) theoretical analytical attacks, the complexity of which is less than the complexity of the attack of the type "full disclosure". This unconditional criterion will be denoted as $W_{\delta 5}$.

6. The absence of weak key pairs, including private keys, for which the complexity of cryptanalytic attacks such as "full disclosure" and "universal disclosure" is less than the complexity of the attack "full disclosure" for other (not weak) private keys. Let us denote this unconditional criterion as $W_{\delta 6}$.

7. The complexity of the direct $I_{пр}$ and reverse $I_{3в}$ cryptographic conversions, as well as the generation or deployment of keys, is polynomial in nature and does not exceed the allowable values $I_{пр}'$, $I_{3в}'$ and $I_{reн}'$. Let us denote this unconditional criterion as well $W_{\delta 7}$.

Since the given partial criteria are unconditional, the selection criterion is a logical change yes/no (1/0), so the unconditional criterion can be written as:

$$(K_{\delta 1}, K_{\delta 2}, K_{\delta 3}, K_{\delta 4}, K_{\delta 5}, K_{\delta 6}, K_{\delta 7}) \in (1, 0). \tag{26}$$

Taking into account the above partial unconditional criteria $W_{\delta 1} - W_{\delta 7}$ the cryptoconversion matching function can be represented as:

$$f_{фв}( ) = W_{\delta 1} \wedge W_{\delta 2} \wedge W_{\delta 3} \wedge W_{\delta 4} \wedge W_{\delta 5} \wedge W_{\delta 6} \wedge W_{\delta 7}. \tag{27}$$

In (26), the symbol «$\wedge$» denotes the conjunction operation of Boolean variables.

Therefore, the quality of cryptographic conversion of ES can be evaluated using an unconditional integral criterion—the function of compliance of cryptographic conversion of ES requirements

$$f_{фв}( ) \in (0;1) \text{ and at } f_{фв}( ) = 1 \tag{28}$$

the cryptographic conversion of the evaluated ES meets the requirements.

The integrated criterion introduced in this way allows to determine whether the cryptoconversions of the ES type meets the considered requirements. If the ES meets the requirements, it can be reasonably recommended for use.

Conditional criteria for evaluating cryptographic conversions of the ES type. Subject to a positive evaluation of the ES on the integral unconditional criterion, further comparison and evaluation can be made on the basis of conditional criteria and the integral conditional criterion.

Studies have shown that qualitative and quantitative comparison of cryptographic conversions of the ES type can be performed using a generalized criterion of superiority [5]. It can be based on conditional criteria of superiority of one ES over another.

As the main components of the generalized criterion of preference, it is proposed to use the partial conditional criteria listed in Table 1 [5].

The choice of the above criteria is not mandatory, depending on the requirements of the applications, the researcher may choose other conditional criteria.

In the future, the assessment of each of the partial unconditional and conditional criteria will be carried out using one or more indicators that support the given criterion.

First, we select and consider the main indicators by which we can assess the standards of cryptographic conversions of the ES type by unconditional criteria [5].

1. The assessment of the reliability of the mathematical base can be carried out on the basis of expert assessments of specialists in the field of cryptology. This should take into account the degree of openness of the design and study of the

Table 1 Conditional criteria for evaluating cryptocurrencies ES

Conditional criteria	Marking
Possibility and conditions of free distribution and application of the international or national standard of cryptographic transformations of the ES type in Ukraine taking into account requirements of regulatory legal acts of Ukraine on export, import and restrictions on its application, including for rendering of electronic trust services	K_{y1}
Level of trust in the international or national standard of cryptographic conversion of the ES type for applications, including for the provision of electronic trust services	K_{y2}
Prospects for the application of the international harmonized or national ES standard in Ukraine, including for the provision of electronic trust services	K_{y3}
Temporal and spatial complexity of software, hardware and software implementations of ES development and testing, including on different platforms	K_{y4}
Possibility and conditions of application of the relevant ES standard with different general parameters and in different modes, including for the provision of existing or permitted electronic trust services	K_{y5}
Flexibility of the standard of cryptographic conversion of type ES from the point of view of application in various applications and at various implementations of means	K_{y6}
The level of security of the ES in the implementation of different types of threats in different conditions of a priori and a posteriori uncertainty, including deviations from the requirements or violation of the properties of general parameters and keys	K_{y7}

standard or in general cryptographic conversion of the ES type. An important factor is the possibility or suspicion of its existence in terms of modeling crypto-conversions and performing cryptanalysis with reduced or significantly reduced complexity.

The main indicator for assessing the reliability of the mathematical base of cryptoconversions in the group of EK points can be used the degree of deterioration of the secure time

$$r_{\text{и}} = \frac{t_6}{t_6'}, \tag{29}$$

where t_6—mathematical expectation of cryptanalysis time for real mathematical conversion, and t_6'—for the mathematical apparatus used in modeling.

In the future, a decision is made on the basis of expert assessments.

If $r_{\text{и}} > 1$, then a decision can be made taking into account expert assessments, that $W_{81} = 0$, otherwise $W_{81} = 1$.

2. We will evaluate the practical protection of cryptographic conversions of ES from force attacks based on the size of the private key. By forceful attack we will understand the directed search of personal keys i and/or session keys (parameters) k_j. As d_i and k_j should be formed randomly, equally probable and independently and

$$0 < d_i, k_j < n, \tag{30}$$

where n—for example the order of the base point G. Under these conditions, the probability of selection from a single attempt P(1) can be estimated as

$$P(1) \geq \frac{1}{n} = \frac{1}{2^m} = 2^{-m}. \tag{31}$$

In k attempts to select the probability $P(k)$ can be calculated as

$$P(k) = \frac{k}{n} = \frac{k}{2^m} = k2^{-m}. \tag{32}$$

Estimation of complexity of force attack can be made by means of an estimation of complexity of execution of attack I_d and secure time t_6. For the condition that $0 < d_i, k_j < n$, complexity can be assessed as

$$I_d = n = 2^m, \tag{33}$$

and the secure time

$$t_6 = \frac{I_d}{\gamma K} P_y, \tag{34}$$

where

γ power of the cryptanalytic system, $K = 3.15 \times 10^7$ s/year,
P_y the probability with which cryptanalysis should or can be successfully
     performed.

Based on the analysis of the state of practical security, it is decided that $W_{82} = 1$
or $W_{82} = 0$.

3.   The real security of a particular cryptoconversion, for example in a group of
     elliptic curve points, is proposed to be assessed by determining the complexity
     I_a and the secure time t_6 implementation of an attack of the type "full disclosure"
     for the i-th method of cryptanalysis. For example, if the solution of a discrete
     logarithmic equation in a group of points of an elliptic curve is carried out by
     Pollard $\rho - 1$ method [7], the complexity and safe time with acceptable accuracy
     can be defined as

$$I_{a_\rho} \approx \sqrt{\frac{\pi n}{4}}, \tag{35}$$

$$t_{6_\rho} = \frac{I_{a_\rho}}{\gamma K} P_y = \frac{\sqrt{\pi n}}{2\gamma K} P_y. \tag{36}$$

When assessing the value I_{a_ρ} and I_{8_ρ} are determined for all methods of cryptanalysis. If they are exponential, then the security of cryptoconversions is assessed as
meeting the requirements. If they are sub-exponential in nature, then the security of
cryptoconversion type ES is assessed as not meeting the requirements.

In addition, even with an exponential nature, in the case of considering different
methods for solving a discrete logarithmic equation in a group of EK points, it is
proposed to evaluate the real security as

$$I_a = \min(I_{a1}, I_{a2}, \ldots, I_{ak}), \tag{37}$$

$$t_6 = \min(t_{61}, t_{62}, \ldots, t_{6k}). \tag{38}$$

That is, it is proposed to evaluate the lowest (worst) value of the complexity of
cryptanalysis. Thus, based on the assessment of real security, it is decided that W_{83}
$= 1$ or $W_{83} = 0$.

For block cryptocurrency conversion, which is, for example, ES-type conversion
in a group of EK points, it is also necessary to assess the probability of collisions.
The peculiarity of cryptocurrency conversion of the ES type is that the result of the
conversion is a point and in fact it is necessary to consider the collision of points.
The analysis showed that this problem of theoretical and practical consideration and
solution in the ES has not yet been found. But it can be solved with the appropriate

justification and determination of the order of the base point n by applying the already standard method based on the paradox of the birthday.

4. Regarding statistical security, it should be noted that regarding cryptoconversions in the group of points of elliptic curves to date, methods and approaches to assessing statistical security are not developed. Their justification and choice are a separate task. Given the block nature of cryptoconversions, it is proposed to perform a preliminary assessment of statistical security by determining the mathematical expectation and variance of the cross-correlation function. First, we determine the cross-correlation function between the Mi input and the Ci output:

$$F_1 = f(M_i, C_i) = \sum_{j=1}^{l} M_{ij} \oplus C_{ij}. \tag{39}$$

In addition, we will define the function of cross-correlation between outputs (cryptograms) for different keys and message blocks:

$$F_2 = f(C_i, C_k) = \sum_{j=1}^{l} C_{ij} \oplus C_{kj}. \tag{40}$$

Next, considering the meaning F_1 and F_2 as random, we calculate the mathematical expectation $m_1(F_1)$ and $m_1(F_2)$, as well as variance $m_2(F_1)$ and $m_2(F_2)$.

Decisions can be made this way. Block of information M_i and block cryptogram C_i are dependent if $m_1(F_1)$ and $m_2(F_1)$ are the parameters of the binomial distribution. Similarly, the decision is made on C_i and C_k. Cryptograms for any M_i, M_k and K_i, K_j can be considered independent if appropriate $m_1(F_2)$ and $m_2(F_2)$ are also parameters of the binomial distribution.

Another valid way to assess statistical security is to determine the redundancy and dependence of cryptographic symbols. In this case, the assessment can be performed as

$$k_2 = \frac{l_c'}{l_c}, \tag{41}$$

where l_c'—the length of the compressed block of the cryptogram, l_c—the length of the cryptogram block.

The decision in this case is made on the basis of analysis of k_2, if k_2 is closer to 1, then statistical security is considered better, because the block cryptogram is increasingly becoming similar to a random sequence.

In general, based on the obtained results and estimates, a conclusion is made about the statistical security of the cryptocurrency algorithm in the group of EK points and $W_{84} = 1$ or $W_{84} = 0$.

5. Assessment of the theoretical security of cryptoconversions ES offers work based on determining the presence, suspicion of the presence or absence of theoretical analytical attacks. Such an assessment can be made on the basis of expert assessments of cryptologists. When evaluating, experts should consider the scope of the standard (eg, electronic digital signature, directional encryption, cryptographic protocol, etc.), possible threats (eg, full disclosure, universal disclosure, etc.). Suspicion of a threat in the form of theoretical attacks, if the complexity of these attacks is subexponential or other vulnerabilities are possible that will not guarantee the required cryptographic stability. The main indicators are the complexity of the theoretical cryptanalytic attack, safe time (time of its implementation) and the probability of analytical attack. The corresponding indicators are given above when assessing the real security of cryptoconversion ES, for example in the group of EK points, Formulas (34)–(38).

Thus, if the complexity of a possible theoretical attack is subexponential in nature and the probability of its implementation $P_{am} > P_{\partial on}$ (i.e. not less than the allowable values), it is decided that $W_{85} = 0$, otherwise $W_{85} = 1$.

6. Regarding weak keys it is necessary to note the following. Finding weak private keys or public keys for which attacks such as "full disclosure" or "universal disclosure" are sub-exponential is a very difficult task. The first cause of weakness may be a tool or system for generating keys. It is clear that personal keys must be formed (generated) randomly, equally and independently. To ensure these properties, it is necessary to use physical sources of "white" noise. But even in this case, the distribution density and its characteristics can be distorted as a result of the presence of a physical generator of ε-asymmetry, when the probabilities $P(1)$ and $P(0)$ differ. Example, $P(1) = 0.5 + \varepsilon$, and $P(0) = 0.5 - \varepsilon$. The presence of sufficient ε-asymmetry reduces the complexity of the attack "brute force", which in this case can be carried out taking into account the shift of the distribution, when the probability of a key with a given ratio between "1" and "0". Therefore, the first step is to submit a valid value requirement to the key source ε, for example, $\varepsilon \leq 10^{-5}$. This requirement can be met through the use of physical noise sources and random sequence generators based on these sources, as well as when $\varepsilon < \varepsilon_{\partial on}$.

That is, when analyzing the keys, first of all it is necessary to analyze the source of the keys and limit the allowable value of ε-asymmetry.

The keys must also be considered weak $d_i(k_i)$, which reduces the complexity of solving a discrete logarithmic equation in a group of points of elliptic curves. Obviously, it is also necessary to consider such keys as weak, the probability of which is too small and which may appear as a result of malfunctions.

Thus, the presence or suspicion of weak keys, as well as equivalent keys, should be established by cryptologists. If weak keys are blocked in the system and the stability is not reduced due to this, then such a standard can be considered as such that its algorithm does not have weak keys. In this case, the unconditional criterion is $W_{86} = 1$, otherwise $W_{86} = 0$.

7. The complexity of the direct I_{np} and reverse I_{3e} of the cryptographic conversions, as well as key generation, can be evaluated both theoretically and experimentally. Experimental evaluation can be carried out using appropriate tools that implement the transformation—software, hardware and combined. In this case, the assessment can be performed by measuring the speed of execution of certain operations. Spatial complexity I_v can be estimated by the amount of memory required to perform forward and reverse transformations in a group of elliptic curve points, placement of keys, tables, parameters and certificates. If at the same time $I_v \leq I_v{}^{\partial on}$, then the expert decides that $W_{87} = 1$, otherwise $W_{87} = 0$.

Next, depending on the requirements for the complexity of the direct I_{np} and reverse I_{3e} conversions, the expert decides that $W_{87} = 1$, if the complexity of the conversions does not exceed the allowable values, and $W_{87} = 0$—if it exceeds.

Thus, based on criteria $W_{81} \div W_{87}$ make decisions according to an unconditional integral criterion.

5 Evaluation Criteria and Indicators to Ensure the Efficiency, Objectivity of Decision-Making, Depersonalization, Reliability, Availability of Decentralized Storage

Consider these properties of decentralized storage, define the criteria and indicators for their evaluation.

By definition, depersonalization is the general name for any process of removing the link between a set of identifying data and a data subject.

One of the main vectors of attacks on information systems in the modern information space is the receipt of personal data (hereinafter—PD) about the users of these systems and their environment. In general, PD includes information or a set of information about an individual by means of which he/she is identified or can be uniquely identified. These include last name, first name, middle name, date of birth, place of residence, telephone number, and so on. However, in the twenty-first century, this list of PD has undergone significant changes. Due to the rapid development of information technology, computerization has penetrated into almost all key areas of infrastructure for human life.

Due to the transfer of human activity to the electronic information space, the above list of PD can be expanded by adding:

- user credentials for access to e-mail, social networks and other information resources on the Internet;
- credit card numbers used to make purchases, pay and various services;
- data on habits, preferences and marks about the person's location;
- photos and videos that a person publishes on the Internet;
- data on information resources visited by the person.

Identifiers / Attributes	Attribute 1	Attribute 2	...	Attribute N
Identifier 1	a_{11}	a_{12}	...	a_{1N}
Identifier 2	a_{21}	a_{22}	...	a_{2N}
...	...	...	...	...
Identifier M	a_{M1}	a_{M2}	...	a_{MN}

$F(x) \rightarrow$

Identifiers / Attributes	Attribute 1	Attribute 2	...	Attribute N
Identifier 1	a_{12}	a_{MN}	...	a_{1N}
Identifier 2	a_{13}	a_{M2}	...	a_{M1}
...	...	...	...	...
Identifier M	a_{2N}	a_{13}	...	a_{12}

Fig. 1 General scheme of implementation of data depersonalization in ISPDP

This data appears and is updated online daily. Their systematic collection and analysis allows for a more accurate identification of a person and the areas of his activity than the traditional list of PD. An additional list of data allows anyone to monitor a person via the Internet, without the use of specialized means of capturing and collecting information.

The significant use of PD in the network, as well as their rather conditional protection, greatly facilitates the task for attackers to collect and analyze information, in the preparation of coordinated cyberattacks on the objects of information activities. In this regard, the security of PD, as one of the vectors of cyber attacks, is becoming an increasingly important issue in the field of information security [8].

In the considered work [8] mechanisms of depersonalization of the information do not provide difficult mathematical transformations of initial PD. The basis of the data depersonalization subsystem is the mixing of data of information systems for personal data processing (hereinafter—ISPDP), in the open, according to a certain algorithm or pattern. This approach is convenient to implement due to the fact that all databases are based on the use of tabular data storage structures (Fig. 1).

Example: perform a primitive depersonalization of the table with PD using a simple algorithm for mixing data. Let table A contain the initial values of PD, and tables π and $\pi \wedge (-1)$ the laws of data movement in the original table A:

$$A = \begin{matrix} 1 & 2 & 3 \\ 4 & 5 & 6 \\ 7 & 8 & 9 \end{matrix} \quad \pi = \begin{matrix} (3,3) & (2,1) & (2,3) \\ (1,1) & (3,2) & (1,2) \\ (2,2) & (1,3) & (3,1) \end{matrix} \quad \pi^{-1} = \begin{matrix} (2,1) & (2,3) & (3,2) \\ (1,2) & (3,1) & (1,3) \\ (3,3) & (2,2) & (1,1) \end{matrix}$$

Then, as a result of depersonalization of data in table A, table C is obtained, and when performing inverse transformations, we obtain the original table. A':

$$C = \pi(A) = \begin{matrix} 9 & 4 & 6 \\ 1 & 8 & 3 \\ 5 & 3 & 7 \end{matrix} \quad A' = \pi(C) = \begin{matrix} 1 & 2 & 3 \\ 4 & 5 & 6 \\ 7 & 8 & 9 \end{matrix}$$

To obtain the source table with open PD, the attacker must access the table of inverse transformations π^{-1} or select $(M \cdot N)!$ combinations, using depersonalized PD. In addition, by analyzing the transformations π, the attacker can understand the inverse algorithm for moving data and thus obtain the original table with PD.

Therefore, all the transformations described above can be generalized to the scheme (Fig. 2), according to which, in both cases, there are schemes of inverse transformations, when obtaining key parameters of which an attacker can obtain PD in the open without obtaining appropriate access.

According to the results of the analysis of the above subsystems of ISPDP protection (encryption and depersonalization of data), in the process of ensuring the confidentiality of data in ISPDP, information about their advantages and disadvantages was obtained (Table 2). The table shows the criteria for assessing depersonalization in relation to encryption.

Objectivity—impartiality, independence of information from the thoughts, will and desires of man. This is an integral criterion that combines the following partial criteria:

- *completeness of information* (availability of information, including contradictory, which is necessary and sufficient for decision-making);
- *accuracy of information* (the degree of conformity of the information to the original);
- *consistency of information* (separate parts of the same information should not contradict each other);
- *persuasiveness of information* (proof, reliability of information);

Efficiency—a property of data, which is that the time of their collection and processing corresponds to the dynamics of the situation;

Performance indicator $P_{оп}$—the probability that the decision will be made in time, which ensures its implementation and the possibility of issuing appropriate orders to subordinates.

The figure depends on the available $t_{расп}$ and the right time $t_{попт}$.

We will describe the process of making recommendations in terms of events.

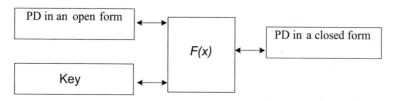

Fig. 2 Generalized scheme of transformations of PD into ISPDP

Table 2 The results of the analysis of subsystems protection ISPDP

Analysis parameters	Protection subsystems	
	Encryption	Depersonalization
Algorithm family	Symmetrical, asymmetrical	Symmetrical
The scheme of conversion	Substitution, permutation, key extension, use of expression a^x (mod n), cyclic transformations	Substitution, permutation
Availability of encryption keys	Yes	Yes
Resistance of keys to compromise	No	No
Resistance to forgery of keys	Yes	No
Resistance to key selection	Yes	Yes
Computational stability	Yes	No
Resistance to coercion	No	No
Resistance to attacks based on open PD	Yes	Yes/No
Resistance to attacks based on encrypted PD	Yes	Yes/No
Resistance to attacks based on reverse engineering algorithm	Yes	Yes/No
Resistance to MITM attacks	No	Yes

Let event A be that the recommendations are made at a time not exceeding $t_{\text{расп}} + \Delta t$; and the event B—that the recommendations are made at an interval that exceeds $t_{\text{расп}}$. Then the values of the random value of the end time of the development of recommendations z satisfy the inequalities

$$t_{\text{расП}} < z \leq t_{\text{расП}} + \Delta t.$$

That is, the value of z is within $(t_{\text{расп}}, t_{\text{расТ}} + \Delta t)$. Using the ratio for conditional probabilities, we obtain:

$$P(A/B) = \frac{P(A \cap B)}{P(B)} = \frac{P(t_{pacn} + \Delta t) - P(t_{pacn})}{1 - P(t_{pacn})} \tag{42}$$

Let there be a limit μ ($\mu = 1/t_{\text{попт}}$), characterizes the probability

$$P(t_{\text{расП}}): = \lim_{\Delta t \to 0} \frac{P(t_{pacn} + \Delta t) - P(t_{pacn})}{\Delta t^* (1 - P(t_{pacn}))} = \frac{P'(t_{pacn})}{1 - P(t_{pacn})} \tag{43}$$

From where after integration we will receive:

$$P\left(t_{pacn}\right) = 1 - \exp\left(\int_0^{t_{pacn}} \partial t_{pacn}\right) \qquad (44)$$

where μ—the intensity of recommendations. Fixing $t_{\text{попр}}$, we will receive:

$$P_{on} = 1 - \exp\left(-\frac{t_{pacm}}{t_{nopm}}\right) \qquad (45)$$

where $t_{\text{попр}}$—average time of information collection and processing, synthesis and decision making.

Next, it is necessary to determine the time interval for deciding on the purpose of the effects of the aircraft on the target [9].

Reliability of information—an indicator of the quality of information, which means its completeness and overall accuracy.

In today's world, due to the ever-increasing importance of information, the industry of its receipt, processing, registration, transmission and dissemination is becoming one of the leading industries of mankind, where every year more and more money is invested. Information becomes the most important strategic resource, the lack of which leads to significant losses in all spheres of life.

Modern information space is a unique opportunity to obtain any information on a particular issue, provided that there are appropriate tools, the use of which allows you to analyze the relationship of possible events that are already happening with the information activity of a number of sources of information.

All types of information that meet the needs of the subject, meet such properties as confidentiality, integrity and availability of information.

Protection of confidentiality and integrity of information is provided by the laws of Ukraine and legal documents: Laws of Ukraine "On Information", "On Personal Data Protection", ND TZI 1.1-002-99 "General provisions on protection of information in computer systems from unauthorized access" etc. There is only an article on reliability (references to sources) in the professional standards of information journalism, but there are no legal norms and regulations for the protection of reliable information.

Actions to disseminate distorted, unreliable and biased information in the information space or negative information influences on the public consciousness in the information space are a violation of the integrity of information [10].

In real life, it is hardly possible to count on the complete reliability of information. There is always some degree of uncertainty. The correctness of decisions made by a person or system depends on the degree of reliability of information to the actual state of the information object or process. Some of the basic properties of information are as follows: quality of information, sufficiency (completeness) of information, relevance of information, adequacy of information, stability of information, timeliness and accuracy of information. Under the reliability of the information understand some function of the probability of error, ie the event, which is that the real information in the system about a parameter does not match within a given accuracy with the true value.

An important stage in the selection, development of methods and mechanisms to ensure the reliability of information is the analysis of its processing. The analysis should investigate the structure of data processing, building a model of errors and their interaction, calculating the probability of occurrence, detection and correction of errors for different variants of data processing structures and the use of mechanisms to ensure the required level of reliability.

Assessment of information reliability is a process of establishing the degree of conformity of information about an object (within the framework of information models adopted in solving problems) to the real state of an object, ie determining the degree of adequacy of representations to the real state of an object (object, phenomenon).

The complexity of the process of assessing the reliability of information is that it is necessary to assess the information as a whole so that the probability of description (or inaccuracy) of insignificant parameters is not a cover (concealment) of the actual state of the object, or even targeted misinformation.

The range of levels of reliability of information is very wide: from complete reliability to complete unreliability of this or that information available in the information space.

However, the assessment of reliability is subject to significant influence of subjective factors (personal and group), which have a significant impact on the assessment of the reliability of information.

It is important to be able to assess the accuracy of the description of the object as a whole, the fundamental possibility of its existence within the limits set by the available information description (even in cases where individual elements of the object description are given with some error) contained in the information describing object and available to the individual.

When analyzing the reliability of information and determining its reliability should focus on the following issues:

- whether this fact or event is possible at all;
- whether the information is inconsistent in itself;
- to what extent the received information corresponds to the available one;
- if the information obtained does not correspond to information obtained from other sources, which of them can be considered the most reliable (i.e. there is a problem of comparative evaluation of information with different levels of reliability of both the information and its sources).

In order to assess the reliability of information in the information space, the following features are proposed:

- doubtfulness of the stated facts, which is determined by the concealment of sources and authors of content, insufficient argumentation, references to the opinion of the general public, the presence of rhetorical questions;

- emotional coloring of content, which is used to reflect the emotional state of the author and is manifested in the oversaturation of content with figurative means, adjectives, comparisons, etc.;
- the tone of the content in relation to an object or event, which reflects the evaluative judgments of the author and may be manifested in the use of images, etc.; sensational content, which aims to attract attention by increasing anxiety, etc.;
- The hidden (implicit) content of the content is related to its deep content.

Monitoring of the reliability of information in the information space will be based on the above features [10].

Criteria for submitting the accuracy of information

The criterion for presenting the reliability of information can be violated in several different ways [10]:

- when the author cites facts in general without reference to their source—in such cases, the facts can be considered neither accurate nor reliable, as their origin is unknown;
- when the author makes vague or generalized references to the source of facts ("law enforcement officers say", "city officials claim", "reported by the Ministry of Defense", etc.)—in these cases the individual is deprived of the opportunity to assess the facts, because it is unknown where the author received the fact;
- when the author gives a subjective opinion without a clear reference to the subject who expressed it ("experts say", "deputies think", "people say", "there is an opinion", etc.)—in all these cases, the opinion is not can be perceived as credible, because any subjective opinion is inseparable from the subject who catches it.

A separate caveat to any facts and opinions submitted by the authors with reference to any of the segments of the information space. Without verification in the original source, such information cannot be presented in the information space. There are many reasons, the main one being the high vulnerability and dynamism of the information space as such, including before hacker attacks, not to mention all possible types of errors.

Another—already purely manipulative—is the violation of the criterion of submission of authenticity, when the author generalizes a certain opinion or attitude to something of one individual to a larger number of individuals. From this point of view, even a vague reference is a gross violation of the criterion of submitting authenticity. In general, generalizations in which a certain opinion or position extends to an immeasurably large group of individuals are wrong.

Here, a certain forecast can be given only by sociological research conducted according to scientific methods. From the same point of view, it would be incorrect for the author to present the opinion of the representative of the collective side of the conflict as the opinion of this whole party.

Availability is an opportunity to receive the necessary information service in a reasonable time.

By definition [10], **availability** is a property of an information resource, which is that a user and/or process that has the appropriate authority can use this resource in accordance with the rules established by the security policy without waiting longer than specified (acceptable) time interval.

Information systems are created to receive certain information services. If, for one reason or another, it becomes impossible to provide these services to users, it obviously harms all parties to the information relationship. Therefore, without contrasting the accessibility of other aspects, it should be highlighted as the most important element of information security.

The main role of accessibility is especially evident in various management systems—production, transport, and so on. Outwardly less dramatic, but also very unpleasant consequences—both material and moral—may have a long-term unavailability of information services used by a large number of people (sale of rail and air tickets, banking services, etc.).

Availability criteria. In order for a system to be assessed for accessibility criteria, the security system of the system being evaluated must provide services to enable the system as a whole, individual functions or processed information to be used for a certain period of time and guarantee the system's ability to function. Availability can be provided in the system by the following services:

- use of resources;
- resistance to failures;
- hot swap;
- recovery after failures.

1. Use of resources. This service allows users to manage the use of services and resources. The levels of this service are ranked based on the completeness of protection and selectivity of managing the availability of system services:

    - Quotas;
    - Preventing the seizure of resources;
    - Priority of resource use.

Quotas—stipulate that the policy of resource use, implemented by a set of means of protection, should determine the set of objects of the system to which it belongs. Preventing the capture of resources and Priority of resource use—provides that the policy of resource use, implemented by a set of means of protection, should apply to all objects of the system.

The policy of use of resources should define restrictions which can be imposed, on quantity of the given objects (volume of resources):

- Quotas and Prevention of resource capture—allocated to an individual user;
- Priority of resource use—allocated to individual users and arbitrary groups of users.

It should be possible to set restrictions in such a way that the security suite is able to prevent actions that could prevent other users from accessing the security features or protected objects. The set of means of protection must control the following actions:

- Preventing the capture of resources—by an individual user;

- Priority of resource use—by individual users and arbitrary groups of users.

2. Resistance to failures. Failure resistance guarantees the availability of the system (the ability to use information, individual functions or the system as a whole) after the failure of its component. Levels of this service are ranked on the basis of ability of complexes of means of protection to provide a possibility of functioning of systems depending on quantity of failures and the services available after refusal:

   - Stability with limited failures;
   - Stability with deterioration of service characteristics;
   - Stability without deterioration of service characteristics.

The developer must analyze the failures of system components. Resilience with limited failures—implies that the policy of resilience to failures, implemented by a set of remedies, should determine the many components of the system to which it belongs, and the types of their failures, after which the system is able to continue to operate.

Stability with deterioration of service characteristics and Stability without deterioration of service characteristics—assumes that the policy of resistance to failures implemented by a complex of means of protection should concern all components of system.

Failure levels must be clearly indicated, above which failures lead to reduced service performance or unavailability of service:

- Resilience in case of limited failures and Resilience with deterioration of service characteristics—failure of one protected component should not lead to unavailability of all services, and should in the worst case be shown in decrease in service characteristics
- Stability without deterioration of service characteristics—failure of one protected component should not lead to unavailability of all services or to decrease in service characteristics

The security suite must be able to notify the administrator of the failure of any protected component.

3. Hot swap. This service allows you to guarantee the availability of the system (the ability to use information, individual functions or the system as a whole) in the process of replacing individual components. The levels of this service are ranked based on the completeness of the implementation.

   - Modernization (upgrade);
   - Limited hot swap;
   - Hot replacement of any component.

Modernization—a policy of hot swapping, implemented by a set of means of protection, should determine the policy of modernization of the system.

Limited hot-swap is a hot-swap policy implemented by a set of protections that must define the set of system components that can be replaced without service interruption.

Hot replacement of any component is a policy of hot replacement implemented by a set of means of protection, must provide the ability to replace any component without interruption of service.

The modernization (upgrade) requires that the administrator or users who are granted the appropriate authority must be able to upgrade the system. System upgrades should not lead to the need to re-install the system or to interrupt the performance of the protection function. Restricted Hot-swap and Hot-swap any component—means that the administrator or users who are authorized to do so must be able to replace any protected component.

4. Disaster recovery. This service ensures that the system returns to a known secure state after a failure or interruption of service. The levels of this service are ranked based on the degree of automation of the recovery process.

The recovery policy implemented by a set of protection tools should define many types of system failures and service interruptions, after which it is possible to return to a known protected state without violating the security policy. Failure levels must be clearly indicated if they need to be reinstalled:

- manual recovery;
- automatic recovery;
- selective recovery.

Manual recovery—implies that after a system failure or service interruption, a set of security measures must bring the system to a state from which it can be returned to normal operation only by the administrator or users who have been granted the appropriate authority.

Automated Recovery—Provides that you must be able to determine if automated procedures can be used to return the system to normal operation in a safe manner. If such procedures can be used, then the set of means of protection must be able to perform them and return the system to normal operation.

Selective recovery—assumes that after any system failure or service interruption that does not necessitate reinstallation of the system, the security suite must be able to perform the required procedures and safely return the system to normal operation or, at worst, operation in mode with degraded service characteristics.

Provided that automated procedures cannot be used, then automated recovery and selective recovery provide that a set of security features must bring the system to a state from which only the administrator or users who have been granted the appropriate authority can return it to normal operation.

6 Summary

The section discusses general approaches that can be used to determine the criteria and performance indicators of cryptographic protection mechanisms for users' personal data and interaction protocols that ensure efficiency, objectivity, decision making, depersonalization, generalized processing, reliability, availability of decentralized privacy storage COVID-19 when tracking contacts with system users. If we talk about COVID-19, it seems that the generalized criterion for the effectiveness of this system should be considered the minimum number of infected per day. Obviously, the considered criteria will solve the problem of collecting information about the state of morbidity in the most objective form. The connection of this criterion with the ones presented in the section is subject to further study.

On the basis of researches the criteria and indicators of efficiency of mechanisms of cryptographic protection of personal data of users and protocols of interaction providing: efficiency are analyzed and generalized; objectivity of decision-making; depersonalization; generalized processing, reliability, availability of decentralized storage. In particular, the criteria and indicators for assessing the security of symmetric and asymmetric transformations are formulated, in particular, it is proposed to use these criteria and indicators for comparative analysis of different methods of cryptographic processing of personal data that can be used in the system. They allow to evaluate the mechanisms of cryptographic protection of personal data of users and protocols of interaction, reasonably using a system of unconditional and conditional criteria. Based on these indicators, it is advisable to evaluate cryptocurrencies such as Gesh functions and ES. In particular for the ES, this assessment is carried out in 2 stages. At the first stage, their compliance with unconditional criteria is checked, and at the second stage, appropriate assessments are obtained using conditional criteria. It is through the use of conditional criteria that it is possible to compare different cryptographic transformations of the ES type according to the integral conditional and general criteria.

References

1. Delfs H, Knebl H (2015) Introduction to cryptography. Springer, Berlin. http://doi.org/10.1007/978-3-662-47974-2
2. Shannon CE (1949) Communication theory of secrecy systems. Bell Syst Tech J 28:656–715. https://doi.org/10.1002/j.1538-7305.1949.tb00928.x
3. Menezes AJ, van Oorschot PC, Vanstone SA, van Oorschot PC, Vanstone SA (2018) Handbook of applied cryptography. CRC Press. https://doi.org/10.1201/9780429466335
4. Hattersley B. NIST SHA-3 competition waterfall hash-algorithm specification and analysis. Режим доступу: https://ehash.iaik.tugraz.at/uploads/1/19/Waterfall_Specification_1.0.pdf
5. Schneier B (1996) Applied cryptography: protocols, algorithms, and source code in C. Wiley, New York
6. UCL: GDPR—glossary of terms and definitions. https://www.ucl.ac.uk/legal-services/gdpr-glossary-terms-and-definitions. Last accessed 2021/11/07

7. Montgomery PL (1987) Speeding the Pollard and elliptic curve methods of factorization. https://doi.org/10.1090/S0025-5718-1987-0866113-7
8. Kuznetsov A, Kalashnikov V, Brumnik R, Kavun S (2020) Editorial "computational aspects of critical infrastructures security", "security and post-quantum cryptography." Int J Comput 19:233–236
9. Isirova K, Kiian A, Rodinko M, Kuznetsov A (2020) Decentralized electronic voting system based on blockchain technology developing principals. In: Subbotin S (ed) Proceedings of the third international workshop on computer modeling and intelligent systems (CMIS-2020), Zaporizhzhia, Ukraine, April 27–May 1, 2020, pp 211–223. CEUR-WS.org
10. Kuznetsov A, Kiian A, Smirnov O, Zamula A, Rudenko S, Hryhorenko V (2019) Variance analysis of networks traffic for intrusion detection in smart grids. In: 2019 IEEE 6th international conference on energy smart systems (ESS), pp 353–358. http://doi.org/10.1109/ESS.2019.8764195

Methods of Evaluation and Comparative Research of Cryptographic Conversions

Alexandr Kuznetsov(ID)**, Yuriy Gorbenko**(ID)**, Anastasiia Kiian**(ID)**,
and Olena Poliakova**(ID)

Abstract The security of decentralized systems built on Blockchain technology is directly determined by the efficiency of the applied cryptographic transformations. Classical blockchain technology mainly uses several cryptographic primitives: hash function, electronic signature, encryption, etc. In particular, the hash function is used for several purposes as the generation of an address or user ID; hashing of transactions and blocks to confirm the indisputable block against network errors, as well as the hash value of the previous block is used as a reference in the formation of the next. The use of digital signatures in the blockchain system introduces services for the integrity and integrity of transactions. In turn, in any blockchain system there is a pool of transactions, each of which must be signed. This ensures that the transaction will not be altered, the signature of the transaction ensures that it is signed by a person who can then be identified, and, in fact, the signature can help determine "whether the user has the right to conduct the transaction." Block ciphers are one of the most common cryptographic primitives, which is also used as a structural element of hash functions, message authentication codes, and so on. Thus, for the operation of monitoring and tracking programs using Blockchain technology, it is necessary to study various cryptographic primitives, in particular, hash functions, electronic signatures, encryption schemes, etc.

Keywords Cryptographic transformations · Blockchain technology · Decentralized systems

A. Kuznetsov (✉) · Y. Gorbenko · O. Poliakova
V. N. Karazin Kharkiv National University, Svobody sq., 4, Kharkiv 61022, Ukraine
e-mail: kuznetsov@karazin.ua

O. Poliakova
e-mail: olena.polyakova@nure.ua

A. Kuznetsov · Y. Gorbenko · A. Kiian
JSC "Institute of Information Technologies", Bakulin St., 12, Kharkiv 61166, Ukraine

© The Author(s), under exclusive license to Springer Nature Switzerland AG 2022
R. Oliynykov et al. (eds.), *Information Security Technologies in the Decentralized Distributed Networks*, Lecture Notes on Data Engineering and Communications Technologies 115, https://doi.org/10.1007/978-3-030-95161-0_7

1   Introduction

Analyzing the already functioning applications and protocols used in contact monitoring applications, it should be noted that for their correct operation it is necessary to use a number of cryptographic primitives and interaction protocols at different stages of work. They are used to maintain the confidentiality of individual user data and increase trust in the applications themselves. In turn, the security of decentralized systems built on Blockchain technology is directly determined by the efficiency of the applied cryptographic transformations. In the following, cryptographic primitives are presented, which are necessary for building contact monitoring systems both according to standard schemes and using Blockchain technology, as well as their main features and criteria, according to which you can choose the necessary algorithm for further effective implementation.

2   Analysis of Contacts Tracking Applications Operating in the World, Regarding the Cryptographic Primitives Used in Them

Summarizing the information on contact tracking applications, we can identify several stages of application operation, for which the use of cryptographic mechanisms is fundamental.

1. Generation of the initial secret key, according to which day keys are generated. This stage involves the use of a pseudo-random sequence generator for further safe operation.
2. Generation of day keys. Typically, this procedure is done by hashing the initial or day key using the hash function.
3. Generation of temporary keys. In most cases, temporary keys are generated using block encryption.
4. In some cases, digital signature and HMAC algorithms are used to generate signatures.

Thus, it is necessary to study, analyze and substantiate the use of digital signature algorithms, encryption, pseudo-random sequence generation and hashing in contact monitoring applications [1].

3   Research of Blockchain Systems on the Used Cryptographic Primitives

In turn, the basis of the operation of the proposed configuration of the contact tracking application is a blockchain system. In October 2018, NIST USA published a report on the analysis of blockchain technology NIST.IR.8202. According to this report, blockchain systems can be defined as follows:

A blockchain is a distributed digital repository of transactions that are cryptographically secure (signed) and that are grouped into blocks. In turn, the blocks are linked by a cryptographic hash function, ie formed as a one-way linked list, where the next block points to the previous one. This provides protection against unauthorized changes to the repository. The next block should be pre-screened and reviewed by consensus protocol. Thus, the more new blocks are added to the repository, the more difficult the task of modifying previous blocks [2].

At its core, the classic blockchain uses two cryptographic primitives: a hash function and an electronic signature. The hash function is used for several purposes:

(1)   generating an address or user ID. In a basic sense, the blockchain system uses asymmetric cryptography, namely an electronic signature, to identify and confirm the transaction of either the owner of the asset who transfers it to another participant, or changes to the document. Of course, the user can be identified by his public key, but the public key as the "destination" of the asset may be too large and not very acceptable to humans. Therefore, as addresses in blockchain systems, it is customary to use hash values from the user's public key;

(2)   hashing of transactions and the block for confirmation of irrefutable block concerning network errors and also hash value of the previous block is used as the reference at formation of the following.

On the other hand, digital signature (DS) in the blockchain system plays a crucial role. The use of securities in the blockchain system implements the services of irrefutability and integrity in relation to transactions. In turn, in any blockchain system there is a pool of transactions, each of which must be signed. This ensures that the transaction will not be altered, the signature of the transaction ensures that it was signed by a person who can then be identified, and, in fact, the signature can help determine "whether the user has the rights to conduct the transaction." Therefore, the set of transactions that fall into the block form a hash value by which, for example, in the Proof of Work model, the hash value of the block is formed. This approach ensures that the list of transactions cannot be changed over time in a block that has already been accepted. Even changing one bit of a transaction will require re-signing it, and changing the signature in a heap with changing transactions will require re-generating hash values for the entire block. Since the next block will have as a pointer the hash value of the previous block, it will also be necessary to change this hash value, and hence the hash value of the next block. Thus, the mechanism that allows you to build a cryptographic system to protect data from modification or destruction, ie guarantees the service of integrity [3].

Thus, for the operation of contact tracking applications using blockchain technology, it is necessary to conduct research and comparative analysis of such cryptographic primitives as the hash function and electronic digital signature.

4   Development of Methods for Evaluation and Comparative Studies of Block Symmetric Ciphers that Can Be Used for the Implementation of Monitoring Systems

Block ciphers are one of the most common cryptographic primitives. In addition to ensuring confidentiality, they are used as a constructive element in the construction of hash functions, message authentication codes, and so on. In order for block symmetric ciphers to function effectively, they must satisfy some important conditions [4].

As you know, the size of the block and the length of the key determine the external parameters of the algorithm and its scope (provided that the internal components of resistance to the respective types of attacks). In turn, the length of the key should provide a practical impossibility of carrying out enumeration attacks and methods of analysis based on pre-calculation tables, with a significant margin of stability.

Also, the block size affects the collision properties of the conversion, which, in turn, determines the level of stability of most modes of operation of the block cipher, aimed at ensuring confidentiality and integrity and other applications (generation of pseudo-random sequences, construction of hash function, etc.).

These parameters of cryptographic conversion determine the limit values of stability and form the general criteria that must satisfy the internal components (the lower limit of the complexity of analytical attacks).

Cryptographic primitives should implement random reflection properties, and provided fixing encryption key cipher endomorphic ideal model is a random permutation corresponding degree.

The requirements of randomness (pseudo-randomness) are necessary to hide the redundancy of input data, when the unequal probability of input symbols or their groups turns into unequal probability of sequences of much greater length, which exponentially increases the number of resources required for cryptanalysis. However, the direct implementation of the substitution requires unattainable amounts of memory in practice, which leads to the need for iterative conversions [5].

In turn, for this type of cipher you need to specify a high-level design. Currently, common conversions use the Feistel chain, SPN structure, Ley-Messi scheme, or modifications thereof. At the same time, the choice of a high-level cipher construction is made on the basis of the developer's preferences and in most cases has no theoretical justification with numerical estimates of efficiency.

For the effective implementation of the scattering and mixing properties in the cyclic function used by the high-level structure, the use of layers of linear and nonlinear mappings is required on modern software and hardware platforms. It should be noted that in the implementation they can be performed in the form of a single substitution table, but in the cryptographic analysis of properties there are two consecutive transformations [4].

In addition to the basic encryption transformation, the block cipher requires a scheme for generating cyclic keys, which are generated based on the encryption key and used at each iteration. This component of cryptographic transformation determines the complexity of meet-in-the-middle attacks and some other methods of analysis, and there are examples where a weak scheme allows even practical attacks on common ciphers that provide a high level of resilience to all other known methods of cryptanalysis. In addition, vulnerable schemes in a number of cases allow the implementation of attacks based on related keys, which jeopardizes other cryptographic primitives built on them (for example, a hash function based on the Davis-Meyer design).

Now there are several different approaches to the construction of schemes for the formation of cyclic keys. The simplest ones provide minimal additional calculations and implement a transposition or a linear combination of bits of the encryption key to obtain the values of the cyclic keys. This approach is implemented, for example, in GOST 28147-89 and DES. The advantage is the simplicity and low computational complexity of the implementation, the disadvantage, as already mentioned, is the potential possibility of implementing cryptanalytic attacks, which require very specific conditions, but in this framework are very effective. Another approach involves the use of a separate nonlinear transformation, which provides high resistance to these attacks, but the computational complexity of the formation of cyclic keys (key agility) can exceed the encryption of even ten blocks of plaintext. Additionally, it should be noted nonlinear complex circuits (Twofish, FOX), which do not provide the property of injectivity and theoretically allow the existence of equivalent keys. However, this design also provides some protection against implementation attacks, which is not possible with an injectable circuit.

Thus, when developing a scheme for the formation of cyclic keys also requires a solution to the problem of finding the optimal solution with limitations on the cryptographic stability, flexibility, speed and compactness of this additional conversion.

On the other hand, one of the most important characteristics of cryptographic conversion is speed, and the comparison of the new standard will be carried out not only with the outdated DSTU GOST 28147: 2009 [6], but also with widespread AES (US standard FIPS-197, introduced in 2002, also included in the international standard ISO/IEC 18033-3: 2010), which has already achieved a number of extreme performance indicators, and some trade-offs have led to some theoretical attacks.

Therefore, cryptographic conversionions that can be applied in monitoring systems must meet the following general requirements:

1. high level of cryptographic stability (the complexity of known cryptanalytic attacks should be higher than the complexity of bust-type attacks, which, in turn, are not feasible in practice, given the prospects for the development of mass semiconductor technologies);
2. comparable performance for effective implementation;
3. simple software and combined (software-hardware) implementation.

These criteria will be used in the future for comparative studies of promising block ciphers for the effective implementation of monitoring systems.

5 Substantiation of Methods of Estimation and Comparative Researches of Hash Functions Which Can Be Applied for Realization of Monitoring Systems

By definition, hashing is the conversion of an input data array of arbitrary length into an output bit string of fixed length. Such transformations are also called hash functions, or convolution functions, and their results are called hash, hash code, hash sum, or message digest. Thus, the hashing function is a function that converts the input data of any usually large size into data of a fixed size [7].

The main international normative document that defines the terms, basic concepts, classification and specification of certain cryptographic hashing algorithms is the international standard ISO/IEC 10118 [8]:

- In the first part of the standard ISO/IEC 10118-1: 2016 "Information technology—Security techniques—Hash-functions—Part 1: General" [9] the basic concepts and definitions on information hashing are given, in particular the general iterative model of hash function (first part standard harmonized in Ukraine in the form of DSTU ISO/IEC 10118-1: 2018 (ISO/IEC 10118-1: 2016, IDT) "Information technology. Methods of protection. Gash functions. Part 1. General provisions" [10]);
- in the second part of ISO/IEC 10118-2: 2010/Cor.1: 2011 "Information technology—Security techniques—Hash-functions—Part 2: Hash-functions using an n-bit block cipher" [11] defining hashing algorithms which use symmetric block ciphers (this part of the standard is harmonized in Ukraine in the form of DSTU ISO/IEC 10118-2: 2015 (ISO/IEC 10118-2: 2010; Cor 1: 2011, IDT) "Information technology. Methods of protection. hash functions. Part 2. hash functions using n-bit block cipher" [12]);
- The third part of ISO/IEC 10118-3: 2018 "IT Security techniques—Hash-functions—Part 3: Dedicated hash-functions" [13] is devoted to the consideration of specialized hashing functions (harmonized in Ukraine in the form of DSTU ISO/IEC 10118-3: 2005 "Information technologies. Methods of protection. Hash functions. Part 3. Specialized hash functions" [14]);
- The fourth part of ISO/IEC 10118-4: 1998/Cor.1: 2014 "Information technology—Security techniques—Hash-functions—Part 4: Hash-functions using modular arithmetic" [15] contains a description of hashing functions based on modular arithmetic (harmonized in Ukraine in the form of DSTU ISO/IEC 10118-4: 2015 (ISO/IEC 10118-4: 1998; Cor 1: 2014; Amd 1: 2014, IDT) "Information technology. Security methods. hash functions. Part 4 hash functions that use modular arithmetic "[16]).

The hashing functions described in a number of ISO/IEC 10118 standards do not use a secret key (ie are keyless hash functions), in particular they can be used to generate manipulation detection code (MDC).

It should be noted that some cryptographic hashing functions can also use a secret key (ie be so-called key hash functions). Such hashing functions are designed to generate message authentication codes (MACs). Another normative document is used for their description and standardization at the international level, namely ISO/IEC9797-1: 2011 Information technology—Security techniques—Message Authentication Codes (MACs).

Analyzing international algorithms, we can identify the following criteria that must meet the cryptographic hash functions [8].

1. Resistance to finding a prototype. This means that the hash function is one-sided, ie from a mathematical point of view it is impossible to calculate the correct input value for a known output value. For example, if a hash value of y is given, then it is computationally difficult to find an x for which $hash(x) = y$.

2. Resistance to finding the second prototype. This means that no one can find the input value that is hashed into a specific result. In more detail, cryptographic hash functions are created in such a way that at a given specific output value it is computationally impossible to find a second input value that gives the same output value. For example, if x is given, it is computationally difficult to find one for which $hash(x) = hash(y)$. The only available approach is to go through the original values throughout the space, but from a computational point of view there is no chance of success.

3. Resistance to collisions. This means that it is not possible to find two input values that would be hashed to the same result. If we consider in more detail, then from a mathematical point of view it is computationally impossible to find two input values that would lead to the same output value. For example, it is computationally difficult to find such x and y, where hash (x) = hash (y).

It should also be noted that an important indicator of the effectiveness of the hashing function is the speed of obtaining a hash and generating key parameters, which should also be explored when comparing alternatives for use in monitoring systems.

6 Development of Methods for Evaluation and Comparative Studies of Asymmetric Cryptoalgorithms and Key Encapsulation Schemes That Can Be Used for the Implementation of Monitoring Systems

Asymmetric encryption algorithms are encryption algorithms that use different keys to encrypt and decrypt data. Asymmetric algorithms do not require the sender and recipient to agree on a secret key on a special secure channel. The encryption procedure is irreversible according to the known encryption key. Thus, knowing the encryption key and the encrypted text, it is impossible to recover the original message. The encryption key in this context is a public key, the decryption key is secret. Encryption and decryption algorithms are built in such a way that it is impossible to recover the secret key with a known public key. Another name for asymmetric cryptosystems is a directional encryption or public key encryption [17].

As asymmetric cryptosystems have a significant advantage over symmetric algorithms due to the lack of need to reliably exchange secret keys internationally during the implementation of the European NESSIE project, special attention was paid to the implementation of the requirements for key encapsulation protocols. Subsequently, based on the obtained results, proposals and recommendations, the international standard ISO/IEC 18033-2 "Information technology—Protection methods—Protection algorithms—Part 2: Asymmetric ciphers" was adopted [18].

In turn, the key encapsulation mechanism in the proposed terminology is intended for encapsulation and decapsulation of keys, as well as calculation (generation) and use of secret keys, when using, for example, modes of operation of symmetric block and stream ciphers.

As already considered, the first thing that is necessary for the evaluation and comparison of cryptographic systems is the choice of criteria and indicators for the evaluation of cryptographic systems.

Studies have shown that the comparison of cryptographic primitives can be performed using two sets of criteria: unconditional and conditional. This approach is based, inter alia, on the consideration or use of expert judgment.

As a first step, it is necessary to make a comparison according to unconditional criteria and calculate an unconditional integral criterion.

Unconditional criteria will include those criteria that are mandatory for cryptographic conversions, ie unconditional. Thus, under the condition of a positive assessment by the integral unconditional criterion, further comparison and evaluation can be made on the basis of definition and comparison of conditional criteria and integral conditional criterion. As unconditional criteria we choose the following [17]:

- W_1—reliability of the mathematical base used in cryptographic conversions;
- W_2—practical protection of cryptographic conversions from known quantum attacks;
- W_3—real protection against all known and potential cryptanalytic attacks;
- W_4—statistical security of cryptographic conversion;
- W_5—theoretical security of cryptographic conversion;
- W_6—the absence of weak private keys of cryptographic conversion or the presence of a proven mechanism for detecting/verifying such keys;
- W_7—the complexity of direct and reverse cryptographic conversions with respect to ES is not higher than polynomial.

The reliability of the mathematical basis means the practical lack of the violator's ability to carry out attacks such as "universal disclosure" due to the imperfections of the mathematical apparatus used, or weaknesses that may be due to the specific properties of general parameters and keys. It should be noted that the criterion of unreliability is the subexponential or polynomial complexity of the attack "universal disclosure", and the criterion of reliability, in turn, is exponential.

If the result of the cryptographic transformation does not depend on the encrypted input block and the private key used, then the transformation will be considered statistically secure.

Practical security of cryptoconversions is protection against force and analytical attacks, which is achieved by choosing the size of general parameters and keys, as well as the means of their generation. That is, the criterion of practical security of cryptoconversions is determined by the dependence of the complexity of the attack on the size of the general parameters and keys.

It is worth noting that weak keys are keys in which the complexity of cryptanalytic attacks such as "full disclosure" and "universal disclosure" is less than the complexity of the attack "full disclosure" for other (not weak) private keys. It is allowed to adopt a mechanism that has weak key pairs, but the probability of their generation is low and there is a proven algorithm for checking the key pair for weakness (or all such key pairs have already been identified).

In the second stage, it is necessary to obtain estimates for cryptoalgorithms according to a set of conditional criteria and calculate the conditional integral criterion accordingly.

In the future, the following conditional criteria for comparison should be identified:

- Ist.—cryptographic stability;
- lp.k—the length of the public key;
- lp.k—the length of the private key;
- lres.—the length of the crypto conversion result;
- Tdir.—the speed of direct cryptoconversion;
- Trev.—reverse conversion rate.

The application of partial conditional criteria, and then on the basis of their integral conditional criterion, allows to obtain a more accurate estimate. Moreover, in the future, under the compliance of a mechanism with unconditional criteria, we will understand the fact that expert assessments of unconditional criteria are positive, ie they are clearly met. A specific description of obtaining a conditional integral criterion is described below.

7 Development of Methods for Evaluation and Comparative Studies of Digital Signature Schemes that Can Be Used for the Implementation of Monitoring Systems

An important area of application of public key cryptography is the authentication of messages using electronic digital signatures. Modern signatures have different properties and are widely used in many areas, which necessitates a comparative analysis to develop recommendations for further appropriate use of selected crypto-primitives.

In order to apply the method of hierarchies, it is necessary to choose a system of conditional criteria. According to which to calculate the conditional integrated criterion and to make a full comparison of electronic digital signature schemes with its use [19].

We choose as conditional criteria for evaluating the ES such partial criteria:

- W_{y1}—possibility and conditions of free distribution and application of the international or national standard of cryptographic transformations of ES in Ukraine taking into account regulatory legal acts of Ukraine on export, import and restrictions on its application, including for rendering of electronic trust services;
- W_{y2}—the level of confidence in the international or national standard of cryptographic conversion, determined by the results of research and the degree of application and recognition in different countries and internationally recognized systems, including for the provision of electronic trust services;
- W_{y3}—prospects for the application of international or national standards in Ukraine, taking into account the recognition and application of advanced information and telecommunications systems, cloud computing and other information technologies, etc.;
- W_{y4}—temporal and spatial complexity of hardware, combined (hardware-software) and software implementations of ES means and management and certification of keys, including for the provision of electronic trust services, etc.
- W_{y5}—the possibility and conditions of application of standards with different values of general system parameters and keys, methods of production and maintenance of public key certificates, including for the provision of electronic trust services, etc.;
- W_{y6}—the degree of flexibility of the ES in terms of use in different applications, under different requirements and restrictions, in different conditions, the degree of unification and standardization, including for the provision of electronic trust services, etc.;
- W_{y7}—the level of security in the implementation of different types of threats, under different conditions of cryptanalytic attacks and deviations of the properties of general parameters from the defined ones, etc.

Expert estimates of alternatives according to certain conditional criteria can be presented as fuzzy sets or numbers expressed by membership functions. To organize fuzzy numbers, there are many methods that differ from each other in the way of convolution and construction of fuzzy relationships. The latter can be defined as the relationship of preference between objects. One such method is the method of pairwise comparison [20].

The method of pairwise comparison of elements can be described as follows. First, it is necessary to construct a set of matrices of pairwise comparisons. It should be noted that pairwise comparisons are made in terms of the dominance of one element over another. The resulting judgments are expressed in integers based on a nine-point scale (Table 1).

Table 1 Degrees of significance

Degree of significance	Definition	Explanation
1	Equal significance	Two actions make the same contribution to achieving the goal
3	Some advantage of the significance of one action over another (weak significance)	There is an understanding in favor of the benefits of one of the actions, but these understandings are not convincing enough
5	Significant or strong significance	There is reliable data or logical reasoning to show the superiority of one of the actions
7	Obvious or very strong significance	Convincing evidence in favor of one action before another
9	Absolute significance	Evidence in favor of the superiority of one action over another is highly convincing
2, 4, 6, 8	Intermediate values between two adjacent judgments	A situation where a compromise solution is needed
The inverse values of the above non-zero values	If the action i when compared with the action j is assigned one of the above non-zero numbers, then the action j when compared with the action i is assigned the inverse value	If the consistency was postulated when obtaining N numerical values for the formation of the matrix

When using this scale, comparing two objects in the sense of achieving a goal located at the highest level of the hierarchy, should put in accordance with this comparison a number in the range from 1 to 9 or the inverse of the number. In cases where it is difficult to distinguish so many intermediate gradations from absolute to weak advantage or it is not required in a particular problem, a scale with a smaller number of gradations can be used. Within the boundary, the scale has two estimates: 1—objects are equivalent; 2—the advantage of one object over another [19].

Filling of square matrices of pair comparisons is carried out by such rule. If the element E_1 dominates the element E_2, then the cell of the matrix corresponding to the row E_1 and column E_2 is filled with an integer, and the cell corresponding to the row E_2 and column E_1 is filled with an inverse number. If the element E_2 dominates E_1, then the integer is placed in the cell corresponding to the row E_2 and column E_1, and the fraction is placed in the cell corresponding to the row E_1 and column E_1. If the elements E_1 are E_2 equally dominant, then in both positions of the matrix are units.

To obtain each matrix, the expert or LPR makes $n(n - 1)/2$ judgments (here n—the order of the matrix of pairwise comparisons).

Consider in general an example of the formation of a matrix of pairwise comparisons.

Let $E_1, E_2, \ldots, E_n$—plural of n elements (alternatives) and $v_1, v_2, \ldots, v_n$—according to their weight or intensity. Let's compare in pairs the weight, or intensity, of each element with the weight, or intensity, of any other element of the set in relation to their common property or purpose. In this case, the matrix of paired comparisons [E] has the following form [21]:

	E_1	E_2	...	E_n
E_1	v_1/v_1	v_1/v_2	...	v_1/v_n
E_2	v_2/v_1	v_2/v_2	...	v_2/v_n
...	...	...	...	...
E_n	v_n/v_1	v_n/v_2	...	v_n/v_n

The following questions should be answered during pairwise comparisons: which of the two elements being compared is more important or more influential, which is more likely and which is better.

When comparing criteria, one usually asks which of the criteria is more important; when comparing alternatives to the criterion—which of the alternatives is better or more likely.

The ranking of the elements analyzed using the matrix of paired comparisons [E] is carried out on the basis of the main eigenvectors obtained as a result of processing the matrices.

The calculation of the principal eigenvector W of the positive quadratic matrix [E] is performed on the basis of equality

$$EW = \lambda_{\max} W,$$

where $\lambda_{\max}$—maximum eigenvalue of the matrix [E].

For a positive square matrix [E] the right eigenvector W corresponding to the maximum eigenvalue $\lambda_{\max}$, up to a constant factor C can be calculated by the formula:

$$\lim_{k \to \infty} \frac{[E]^k e}{e^T [E]^k e} = CW,$$

where

$e = \{1,1, \ldots,1\}^T$ single vector;
$k = 1, 2, 3, \ldots$ degree indicator;
C constant;
T transposition sign.

The calculation of the eigenvector W according to (6.49 3.20) is carried out until the specified accuracy is reached:

$$e^T \left| W^{(l)} - W^{(l+1)} \right| \leq \xi,$$

where l—iteration numbe—such that $l = 1$ meets $k = 1$; $l = 2$, $k = 2$; $l = 3$, $k = 4$ and etc.;

ξ—permissible error (with sufficient accuracy for practice can be accepted $\xi = 0, 01$ regardless of the order of the matrix).

The maximum eigenvalue of the matrix is calculated by the formula:

$$\lambda_{max} = e^T [E] W.$$

This approach allows you to compare different algorithms according to certain conditional criteria and obtain a quantitative assessment of their effectiveness.

8 Substantiation of Evaluation Methods and Comparative Studies of Streaming Encryption Algorithms That Can Be Used for the Implementation of Monitoring Systems

Among symmetric cryptocurrencies, streaming algorithms occupy a special place, in which information is presented and processed in the form of an infinite stream, ie a sequence that can hypothetically be of infinite length. The main advantage of this conversion is the establishment of a certain relationship between the individual symbols of the data stream, which provides additional protection against the imposition of false information or false modes of protection equipment or terminal equipment of telecommunications systems and networks. Accordingly, cryptographic streaming is usually more trusted by users because the data streams protected by the streaming algorithm cannot be distorted in any way as a result of intentional or unintentional actions of users and attackers, or any random natural factors or factors [22].

Today, the current code of the European standard must meet a fairly high figure— hundreds of Mbit/sec and even a few Gbit/sec, if you look to the future and make a stock for several decades. An effective solution, in addition to high performance, must have validity, proven reliability, simplicity and scalability, completeness and clarity of the algorithm, to ensure confidentiality in information transmission channels.

The criteria for selecting the current cipher are usually indicators of the encryption speed of long sequences and the time of initialization/generation of key parameters. In turn, the following method was used during the international eSTREAM competition. According to this method, the current ciphers are compared according to the following criteria [23]:

- criterion for encrypting long streams, current ciphers have the most potential advantage over block ciphers when encrypting long streams. Therefore, this indicator is an important evaluation criterion;

- criterion for encrypting short streams, this indicator reflects the encryption speed of packets of different lengths. Each function call includes a separate initialization vector (IV) setting, packet lengths (40, 576, and 1500 bytes) were selected to be representative of telecommunication traffic;
- initialization criterion / generation of key parameters. Separately displays the efficiency of setting the key and the initialization vector. These two parameters are the least critical for displaying packet encryption speeds so negligibly small compared to the key creation and recovery process.

9   Development of Methods for Comparing Algorithms for the Synthesis of Pseudo-random Sequences

It is worth noting that stream ciphers are also used to generate pseudo-random sequences on a par with typical pseudo-random sequence generators. In this context, the statistical properties of the resulting generated sequences are important.

Informally speaking, an algorithm is statistically safe if the sequence it generates is not inferior in its properties to a "random" sequence. Statistical tests are used to experimentally assess how closely cryptoalgorithms approximate "random" sequence generators. One such common and proven test suite is the NIST STS test suite, developed by the National Institute of Standards and Technology [24].

The procedure for testing a single binary sequence is as follows:

(1)   The null hypothesis H_0 is put forward, which defines the sequence as random;
(2)   According to the sequence, the statistical index of the test c (S) is calculated;
(3)   Using a special function and statistics, the probability value is calculated $P = f(c(S))$, $P \in [0, 1][0, 1]$ [25];
(4)   The obtained value of P is compared with the level of significance $\alpha \in [0.001, 0.01]$. If $P \geq \alpha$ -then the hypothesis H_0 is accepted, otherwise the hypothesis will be rejected.

NIST STS consists of 16 statistical tests, but actually calculates 189 probability values, which can be considered as separate test results [26].

Thus, as a result of testing the binary sequence, a vector of probability values is formed.

According to the described method the decision on passing of statistical testing is accepted according to the following rules:

(1)   Testing was passed on all q tests and the value of the coefficient is in the confidence interval [0.96, 1.00];
(2)   Testing passed on all q tests and if for all tests by Pearson's criterion χ^2 the condition is fulfilled $P(\chi^2) > 0,0001$.

In the future, the choice of an efficient algorithm for generating a pseudo-random sequence is proposed to be carried out according to the described method.

References

1. Contact Tracing Apps. [Online]: https://www.coe.int/en/web/data-protection/contact-tracing-apps
2. Isirova K, Kiian A, Rodinko M, Kuznetsov A (2020) Decentralized electronic voting system based on blockchain technology developing principals. In: Subbotin S (ed) Proceedings of the third international workshop on computer modeling and intelligent systems (CMIS-2020), Zaporizhzhia, Ukraine, April 27–May 1, 2020. CEUR-WS.org, pp 211–223
3. Fitzi M, Gaži P, Kiayias A, Russell A (2018) Parallel chains: improving throughput and latency of blockchain protocols via parallel composition. Cryptology eprint Archive, Report 2018/1119
4. Gorbenko I, Kuznetsov A, Lutsenko M, Ivanenko D (2017) The research of modern stream ciphers. In: 2017 4th international scientific-practical conference problems of infocommunications. Science and TECHNOLOGY (PIC S T), pp 207–210. https://doi.org/10.1109/INFOCO MMST.2017.8246381
5. Gorbenko I, Kuznetsov A, Tymchenko V, Gorbenko Y, Kachko O (2018) Experimental studies of the modern symmetric stream ciphers. In: 2018 international scientific-practical conference problems of infocommunications. Science and technology (PIC S T), pp 125–128. https://doi.org/10.1109/INFOCOMMST.2018.8632058
6. Kuznetsov AA, Potii OV, Poluyanenko NA, Gorbenko YI, Kryvinska N (2022) Stream ciphers in modern real-time it systems: analysis, design and comparative studies. Springer International Publishing. https://doi.org/10.1007/978-3-030-79770-6
7. Kuznetsov A, Lutsenko M, Kuznetsova K, Martyniuk O, Babenko V, Perevozova I (2019) Statistical testing of blockchain hash algorithms. In: Fedushko S, Gnatyuk S, Peleshchyshyn A, Hu Z, Odarchenko R, Korobiichuk I (eds) Proceedings of the international workshop on conflict management in global information networks (CMiGIN 2019) co-located with 1st international conference on cyber hygiene and conflict management in global information networks (CyberConf 2019), Lviv, Ukraine, November 29, 2019. CEUR-WS.org (2019), pp 67–79
8. Menezes AJ, van Oorschot PC, Vanstone SA, van Oorschot PC, Vanstone SA (2018) Handbook of applied cryptography. CRC Press (2018). https://doi.org/10.1201/9780429466335
9. ISO/IEC 10118-1:2016 «Information technology—Security techniques—Hash-functions—Part 1: General». [Online]: https://www.iso.org/ru/standard/64213.html
10. DSTU ISO/IEC 10118-1:2018 (ISO/IEC 10118-1:2016, IDT) [Online]: http://online.budstandart.com/ru/catalog/doc-page?id_doc=80113
11. ISO/IEC 10118-2:2010/Cor.1:2011 «Information technology—Security techniques—Hash-functions—Part 2: Hash-functions using an n-bit block cipher». [Online]: https://www.iso.org/standard/44737.html
12. DSTU ISO/IEC 10118-2:2015 (ISO/IEC 10118-2:2010; Cor 1:2011, IDT) [Online]: http://online.budstandart.com/ua/catalog/doc-page?id_doc=66931
13. ISO/IEC 10118-3:2018 «IT Security techniques—Hash-functions—Part 3: Dedicated hash-functions». [Online]: https://www.iso.org/standard/67116.html
14. DSTU ISO/IEC 10118-3:2005 [Online]: https://metrology.com.ua/ntd/skachat-iso-iec-ohsas/eea/dstu-iso-ies-10118-3-2005/
15. ISO/IEC 10118-4:1998/Cor.1:2014 «Information technology—Security techniques—Hash-functions—Part 4: Hash-functions using modular arithmetic». [Online]: https://www.iso.org/ru/standard/65665.html
16. DSTU ISO/IEC 10118-4:2015 (ISO/IEC 10118-4:1998; Cor 1:2014; Amd 1:2014, IDT) [Online]: http://online.budstandart.com/ua/catalog/doc-page?id_doc=66932
17. Computer Security Division, I.T.L.: Post-Quantum Cryptography: Proposed Requirements & Eval Criteria|CSRC. https://content.csrc.e1c.nist.gov/News/2016/Post-Quantum-Cryptography-Proposed-Requirements. Last accessed 30 Nov 2020
18. DSTU ISO/IEC 18033-2 [Online]: http://online.budstandart.com/ua/catalog/doc-page?id_doc=66929

19. Kuznetsov AA, Gorbenko YI, Prokopovych-Tkachenko DI, Lutsenko MS, Pastukhov MV (2019) NIST PQC: CODE-BASED CRYPTOSYSTEMS. TRE. 78 (2019). https://doi.org/10.1615/TelecomRadEng.v78.i5.50

20. Kuznetsov A, Girzheva O, Kiian A, Nakisko O, Smirnov O, Kuznetsova T (2020) Advanced code-based electronic digital signature scheme. In: 2020 IEEE international conference on problems of infocommunications. Science and technology (PIC S T), pp 358–362 (2020). https://doi.org/10.1109/PICST51311.2020.9467895.

21. Kuznetsov A, Kiian A, Babenko V, Perevozova I, Chepurko I, Smirnov O (2020) New approach to the implementation of post-quantum digital signature scheme. In: 2020 IEEE 11th international conference on dependable systems, services and technologies (DESSERT), pp 166–171 (2020). https://doi.org/10.1109/DESSERT50317.2020.9125053

22. Gorbenko I, Kuznetsov A, Gorbenko Y, Vdovenko S, Tymchenko V, Lutsenko M (2019) Studies on statistical analysis and performance evaluation for some stream ciphers. Int J Comput 18:82–88

23. Gorbenko I, Kuznetsov O, Gorbenko Y, Alekseychuk A, Tymchenko V (2018) Strumok keystream generator. In: 2018 IEEE 9th international conference on dependable systems, services and technologies (DESSERT). IEEE, Kyiv, Ukraine, pp 294–299. https://doi.org/10.1109/DESSERT.2018.8409147

24. Blum M (1986) How to generate cryptographycally strong sequences of pseudorandom bits/Blum M, Micali S//SIAM J Comput 1:850–864

25. Kuznetsov A, Kavun S, Panchenko V, Prokopovych-Tkachenko D, Kurinniy F, Shoiko V (2018) Periodic properties of cryptographically strong pseudorandom sequences. In: 2018 international scientific-practical conference problems of infocommunications. Science and technology (PIC S T), pp 129–134. https://doi.org/10.1109/INFOCOMMST.2018.8632021

26. A statistical test suite for random and pseudorandom number generators for cryptographic applications national institute of standards and technology special publication 800-22 revision 1a//Natl Inst Stand Technol Spec Publ 800-22 rev1a, 131 p

Cryptographic Mechanisms that Ensure the Efficiency of SNARK-Systems

Lyudmila Kovalchuk⊙, Roman Oliynykov⊙, Yurii Bespalov⊙, and Mariia Rodinko⊙

Abstract The section provides, strictly substantiated, estimates of the resistance of the block encryption algorithm HadesMiMC and the hash function Poseidon to linear and differential attacks. Based on the obtained results, specific parameters for the HadesMiMC encryption algorithm and the Poseidon hash function were determined, which allow them to be used for recurrent SNARK-proofs based on MNT-4 and MNT-6 triplets. It is also shown how better to choose S-boxes for these algorithms, so that this choice was optimal both in terms of stability and in terms of complexity of calculations in SNARK-proofs. It is determined how many rounds are sufficient to guarantee the stability of algorithms for different options. The number of constraints per bit is calculated for different options, and it is shown that it is significantly smaller than for the Pedersen function, which is currently used in the cryptocurrency Zcash. The significant advantages of the considered HadesMiMC block encryption algorithm and the corresponding Poseidon hash function when using these cryptographic primitives in the blockchain as a decentralized environment, an additional function of which is to ensure the anonymity of personal information, are substantiated. These advantages are both a reasonably high level of resistance to known attacks on such cryptographic primitives, and a significant increase in speed (up to 15 times) in the construction of appropriate SNARK-proofs, which are an integral part of the mechanisms of anonymity in the blockchain. The results showed that the HadesMiMC block encryption algorithm, based on operations over a simple finite field, is secure against linear and differential cryptanalysis, with the correct choice of parameters of this algorithm. As a result, the Poseidon hash function, built using the SPONGE design and the HadesMiMC permutation, will be stable. It was also shown how the parameters of the algorithm should be selected to ensure its specified level of security.

L. Kovalchuk
National Technical University of Ukraine "Igor Sikorsky Kyiv Polytechnic Institute", 37, Prosp. Peremohy, Kyiv 03056, Ukraine

R. Oliynykov (✉) · M. Rodinko
V. N. Karazin Kharkiv National University, 4 Svobody Sq., Kharkiv 61022, Ukraine

Y. Bespalov
M.M. Bogolyubov Institute for Theoretical Physics of the National Academy of Sciences of Ukraine, 14-b Metrolohichna str., Kyiv 03143, Ukraine

© The Author(s), under exclusive license to Springer Nature Switzerland AG 2022
R. Oliynykov et al. (eds.), *Information Security Technologies in the Decentralized Distributed Networks*, Lecture Notes on Data Engineering and Communications Technologies 115, https://doi.org/10.1007/978-3-030-95161-0_8

185

In addition, the number of constraints per bit of information was calculated and it was shown that the HadesMiMC algorithm and the Poseidon hash function currently have no competitors when used in SNARK-proofs.

Keywords SNARK-proof · Poseidon hash function · Differential cryptanalysis · Linear cryptanalysis · Cryptocurrency

1   Introduction

Consider in detail the hash function and block encryption algorithm, which are particularly interesting because they are very suitable for use in a mathematical apparatus that provides complete anonymity of processed information, which, in particular, can be effectively used to protect personal information of patients in a pandemic.

Block encryption algorithms, which use simple finite field operations instead of binary ones, have emerged with the advent of new non-disclosure methods used in some blockchain systems. Here we mean the so-called SNARK- and STARK-proofs [1–3]. The construction of such proofs begins with the fact that a certain transformation (for example, a hash function) should be described as a system of certain equations of many variables over a finite field, the left part of which contains a polynomial of many second-degree variables, and the right part of a polynomial of many variables of the first degree. These equations are called constraints, and the complexity of constructing the corresponding SNARK-proof depends on their number. SNARK-proofs are most often used to prove knowledge of a prototype of some hash function. Therefore, the hash functions used in such blockchains should be designed so that they can be described by as few constraints as possible.

One of the first hash functions convenient for constructing SNARK-proofs was the Pedersen hash function [4], p. 134]. It is based on operations in a group of points of an elliptic curve, which, in turn, can be reduced to operations in the corresponding finite field. Since constraints are polynomials just above such a field, the number of constraints required to specify such a hash function is ten times less than for "classical" hash functions that operate with byte and bit operations (about 1.68 constraints per 1 bit of input). This number of constraints is quite acceptable, but the question of reducing it still remains relevant.

The new Poseidon hash function proposed in [5] allows reducing the number of constraints up to 15 times, compared to the Pedersen hash function. It uses two constructs—the so-called SPONGE construction [6] and a new block encryption algorithm HadesMiMC [7], which is also adapted for use in SNARK-proofs and is therefore based on operations in simple finite fields.

To construct the Poseidon hash function, the HadesMiMC algorithm is used in the SPONGE construction as an internal random permutation. Therefore, the cryptographic security of the Poseidon hash function is largely determined by the resistance of this algorithm to the main cryptographic attacks that can be applied to such algorithms.

The authors of the HadesMiMC algorithm, and later the authors of the Poseidon hash function, performed a fairly detailed analysis of the resistance of this algorithm to a certain class of attacks, which they called "algebraic attacks". However, their estimates of resistance to statistical attacks, such as linear and differential, are mostly empirical, and, moreover, are significantly based on false statements. In particular:

it is not specified on which parameters the resistance of the algorithm to linear attacks depends;

it is erroneously assumed that the coordinate functions of S-boxes determine the resistance of a nonbinary algorithm to statistical attacks;

the number of active S-boxes of the algorithm is incorrectly calculated (without taking into account its peculiarity, which consists in the presence of a large number of rounds with only one S-box), etc.

In addition, the authors argue that estimates of resistance to linear cryptanalysis are obtained similarly to the differential one. However, as proved in [8, 9], for non-binary algorithms, i.e. those in which the key adder, substitution unit and linear operator use operations in a finite simple field, the methods for estimating resistance to statistical attacks, especially linear ones, differ significantly from classical, as well as methods for assessing resistance to linear cryptanalysis are very different from methods for assessing resistance to differential one.

Based on the above, it can be argued that the issue of resistance to statistical attacks of both the HadesMiMC algorithm and the Poseidon hash function remains open.

In this section, we provide strictly proven security estimates of the HadesMiMC block encryption algorithm (and, consequently, the Poseidon hash function) against linear and differential attacks. As an intermediate result, S-box-dependent parameters will be obtained that determine the resistance of the algorithm to linear attacks. Also, for certain parameters of this algorithm, in which it is compatible with the MNT-4 and MNT-6 triplets used for recursive SNARK-proofs, specific values of such estimates will be given and how to determine the number of rounds with a full layer of S-boxes, guaranteeing a sufficient level of security of this algorithm.

2 Mathematical Models of HadesMiMC Block Encryption Algorithm and Poseidon Hash Function

The Poseidon hash function [5] uses the SPONGE [6] construction with an internal permutation specified by the HadesMiMC block algorithm [7] (see Fig. 1).

This design is set by three parameters: capacity c, speed r and the length of the permutation N, where $N = c + r$. As a partial case, for some practical reasons, we will be interested in the parameters $N = 3\lceil \log p \rceil$, $c = 2\lceil \log p \rceil$ and $r = \lceil \log p \rceil$. As already mentioned, the authors [1] proposed HadesMiMC as an internal permutation in order to reduce the number of constraints per input bit.

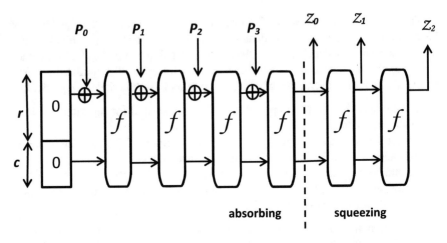

Fig. 1 SPONGE design

To construct estimates of resistance to statistical attacks, we will consider HadesMiMC as a block encryption algorithm, the specific feature of which is that its round functions differ from each other. The main idea of building this algorithm is to use round functions of two types: functions with a full layer of S-boxes and functions with a partial layer of S-boxes (for example, with only one S-box). The general scheme of the HadesMiMC algorithm is shown in Fig. 2. As will be shown below, the use of such an algorithm allows to significantly reduce the number of constraints, while maintaining an acceptable level of resistance to statistical (as well as algebraic) attacks.

Let's move on to building a mathematical model of HadesMiMC, which is necessary to analyze its resistance to these attacks.

Let p is a large prime number, l its bit length, $l \approx \log p$.

Let's define bijective reflections $s : F_p \to F_p$ as $s(x) = x^u \mod p$, where $(u, p - 1) = 1$.

For some $t \in N$ determine the values $x, C \in (F_p)^t$ as $x = (x_t, \ldots, x_1)$, $C = (c_t, \ldots, c_1)$, where $x_i, c_i \in F_p, i = \overline{1, t}$. For $x \in (F_p)^t$ we also define two mappings, $S^{full} : (F_p)^t \to (F_p)^t$ and $S^{part} : (F_p)^t \to (F_p)^t$, as

$$S^{full}(x) = (s(x_t), \ldots, s(x_1)), \quad S^{part}(x) = (x_t, \ldots x_2, s(x_1)). \qquad (1)$$

Finally, define MDS-matrix $A : (F_p)^t \to (F_p)^t$ of the size $t \times t$.

Now define the round functions for the HadesMiMC algorithm. As mentioned, round functions in this algorithm are of two types: round functions with a full layer of S-boxes, which are defined as $f_C^{full} : (F_p)^t \to (F_p)^t$, where for arbitrary $C \in (F_p)^t$:

$$f_C^{full}(x) = A \circ S^{full}(x * C), \qquad (2)$$

Fig. 2 HadesMiMC

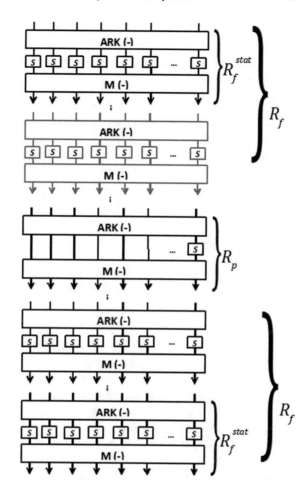

and round functions with a partial layer of S-boxes, which are defined as f_C^{part} : $\left(F_p\right)^t \to \left(F_p\right)^t$, where for arbitrary $C \in \left(F_p\right)^t$:

$$f_C^{part}(x) = A \circ S^{part}(x * C), \tag{3}$$

where $x * C = (x_t + c_t, \ldots, x_1 + c_1)$, and " $+$ " is an addition operation in a finite field F_p (that is, the addition operation modulo p).

Definition 1 HadesMiMC-like permutation with parameters p, t, u, r_{full} and r_{part} we will call the family of permutations $H_C^{(p,t,u,r_{full},r_{part})}$: $\left(F_p\right)^t \to \left(F_p\right)^t$, parameterized by a set of round constants $C = \left(C_1, \ldots, C_{2r_{full}+r_{part}}\right)$, $C_i \in \left(F_p\right)^t$, and determined as.

$$H_C^{(p,t,u,r_{full},r_{part})}(x) = f_{C_{2r_{full}+r_{part}}}^{full} \circ \ldots \circ f_{C_{r_{full}+r_{part}+1}}^{full} \circ f_{C_{2r_{full}+r_{part}}}^{part}$$

$$\circ \ldots \circ f_{C_{r_{full}+1}}^{part} \circ f_{C_{r_{full}}}^{full} \circ f_{C_1}^{full}(x). \tag{4}$$

If the parametres p, t, u, r_{full} and r_{part} are fixed, we will use notation to simplify writing H_C.

Remark Transformation (4) is defined as follows. For fixed constants $C = \left(C_1, \ldots, C_{2r_{full}+r_{part}}\right)$ we first apply to the argument x function $f_{C_1}^{full}, \ldots, f_{C_{r_{full}}}^{full}$, defined in (2) (with a full layer of S-boxes), with constants $C_1, \ldots, C_{r_{full}}$ as the appropriate "round keys". This is how the first rounds of this conversion r_{full} are performed. Then we apply the functions defined in (3) (i.e. functions with a partial layer of S-boxes), during r_{part} rounds, on the corresponding "round keys". Finally, we apply r_{full} rounds with functions (2) again.

3   Estimates of Cryptographic Security of HadesMiMC Block Encryption Algorithm and Poseidon Hash Function

As noted, the cryptographic security of the SPONGE construction is largely determined by the stability of its internal permutation, in our case—the permutation of HadesMiMC. Thus, it was proved[1] that if the internal permutation of the SPONGE construction is indistinguishable from a random permutation, then the corresponding hash function is indistinguishable from a random oracle, provided that the number of requests to the oracle is not more than $2^{\frac{c}{2}}$.

For practical applications, it is convenient to choose the capacity as $c = 2l(p)$. With this choice of parameter, i.e. the level of the hash function is equal to $l(p)$ (of course, subject to "stable" internal permutation). Therefore, we need to prove that the level of cryptographic security of the internal permutation is not less than $l(p)$. Next, we will use this requirement to determine the number of rounds with a full layer of S-boxes in the HadesMiMC permutation.

Next, we construct, strictly substantiated, estimates of the cryptographic strength of the HadesMiMC permutation to two types of statistical attacks—linear and differential. Note that when constructing estimates of resistance to differential cryptanalysis, we will mostly use known results or their generalizations. But to build estimates of resistance to linear attacks, it was necessary to prove a number of non-trivial statements about the properties of the characters of the finite field additive group and the sums of such characters.

---

[1] Bertoni et al. [10].

3.1 Security Estimates of Non-binary HadesMiMC Block Encryption Algorithm Against Differential Cryptanalysis

In what follows, we will use the following ancillary results.

Definition 2[2]:
Block cipher with round function.

$$f : M \times K \to M$$

(here M is an Abelian group with some operation "$*$" and a neutral element 0) is called a *Markov cipher* in relation to the operation "$*$", if $\forall x, \alpha, \beta \in M$:

$$\frac{1}{|K|} \sum_{k \in K} \delta\big(f(k, x * \alpha) * f(k, x)^{-1}, \beta\big) = \frac{1}{|K|} \sum_{k \in K} \delta\big(f(k, \alpha) * f(k, 0)^{-1}, \beta\big), \quad (5)$$

where δ is a symbol of Kronecker: $\delta(x, y) = \begin{cases} 1, & \text{if } x = y, \\ 0, & \text{else.} \end{cases}$

Note: this definition can be easily summarized in the case of a cipher with different round functions.

Definition 3[3]:
Matrix branching index $A : (F_p)^t \to (F_p)^t$ dimension $t \times t$ defined as.

$$br(A) = \min_{x \in (F_p)^t \setminus (0, \dots, 0)} \{wt(Ax) + wt(x)\},$$

where wt is the Hamming weight.

Note that if A is MDS matrix, then its branching index is the maximum possible (for its size) and is equal to $br(A) = t + 1$. For example, if $t = 3$, then $br(A) = 4$.

Statement 1: block cipher (4) is a Markov cipher.

This statement can be proved directly by checking property (5) for the round functions of this cipher.

Statement 2: (easily obtained from work Vaudenay[4]): for the Markov cipher (4), the probability of its differential characteristic is limited from above by the value Δ^b, where

$$\Delta = \max_{\alpha, \beta \in F_p^*} \frac{1}{p} \sum_{x \in F_p} \delta(s(x + \alpha) - s(x), \beta), \quad (6)$$

---

[2] Lai et al. [11].

[3] Youssef et al. [12].

[4] Vaudenay [13].

" + " is an addition operation in the field, and b is the number of active S-boxes in all rounds together.

Statement 3[5]: the number of active S-boxes in two consecutive rounds with round functions determined according to (2) is not less than the value $br(A)$.

In particular, the matrix$A$ is an MDS-matrix of dimension $t \times t$, then $br(A) = t+1$ and, according to Proposition 3, the number of active S-boxes in two consecutive rounds with a full layer of S-boxes is not less than $t + 1$. But if several rounds with an incomplete layer of S-boxes are placed between two rounds with a full layer of S-boxes, it is impossible to say anything about the number of active S-boxes in these rounds, except that they are not less than the number of rounds with a full layer S-boxes. It should be noted that the authors [5] did not take this fact into account, and as a result incorrectly determined the number of active S-boxes in the algorithm, which led to incorrect estimates of stability.

Statement 4 (obvious): the number of active S-boxes in all rounds of algorithm (4) is not less than the number of active S-boxes in rounds with a full layer of S-boxes.

Statement 5 (consequence of Statements 2 and 3): a number of b active S-boxes in (4) is not less than

$$b \geq 2(t + 1) \cdot \left[\frac{r_{full}}{2}\right], \tag{7}$$

and, if r_{full} is even, is not less than

$$b \geq (t + 1)r_{full}.$$

Note: it follows from (7) that the use of an even value of a quantity r_{full} is more efficient, since in this case the addition of another round, which will make this quantity odd, will not increase the quantity (7). Therefore, in what follows, we believe that r_{full} takes even values.

In this section, we consider two cases of S-box selection for a HadesMiMC-like algorithm: inverse S-boxes and S-boxes given by power functions. Before obtaining the main results regarding the stability of such an algorithm, we formulate the following auxiliary statement.

Statement 6[6]:

1.   Let $s(x) = x^u \bmod p$, where $(u, p - 1) = 1$. Then $\Delta \leq \frac{(u-1)}{p}$.

2.   If $s(x) = \begin{cases} x^{-1} \bmod p, & if \ x \neq 0; \\ 0, & else. \end{cases}$. Then $\Delta \leq \frac{4}{p}$.

Theorem 1

---

[5] Youssef et al. [12].
[6] Nyberg [14].

1. Let r_{full} be even, $s(x) = x^u \bmod p$, $A : (F_p)^t \rightarrow (F_p)^t$ is MDS matrix of the dimension $t \times t$. Then the average probability of the differential characteristic of the block cipher (4) does not exceed the value

$$\left(\frac{u-1}{p}\right)^{(t+1)r_{full}}. \tag{8}$$

2. Let r_{full} be even, $s(x) = x^{-1} \bmod p$, $A : (F_p)^t \rightarrow (F_p)^t$ is MDS matrix of the dimension $t \times t$. Then the average probability of the differential characteristic of the block cipher (4) does not exceed the value

$$\left(\frac{4}{p}\right)^{(t+1)r_{full}}. \tag{9}$$

The proof of this theorem follows immediately from Statements 1–6 and the fact that in this case $\left[\frac{r_{full}}{2}\right] + \left[\frac{r_{full}}{2}\right] = r_{full}$. $\square$

Usually a block cipher is considered practically resistant to differential cryptanalysis, if the probability of its differential characteristic does not exceed 2^{-N}, where N is the size of the block. But in our case, as mentioned at the beginning of 2, the maximum level of security of the SPONGE structure is equal to $l(p) \approx \log p$. Therefore, we can formulate a weaker requirement as

$$\Delta^b < 2^{-\log p}.$$

But to increase stability and bring it closer to theoretical, we will require inequality

$$\Delta^b < 2^{-2\log p}, \tag{10}$$

that means

$$\left(\frac{u-1}{p}\right)^{(t+1)r_{full}} < 2^{-2\log p} \quad \text{or} \quad \left(\frac{4}{p}\right)^{(t+1)r_{full}} < 2^{-2\log p}, \tag{11}$$

for degree and inverse S-boxes, respectively.

In addition, in [5] the authors propose to add two additional rounds "just in case". But adding two paddle rounds (one at the beginning, one at the end) will do r_{full} odd and, as shown earlier, will not increase the stability of the algorithm. Therefore, if you add additional rounds, then two rounds at the beginning and end to maintain parity r_{full}.

As can be seen from (11), permutation with degree S-boxes $u > 5$ requires more rounds than permutation with inverse, for the same level of security. Next, we will discuss which type of S-boxes has advantages from different points of view.

3.2 Estimates of the Stability of the Nonbinary Block Encryption Algorithm HadesMiMC to Linear Cryptanalysis

According to [9], the parameters that characterize the practical stability of the block cipher to linear cryptanalysis, significantly depend on the structure of this cipher, in particular on the operation in the key adder. Thus, the estimate of the average (by keys) probability of the linear characteristic (in relation to the operation of adding in the field F_p) of the cipher E does not exceed the value

$$\max_{\chi, \rho \in \hat{F}_p} ELP^E(\chi, \rho) = L^b,$$

where the value b is equal to the number of active S-boxes, and the parameter L depends on the S-box (see Definition 15 and explanation on p. 25 in [9]):

$$L = L(s) = \max_{\chi, \rho \in \hat{F}_p} \left| \frac{1}{p} \sum_{x \in F_p} (\overline{\chi}(x), \rho(s(x))) \right|^2, \tag{12}$$

where χ and ρ are additive characters of the F_p field (that is, the characters of the additive group of this field).

The value b is completely determined by the linear (in relation to the addition operation in the field) operator A and is calculated in the same way as when constructing the probability of the differential characteristic, i.e. b equals to the smallest possible number of active S-boxes of the cipher. Applying the same considerations as in 2.1, we obtain $b = (t + 1)r_{full}$, if r_{full} is an even number (and we consider just such a case).

Statement 7:

1. If $s(x) = x^u \bmod p$, where $(u, p - 1) = 1$. Then $L(s) \leq \frac{(u-1)^2}{p}$.

2. If $s(x) = \begin{cases} x^{-1} \bmod p, & if \ x \neq 0; \\ 0, & else. \end{cases}$. Then $L(s) \leq \frac{16}{p}$.

Proof

1. First, estimate the value

$$\sum_{x \in F_p} (\overline{\chi}(x)\rho(s(x))) = \sum_{x \in F_p} (\overline{\chi}(x)\rho(x^{13})). \tag{13}$$

Note that the group $(F_p, +)$ is cyclical (with a creative element $g = 1$), so the corresponding group of characters $(\hat{F}_p, \times)$ ((which is isomorphic to it) is also cyclic.

Let ψ be a creative element of the group $\left(\hat{F}_p, \times\right)$. Then any element of this group, in particular the characters $\overline{\chi}$ and ρ, can be represented as

$$\overline{\chi} = \psi^\alpha, \rho = \psi^\beta,$$

for some $0 \le \alpha, \beta \le p - 1$.

Then we get

$$\overline{\chi}(x)\rho(x^{13}) = \psi(x)^\alpha \cdot \psi(x^{13})^\beta = \psi(\alpha x) \cdot \psi(\beta x^{13}) = \psi(\alpha x + \beta x^{13}), \quad (14)$$

using the fact that ψ is a homomorphism.

We can now rewrite (13) using (14) as

$$\sum_{x \in F_p} \left(\overline{\chi}(x)\rho(x^{13})\right) = \sum_{x \in F_p} \psi(\alpha x + \beta x^{13}). \quad (15)$$

Applying Weil's Theorem ([15], Theorem 5.38) to (15), we obtain:

$$\sum_{x \in F_p} \psi(\alpha x + \beta x^{13}) \le \left(\deg(\alpha x + \beta x^{13}) - 1\right) \cdot \sqrt{p} = 12\sqrt{p}. \quad (16)$$

After applying (13)–(16) to (12), we obtain

$$L = L(s) = \max_{\chi, \rho \in \hat{F}_p} \left| \frac{1}{p} \sum_{x \in F_p} (\overline{\chi}(x), \rho(s(x))) \right|^2$$

$$= \max_{\alpha, \beta} \left| \frac{1}{p} \sum_{x \in F_p} \psi(\alpha x + \beta x^{13}) \right|^2 \le \left| \frac{1}{p} \cdot 12 \cdot \sqrt{p} \right|^2 = \frac{144}{p}.$$

2. First, estimate the value

$$\sum_{x \in F_p} (\overline{\chi}(x)\rho(s(x))) = \sum_{x \in F_p^*} \left(\overline{\chi}(x)\rho(x^{-1})\right) + 1. \quad (17)$$

Since the group $(F_p, +)$ is cyclic (with a generating element $g = 1$), the corresponding group of characters $\left(\hat{F}_p, \times\right)$ is also cyclic. Let ψ be a creative element of the group $\left(\hat{F}_p, \times\right)$. Then any element of this group, in particular the characters $\overline{\chi}$ and ρ, can be represented as

$$\overline{\chi} = \psi^\alpha, \rho = \psi^\beta,$$

for some $0 \leq \alpha, \beta \leq p - 1$.

Then

$$\overline{\chi}(x)\rho(x^{-1}) = \psi(x)^{\alpha} \cdot \psi(x^{-1})^{\beta} = \psi(\alpha x) \cdot \psi(\beta x^{-1}) = \psi(\alpha x + \beta x^{-1}), \quad (18)$$

using the fact that ψ is a homomorphism.

We can now rewrite (17) using (18) as

$$\sum_{x \in F_p^*} \left(\overline{\chi}(x)\rho(x^{-1}) \right) = \sum_{x \in F_p^*} \psi(\alpha x + \beta x^{-1}). \quad (19)$$

Applying the Kloosterman sum theorem ([15], Theorem 1.5], Kloosterman sum) to (19), we will receive:

$$\sum_{x \in F_p^*} \psi(\alpha x + \beta x^{-1}) = 2 \sum_{x_1, x_2 \in F_p: \, x_1 x_2 = \alpha \beta} \psi(x_1 + x_2) \leq 2 \cdot 2 \cdot \sqrt{p} = 4\sqrt{p}. \quad (20)$$

Next, after applying (17)–(20) to (12), we obtain:

$$L = L(s) = \max_{\chi, \rho \in \hat{F}_p} \left| \frac{1}{p} \sum_{x \in F_p} (\overline{\chi}(x), \rho(s(x))) \right|^2$$

$$= \max_{\alpha, \beta} \left| \frac{1}{p} \left(\sum_{x \in F_p^*} \psi(\alpha x + \beta x^{-1}) + 1 \right) \right|^2 \approx \left| \frac{1}{p} \cdot 4 \cdot \sqrt{p} \right|^2$$

$$= \frac{16}{p}. \quad \square$$

Theorem 2

1. *Let r_{full} be even, $s(x) = x^u \bmod p$, $A : (F_p)^t \to (F_p)^t$ is MDS matrix dimension $t \times t$. Then the average probability of the linear characteristic of the cipher (4) does not exceed the value*

$$\left(\frac{(u-1)^2}{p} \right)^{(t+1)r_{full}}.$$

2. *Let r_{full} be even, $s(x) = x^{-1} \bmod p$, $A : (F_p)^t \to (F_p)^t$ is MDS matrix dimention $t \times t$. Then the average probability of the linear characteristic of the cipher (4) does not exceed the value*

$$\left(\frac{16}{p} \right)^{(t+1)r_{full}}.$$

Proof Follows from Statements 1–5, Statement 7 and the fact that in this case

$$\left[\frac{r_{full}}{2}\right] + \left[\frac{r_{full}}{2}\right] = r_{full}.$$

4 S-Boxes Selection

When choosing S-boxes you need to consider the following points:

- Reflection $s : F_p \rightarrow F_p$ must be bijective, i.e. the requirement must be met for degree S-boxes $(p-1, u) = 1$;
- For reasons of resistance to linear and differential cryptanalysis, the parameters Δ and L (which are completely determined by S-boxes) should be as small as possible;
- In terms of the complexity of the implementation of SNARK-proofs, the number of constraints (which depends only on the number and type of S-boxes) should be as small as possible.

Recall that for inverse S-boxes parameters Δ and L are limited at the top values $\Delta \leq \frac{4}{p}$ and $L \leq \frac{16}{p}$, and for the degree S-boxes—are limited at the top values $\Delta \leq \frac{(u-1)}{p}$ and $L \leq \frac{(u-1)^2}{p}$ (Statements 6 and 7). Therefore for the degree S-boxes it makes sense to choose a parameter u as $u = \min\{v \in N : (v, p-1) = 1\}$.

Parameters for inverse S-boxes Δ and L will be guaranteed to be smaller than for degree S-boxes if $u \neq 3$. Therefore, from the point of view of cryptographic security, the most attractive are inverse and cubic S-boxes (if only the cubic ones give a bijective reflection). If $p-1$ is divided into 3, then inverse S-boxes have no competitors (in terms of cryptographic requirements).

In terms of minimizing the number of constraints, inverse and degree S-boxes with a small value of the parameter u is also the most attractive. Indeed, to specify an inverse S-box, 3 counters are required; to specify a cubic S-box—2 constraints:

$$\begin{cases} x_1 x_1 = x_2; \\ x_1 x_2 = x_3, \end{cases}$$

to set the S-box $s(x) = x^5 \bmod p$, 3 constraints are required:

$$\begin{cases} x_1 x_1 = x_2; \\ x_2 x_2 = x_3; \\ x_1 x_3 = x_4, \end{cases}$$

etc., in the general case, as the indicator increases, the number of constraints will increase approximately as $2 \log u$. So if $p-1$ can be divided into all relatively small simple 3, 5, 7, 11,…, then both from the point of view of stability, and from the point of view of complexity of realization of SNARK-proofs, the inverse S-boxes are the best.

However, on the other hand, the implementation of the inversion transformation is quite time consuming, as it requires the use of the Euclidean algorithm, which, in turn, requires the order $O(\log p)$ of divisions with the remainder. Thus, when choosing S-boxes, all these factors must be taken into account, and some compromise must be made.

Next, when selecting the parameters of the encryption algorithm for a given value of the field characteristics, we consider two options for selecting S-boxes - inverse S-boxes and power S-boxes, with the smallest value of the degree at which the S-box specifies bijective mappings.

5 The Number of Rounds with a Full and Partial Layer of S-Boxes and the Number of Constraints Required to Specify the Cipher

According to [5, 7], we will determine the number of rounds with a full layer of S-boxes, r_{full}, as a minimum number of rounds, which guarantees resistance to differential and linear cryptanalysis (given that crypto analysis can be performed in one direction and vice versa). Then determine the number of rounds with a partial layer of S-boxes, based on considerations of resistance to algebraic attacks.

As noted, the authors [5] also recommended adding two rounds just in case. However, after adding two rounds (one at the beginning, one at the end) the value r_{full} will become odd, ie adding two rounds will not increase the resistance to statistical attacks. If you add two rounds at the beginning and end, it will significantly increase the number of constraints. We do not see a reasonable need to increase the number of rounds with a full layer of S-boxes, especially in a situation where the number of constraints is critical.

The number of constraints per bit is determined as follows. The number of constraints required to specify one S-box must be multiplied by the number of all S-boxes, which is completely determined by the number of rounds of both types and their structure. Then the resulting value must be divided by the value r that determines the length of the output on one iteration of the SPONGE design.

6 Numerical Results for MNT-Compliant Parameters

We calculated the number of rounds with a full layer of S-boxes for a simple characteristic field p, where the bit length of the characteristic is equal to 753. As a simple field, we chose one of the fields for which the triplets MNT-4 and MNT-6[7] are defined. In rounds with a full layer of S-boxes, we will place 3 S-boxes, in rounds with a partial layer—one S-box. We define a linear operator A with an MDS matrix of dimension 3×3. In this case, the capacity $c = 2 \cdot 752 = 1504$ and the speed $r = 752$, if you use a byte representation. Degree S-boxes were chosen for reasons $s(x) = x^u \bmod p$, where $u = \min\{v \in N : (v, p - 1) = 1\} = 13$. For inverse S-boxes and power S-boxes of the form $s(x) = x^{13} \bmod p$ the number of rounds with a full layer of S-boxes is 4 (2 rounds at the beginning, 2 at the end to eliminate the possibility of both attacks with the selected plaintext and attacks with the selected ciphertext). The number of rounds with a partial layer of S-boxes, according to (1) and (2) in [5], will be nearly 60. In this case, the number of constraints per bit is equal to 0.48 for the degree S-boxes and 0.29 for inverse S-boxes, which is 3.5–5.8 times less than the same for the Pedersen function.

The number of rounds with a full layer of S-boxes was determined from the inequality

$$\left(\frac{144}{p}\right)^{4r_{full}} \le 2^{-2N} = 2^{-6p},$$

which is a sufficient number for both power and inverse S-boxes. From the inequality got the value

$$r_{full} = \left\lceil \frac{6 \cdot 753}{4 \cdot 745} \right\rceil = 2,$$

that is, two rounds with a full layer of S-boxes at the beginning of the algorithm and at the end.

If, on the recommendation of the authors, add two additional rounds, we obtain (maintaining the parity of the value r_{full}): $r_{full} = 4$, that is, the total number of such rounds is equal to 8.

7 Summary

1. Summing up the information discussed in this section, it should be noted that for the proper functioning of monitoring systems, it is necessary to comprehensively use cryptographic mechanisms such as block symmetric ciphers, hash functions, electronic digital signature schemes, streaming ciphers and pseudo-random sequence generators.
2. Block symmetric ciphers must meet the following indicators: high level of cryptographic security, speed and simplicity of software and hardware-software implementation. The hash functions must provide the conversion rate, the

---

[7] https://coinlist.co/build/coda/pages/theory.

optimal amount of key data, be resistant to the first prototype, to the second prototype, and must be resistant to collisions.

3. Electronic digital signature schemes should provide speed in terms of generating parameters and keys, signing and verifying the signature, the amount of key data and the resulting signature, as well as have a reliable mathematical basis and provide protection against known and potential attacks.

4. In turn, stream ciphers should be compared according to the length of short packets, long packets, as well as according to the criterion of initialization of key parameters. In the case of using stream ciphers in pseudo-random sequence generation mode and directly pseudo-random sequence generators to determine the efficiency of the mechanism, the generated sequences should be investigated for their statistical properties according to the described method using the NIST STS statistical test package.

5. Strictly substantiated estimates of the resistance of the HadesMiMC block encryption algorithm and the Poseidon hash function to linear and differential attacks are obtained. Note that the construction of such estimates, especially for linear cryptanalysis, required the proof of non-trivial statements about the properties of the additive characters of a finite field.

6. Specific parameters for the HadesMiMC encryption algorithm and the Poseidon hash function have been identified, which allow them to be used for recurrent SNARK-proofs based on the MNT-4 and MNT-6 triplets.

7. It is shown how best to choose S-boxes for these algorithms, so that this choice was optimal both in terms of stability and in terms of complexity of calculations in SNARK-proofs.

8. It is shown how many rounds are enough to guarantee the stability of algorithms for different options.

9. The number of constraints per bit is calculated for different options, and it is shown that it is significantly smaller than for the Pedersen function, which is currently used in the cryptocurrency Zcash.

All these results are strictly justified.

The results showed that the HadesMiMC block encryption algorithm, based on operations over a simple finite field, is secure against linear and differential cryptanalysis, with the correct choice of parameters of this algorithm. As a result, the Poseidon hash function, built using the SPONGE design and the HadesMiMC permutation, will be stable. It was also shown how the parameters of the algorithm should be selected to ensure its specified level of security. In addition, the number of constraints per bit of information was calculated and it was shown that the HadesMiMC algorithm and the Poseidon hash function currently have no competitors when used in SNARK-proofs.

References

1. Ben-Sasson E, Bentov I, Horesh Y, Riabzev M (2018) Scalable, transparent, and post-quantum secure computational integrity. IACR Cryptol 46. ePrint Archive 2018. Available at https://eprint.iacr.org/2018/046.pdf
2. Groth J (2016) On the size of pairing-based non-interactive arguments. In: EUROCRYPT 2016. LNCS, vol 9666. Springer, Berlin, pp 305–326. https://doi.org/10.1007/978-3-662-49896-5_11
3. Parno B, Howell J, Gentry C, Raykova M (2013) Pinocchio: nearly practical verifiable computation. In: IEEE symposium on security and privacy. IEEE Computer Society, pp 238–252. https://doi.org/10.1109/SP.2013.47
4. Hopwood D, Bowe S, Hornby T, Wilcox N (2019) Zcash protocol specifcation: Version 2019.0-beta-37 [overwinter+sapling]. Technical Report, Zerocoin Electric Coin Company. Available at https://github.com/zcash/zips/blob/master/protocol/protocol.pdf
5. Grassi L, Kales D, Khovratovich D, Roy A, Rechberger C, Schofnegger M (2019) Starkad and poseidon: new hash functions for zero knowledge proof systems. IACR Cryptol 458. ePrint Archive 2019. Available at https://eprint.iacr.org/2019/458.pdf
6. Guido B, Joan D, Michaël P, Gilles VA (2011) Cryptographic sponge functions, pp 1–93. Available at https://keccak.team/files/CSF-0.1.pdf
7. Grassi L, Lüftenegger R, Rechberger C, Rotaru D, Schofnegger M (2019) On a generalization of substitution-permutation networks. The HADES design strategy. IACR Cryptol 1107. ePrint Archive, 2019. Available at https://eprint.iacr.org/2019/1107.pdf
8. Abdelraheem MA, Ågren M, Beelen P, Leander G (2012) On the distribution of linear biases: three instructive examples. In: Advances in cryptology—Crypto 2012. LNCS, vol 7417. Springer, Berlin, pp 50–67. https://doi.org/10.1007/978-3-642-32009-5_4
9. Baignères T, Stern J, Vaudenay S (2007) Linear cryptanalysis of non binary ciphers. In: Adams C, Miri A, Wiener M (eds) Selected areas in cryptography. SAC 2007. Lecture notes in computer science, vol 4876. Springer, Berlin. https://doi.org/10.1007/978-3-540-77360-3_13
10. Bertoni, G., Daemen, J., Peeters, M., Assche, G.V.: On the Indifferentiability of the Sponge Construction. In: EUROCRYPT 2008. LNCS, vol. 4965, pp. 181–197 (2008). https://doi.org/10.1007/978-3-540-78967-3_11.
11. Lai X., Massey J. L., Murphy S. Markov ciphers and differential cryptanalysis // Advances in Cryptology–EUROCRYPT'91: Proceedings of Workshop on the Theory and Application of of Cryptographic Techniques, 8–11 Apr., 1991, Brighton, UK. Vol. 547. Berlin, Heidelberg: Springer, 1991. P. 17–38. https://doi.org/10.1007/3-540-46416-6_2
12. Youssef, A.M., Mister, S., Tavares, S.E.: On the Design of Linear Transformations for Substitution Permutation Encryption Networks. In: School of Computer Science, Carleton University. pp. 40–48 (1997), available at http://citeseerx.ist.psu.edu/viewdoc/download?doi=10.1.1.17.1208&rep=rep1&type=pdf
13. Vaudenay, S. On the security of CS-cipher. In: International Workshop on Fast Software Encryption. Springer, Berlin, Heidelberg, 1999. p. 260–274. https://doi.org/10.1007/3-540-48519-8_19
14. Nyberg, K.: Differentially uniform mappings for cryptography. In: EUROCRYPT 1993. LNCS, vol. 765, pp. 55–64 (1994). https://doi.org/10.1007/3-540-48285-7_6
15. Lidl R, Niederreiter H (1996) Finite fields and their applications. https://doi.org/10.1017/CBO9781139172769

Comparative Analysis of Consensus Algorithms Using a Directed Acyclic Graph Instead of a Blockchain, and the Construction of Security Estimates of Spectre Protocol Against Double Spend Attack

Lyudmila Kovalchuk⬥, Roman Oliynykov⬥, Yurii Bespalov⬥, and Mariia Rodinko⬥

Abstract In recent years, the possibility of moving from a blockchain to a more general structure, which has additional capabilities, but remains resistant to major attacks, has been very actively explored. Basically, the new structure is a blockgraph (DAG), which has certain features. Various ways of such generalizations and partial analysis of their properties are offered. At the same time, the results of such research raise a large number of questions, so the field for research in this area remains extremely large, and the task of scaling the blockchain is very important. Over the last 5–7 years, many consensus protocols have been proposed on blockgraphs. It should be noted that the use of such protocols, although it leads to a significant increase in the speed of transaction processing, but carries certain risks. At present, none of these protocols have rigorous evidence of resistance to major attacks, such as double spend attack, network splitting attack, and so on. The construction of such proofs requires overcoming great analytical difficulties, even if you use a very simplified mathematical model of the network. In this section, we took the first step toward constructing consensus protocol stability estimates on block graphs by constructing upper estimates for the probability of a double spend attack on the SPECTRE consensus protocol. To increase its security, we have proposed a somewhat non-standard algorithm for accepting a transaction by a vendor: it must wait for a certain number of confirmation blocks to be created, so to speak, under its supervision. Methods have been developed to calculate the upper estimates of this probability depending on the network parameters and the number of confirmation blocks that the

L. Kovalchuk
National Technical University of Ukraine "Igor Sikorsky Kyiv Polytechnic Institute", 37, Prosp. Peremohy, Kyiv 03056, Ukraine

R. Oliynykov (✉) · M. Rodinko
V. N. Karazin Kharkiv National University, 4 Svobody Sq., Kharkiv 61022, Ukraine

Y. Bespalov
M.M. Bogolyubov Institute for Theoretical Physics of the National Academy of Sciences of Ukraine, 14-b Metrolohichna str., Kyiv 03143, Ukraine
e-mail: yu.n.bespalov@gmail.com

© The Author(s), under exclusive license to Springer Nature Switzerland AG 2022
R. Oliynykov et al. (eds.), *Information Security Technologies in the Decentralized Distributed Networks*, Lecture Notes on Data Engineering and Communications Technologies 115, https://doi.org/10.1007/978-3-030-95161-0_9

vendor must see before considering the transaction irreversible. For greater clarity, the corresponding numerical results are also given. An important property of the results obtained is that, in addition to the probability estimate itself, we can also calculate the required number of confirmation blocks sufficient to guarantee the irreversibility of the transaction with a probability as close to 1 as desired.

Keywords Consensus algorithm · Directional acyclic graph · Blockchain technology · Double spend attack

1 Introduction

In this section, we will consider and analyze the use of monitoring systems in a decentralized environment, which has significant conveniences and advantages over a centralized one. We will show exactly what characteristics such an environment should have and how you can try to provide them. As an example of a protocol that can be used in decentralized environments in monitoring for rapid information processing, a modification of the SPECTRE protocol is considered. Estimates of the security of this modified protocol are constructed and some numerical results are given.

The most common modern example of a decentralized environment is blockchain technology, which was first used primarily in cryptocurrencies. Modern cryptocurrencies ensure the circulation of decentralized payments, using a publicly available transaction ledger, without the involvement of a trusted party. The transaction log is the result of the joint work of all participants in the decentralized network. It is created on the basis of a consensus protocol, which is protected from the actions of an attacker (dishonest miners) with limited resources.

However, over the 10 years of existence and rapid spread of blockchain technology has found many other, including potential, applications in addition to cryptocurrency. These are smart-contracts, voting, electronic document management, and much more. Summarizing the scope of its application, we can say that it is extremely convenient and virtually unalterable there, it takes a long time to store important, including confidential information. The method of storing information must meet the following requirements:

- the information network must function correctly in conditions of incomplete trust in the parties who work in it, and sometimes in conditions of complete distrust;
- an attacker who tries to distort information has a chance to attack only when he has dominant resources on that network;
- the longer the information is stored, the more difficult it is to falsify, change or corrupt it;
- stored information is somewhat anonymous, but only users with certain access rights can add and process new information;
- each user can verify their personal information, its correctness and relevance, without disclosing their identity;

- in order to protect information from forgery, it is enough to store not all information on each local device, but only some, very small, part of it;
- when synchronization is violated, the network synchronizes itself for a certain (small) period of time.

Here we have listed only the most important properties of storing information in the blockchain, but in fact there are many more such useful properties.

The transaction information storage protocol proposed by Nakamoto, which uses the PoW consensus [1] with the parameters set in the Bitcoin network, satisfies almost all of these properties. More precisely, it can be modified so that these properties are met. It is as a result of such "useful" modifications that new cryptocurrencies appear, such as forks from the BTC.

However, one of the disadvantages of the Nakamoto protocol is its low bandwidth. The average bitcoin transaction processing speed is only about 7 transactions per second, i.e. less than 500,000 transactions per day.[1] For some other decentralized cryptocurrencies, these values are slightly higher, but still remain very small compared to centralized financial services. Such constraints for the consensus proposed by Nakamoto are determined by constraints on block size and block generation intensity. A direct change in these parameters immediately leads to such undesirable phenomena as partial centralization or a significant reduction in resistance to a double spend attack or a branching attack. For example, increasing the size of the block leads to an increase in the delay time of the block, which significantly increases the probability of a double spend attack and a branching attack. For example, increasing the size of the block leads to an increase in the delay time of the block, which significantly increases the probability of a double spend attack and a branching attack. Reducing the output time of the block (and, consequently, reducing the amount of work to create a block) leads to various negative phenomena, namely: to increase the probability of unintentional fork, and hence to the loss of a large amount of "work"; to increase the likelihood of attacks—provided there is a well-synchronized attacker. Currently, the fork problem is solved, according to the Nakamoto protocol, by the rule of a longer chain. However, when you change one of these parameters, this method of consensus, as we see, does not protect the network from various vulnerabilities and problems. This means that with increasing block size or block output intensity, part of the computing power of honest miners seems to be "lost", and the relative share of the attacker's computing power increases. And then there is a situation when in order for the probability of a successful attack to be equal to 1, it is enough that the attacker had not 50% of the power, but significantly less. For example, in a network with a sufficiently long delay time and high intensity of block output, a double spend attack is possible in the presence of 30% of the attacker's hashrate. Therefore, the issue of modifying the consensus protocol, which would allow faster processing of transactions while maintaining resistance to major attacks, is very relevant. But a company that can achieve this and create a cryptocurrency with the right parameters will have a huge advantage over its competitors.

---

[1] Bitcoin confirmed transactions per day stats, 2018. URL: https://www.blockchain.com/charts/n-transactions.

Therefore, the topic of scaling the blockchain while preserving its useful properties is actively discussed and developed.

The new generation of consensus protocols proposed by many blockchain experts in recent years is based on a somewhat new concept—the use of a blockgraph instead of a blockchain. All proposed blockgraphs have the structure of a directed acyclic graph (DAG). Such protocols include GHOST [2] (the first of them), SPECTRE [3], PHANTOM [4], Graphchain [5], Tangle [6] and many others. In such protocols, the blocks form the DAG as a distributed financial book. The main reason for such persistent attempts to create an alternative block structure is the extremely low bandwidth of the blockchain. For example, the Bitcoin network, according to various estimates, processes from 3 to 7 transactions per second, which is just a negligible amount compared to the speed of banking operations. Given that one block in Bitcoin is issued on average every 10 min, and the number of confirmation blocks is considered equal to 6 blocks, then, even if the transaction was processed instantly, you need to wait about an hour for its final confirmation. This situation makes it impossible to use cryptocurrency for various household transactions—no one will wait an hour to pay for a cup of coffee with bitcoins.

Therefore, the main purpose of developing new consensus protocols is, above all, to maximize network bandwidth, or, almost the same thing, to increase the processing speed of transactions. In this case, the main characteristics of the network, ensuring its resistance to attacks of various types, must be preserved.

The simplest ways to solve this problem are the following modifications:

- increase the size of the block, i.e. the maximum possible number of transactions in it;
- reducing the interval between the output blocks.

However, unfortunately, both such approaches significantly reduce the resilience of the network, especially to double spend attack. As shown in [7, 8], with a significant change in these parameters, an attacker can attack a double spend with a probability of 1, even when its computing power is significantly less than 50%.

Some of the proposed consensus protocols on graphs do improve certain characteristics, compared to the classic "blockchain" consensus protocol proposed by Nakamoto in [1]. Yes, by using different types of blockers, you can really increase the speed of transaction processing. However, the "payment" for such an improvement is the occurrence of significant shortcomings in such protocols, the main of which are the following:

- reduction of resistance to attacks (first of all to attack of a double expense and to attack of splitting of a network);
- reducing the "stability" of the formed blockgraph;
- lack of linear order on the blocks (it is possible to specify only a partial order), etc.

In addition, the complexity of constructing sound stability assessments of such protocols increases significantly. At present, in all works [2–6] significant mathematical errors were found in the statements concerning the estimates of the probability of attacks. A comparative analysis of consensus protocols on blockgraphs is presented in Sect. 2.

In Sect. 3, we consider the SPECTRE consensus protocol [3]. This is one of the first DAG protocols. According to the protocol, the new block must contain links to all "fresh" blocks (i.e. those that are not yet referenced), which are called sheets. Miner creates a new block, including the headers of all the leaves, and immediately sends it to all other nodes. Also, this protocol has a very unusual algorithm for choosing between conflicting transactions in the case of a malicious or accidental fork. According to this algorithm, the blocks of the graph seem to "vote", depending on their location in the graph, for one of the conflicting transactions. Next, we will analyze in detail this algorithm.

The SPECTRE protocol provides fast processing of transactions and their fast confirmation, and also can be rather simply scalable. However, it guarantees only a weak property of liveness. The authors argue that this protocol is resistant to various attacks, such as double spend attacks and censorship attacks, but do not provide any evidence for these claims, only questionable empirical considerations.

We will consider a somewhat amplified double spend attack that uses censorship at some point. During this attack, the attacker not only builds alternative subgraphs, but also chooses which blocks he will confirm. Such an attack is more effective for each of these methods separately and its probability can be significant. We will also show that the probability of success of this attack can be significantly reduced if you slightly change the rules of acceptance of the transaction by the vendor.

We will also provide well-founded formulas for calculating the upper estimates of this probability, depending on the network parameters and the number of confirmation blocks that the vendor must see before considering the transaction irreversible. For greater clarity, the corresponding numerical results will also be given.

2 Comparative Analysis of Consensus Protocols on Blockgraphs

2.1 An Overview of the Main Problems that Require Blockchain Scaling

To date, none of the methods of scaling the blockchain proposed in various works are perfect. Significant vulnerabilities to attacks have been found in some; others offer only a partial solution to these problems. After careful study of a huge number of all such proposals, it is even suspected that it is impossible to scale the blockchain so that the resulting consensus protocol on the blockgraph has, at least in part, all the desired properties.

First, let's name all the problems that exist today in the blockchain network and which the researchers wanted to solve in their proposed consensus protocols on the blockgraph.

1. **Fast transaction processing**. This can be achieved by increasing the volume of the blocks, as well as by increasing the intensity of their output.
2. **Increasing real decentralization**. The only solution is to reduce the amount of resources needed to create the unit. A resource is either a computing power (for the PoW protocol), or a deposit in the appropriate cryptocurrency (for the PoS protocol), or another similar resource, depending on the consensus protocol. Today, blockchain decentralization is only relative, as the need to have a large enough resource to generate blocks forces miners to unite in huge mining pools.
3. **Reduce transaction confirmation timeout**. For example, in the Bitcoin network, a transaction is considered confirmed if at least six blocks are issued after the block containing it. These blocks are called "confirmation blocks". Since the average block is issued once every 10 min, the confirmation of the transaction has to wait an average of 1 h. And this is only the average waiting time; real time can vary significantly due to the large variance.
4. **Linear order of blocks**. When switching to a graphchain, you have to spend a lot of effort on the property that goes to the blockchain "for free"—maintaining the linear order of the blocks. This feature, which does not seem to play a significant role in transaction processing, is extremely important in protocols that work with smart-contracts (for example, in the cryptocurrency blockchain DASH). In these cases, the existence of a linear order of blocks is necessary for the correctness of the procedure for voting for contracts.
5. **Maintaining the resistance of the blockgraph to the main types of attacks**. For a blockchain, there are different ways to bring resilience to basic attacks: double spend attacks, branching attacks, censorship attacks (failures to process a transaction or block), linear-order attacks (for a blockchain, this is just a partial case of a double spend attack). For some attacks, such as the double spend attack, analytical upper probability estimates are obtained in the most general assumptions; for some other attacks, there are only asymptotic estimates under the assumption that time is a continuous quantity. Building a linear order on a blockgraph that will be resistant to attacks is an extremely time consuming task.

2.2 Review of the Main Consensus Protocols on Blockgraphs

The term "blockgraph" most accurately reflects the design by which researchers try to generalize the traditional blockchain. Although in different publications the authors use different names: graphchain, hashgraph, DAG (direted asyclic graph), others. Each name reflects some features of the offered blockgraph. Today, the most commonly used abbreviation is DAG—directional acyclic graph. We will also use this term as a synonym for the term "graphchain".

One of the first works to generalize the blockchain is [2], which proposed a structure called GHOST. In this work, instead of the linear structure introduced by Nakamoto, another, more general structure was constructed, which was a "tree" of blocks. In contrast to the Nakamoto protocol, where each block has only one "descendant" that refers to it, the GHOST structure allows the existence of, generally speaking, an arbitrary number of such descendants. In other words, this structure allows the existence of forks. Forks can occur for two reasons: poor network synchronization or during an attack by an attacker. Two blocks built at the same "height" can both be valid. Each linear chain in this tree that has a consistent transaction history is valid.

This approach to consensus allows you to increase the bandwidth of the protocol, although some number of blocks created may still be "lost".

According to the protocol, each miner, if there are several chains, can choose the one that seems more "correct". In the event of a conflict between blocks, the block that has the largest number of "descendants" (including forks) is considered valid.

Although the description of the protocol and its advantages in the performance of the authors looks very tempting, but in their work on this protocol, firstly, there is no exhaustive justification for its resistance to major attacks, and secondly, there are significant mathematical errors that can not be overlooked (for example, Lemma 8 in the mentioned work).

The next two works by the same authors and on the same subject are called SPECTRE [3] and PHANTOM [4]. They also (not very successfully) promote the idea of building a graphchain, and the protocols proposed in the works differ significantly. The only common feature of these works is the presence of gross mathematical errors and the absence of strict mathematical justifications for their loud statements.

In addition to the works described above, the results of other authors published in Graphchain [5], Tangle [6], and some others are also interesting. They all suggest using the blockchain as a generalization of the blockchain. The main purpose of such work is to increase the capacity of the network while maintaining its resistance to attacks. These protocols, on the one hand, do improve the characteristics of the Nakamoto protocol, such as high bandwidth, block confirmation time, line ordering, etc. The approaches proposed in [9, 10] seem interesting and promising. The protocol described in [9] suggests the use of "parallel" blockchains for transaction processing. This approach really improves the bandwidth of the blockchain, while maintaining resistance to attacks, but the time to confirm the block still remains large. The design in [10] proposes to divide the blocks according to their basic functions—transaction processing, log support, confirmation of the block with the transaction, etc. This allows you to scale each type of block separately.

Consider in more detail these and other works that offer different types of blockchain scaling. The first work to describe a cryptocurrency based on a graphchain (it uses the term DAG) is a 2012 blog post.[2] This paper describes the DagCoin cryptocurrency,[3] built without blocks, which was presented in 2015. The role of blocks here is played by individual transactions. This approach did not show any significant advantages, but introduced new types of consensus protocols based on the DAG design.

---

[2] J. Dilley, A. Poelstra, and J. Wilkins. Unfreezable blockchain. Bitcoin forum, 2016. URL: https://bitcointalk.org/index.php?topic=57647.msg686497#msg68649.

[3] Sergio Demian. Lerner. "Dagcoin: a cryptocurrency without blocks", 2015.

The first real cryptocurrency to use DAG instead of blockchain is IOTA cryptocurrency [6]. The consensus protocol used for this currency is called Tangle. According to the mining protocol, each subsequent block refers exactly to the previous two. To resolve a conflict situation in the presence of conflicting transactions, it is suggested to calculate the "height" of the conflicting blocks and choose the one that is at a higher height. This approach provides excellent opportunities for a double spend attack, as the developers of the protocol were informed in private correspondence. Moreover, at the CryBlock-19 conference, one of the speakers on the Tangle protocol confirmed that there is currently no mathematically sound proof of the resistance of this protocol to attacks, in particular to a double spend attack. Therefore, it seems that this protocol is not stable, just the attackers have not yet shown sufficient interest in attacks on it.

2.3   Access Control Protocols

Hedera Hashgraph[4] is a blockgraph-based consensus protocol. This protocol is intended for use in access control environments. Hashgraph is a completely asynchronous protocol, which means that it has persistence and liveness properties without any additional assumptions at the time of synchronization. Hashgraph guarantees a complete solution to the problem of the Byzantine agreement under the following conditions: no more than 1/3 of the resources are controlled by attackers; an attacker can delete messages between honest participants or detain them indefinitely. At the same time, in order to achieve sustainability, it is necessary that at least 2/3 of all participants must be online at all times; all these participants must be honest. Transaction processing is based on the continuous synchronization of participants' states when receiving new information about transactions or new information about the update of information by some participants ("gossip about gossip"). Hashgraph has an extremely high bandwidth (about 250,000 transactions per second[5]), but keep in mind that it only works in an access control environment. Such high bandwidth is possible due to the lack of significant network congestion: about $nlogn$ for the transaction to be "seen" by all network members.

The weakness of this protocol is that it focuses on environments with access control. This makes it impossible to use this protocol (at least the existing version of this protocol) in various applications that require full decentralization.

The recently introduced Casanova[6] protocol is also targeted for use in access control environments. Like most such protocols, Casanova is partially synchronous,

---

[4] Leemon Baird. Hashgraph consensus: fair, fast, Byzantine fault tolerance. swirlds tech report, 2016. http://www.swirlds.com/wp-content/uploads/2016/06/2016-05-31-Swirlds-Consensus-Algorithm-TR-2016-01.pdf.

[5] Yaoqi. Jia. Demystifying hashgraph: Benefits and challenges. URL: https://hackernoon.com/demystifying-hashgraph-benefits-and-challenges-d605e5c0cee5.

[6] Kyle Butt, Derek Sorensen, and Michael Stay. Casanova. URL: https://arxiv.org/pdf/1812.02232.pdf.

which requires guarantees of its liveness. One of the important features of this protocol is the support of partial order of transactions instead of full, or linear, order. That is, in this protocol, transactions are not fully ordered. In the Casanova protocol, blocks are created by ulidators at regular intervals. Consensus is reached in a few rounds. It is based on validator voting. To optimize the process of reaching consensus, Casanova supports several protocols, the use of which depends on the situation. That is, different protocols are used in different conditions, both in the presence of conflicting transactions and in their absence. To date, there is no information on the practical application of the Casanova protocol.

2.4 Protocols Without Access Control

The SPECTRE protocol [3] was introduced in 2016. In this protocol, a new block is sent immediately after creation. Each new block refers to all valid blocks in the column that it sees at the time of creation, and to which there are no links yet. In the event of a conflict, the protocol provides for a "voting" procedure. Unexpectedly, it is not the participants who vote in this procedure, but the blocks themselves. Each block in the blockgraph votes for one of the blocks in which transactions conflict. Legitimate is the block (and, accordingly, from conflicting transactions) for which more blocks voted. Exactly how a blockshould vote is determined by the position of that blockin the graphchain. Such a voting protocol in SPECTRE provides fast block confirmation and good scaling, but only a weak liveness property. An attacker has the ability to maintain a balance between two conflicting transactions, i.e. the protocol is vulnerable to a generalized branching attack. In addition, the voting protocol in SPECTRE is vague, allowing for ambiguous interpretation. In particular, it provides opportunities to double spend attack, subject to certain additional conditions that are likely. The paper presenting this protocol also does not contain mathematically sound statements regarding estimates of the probabilities of major blockchain attacks. Instead, certain empirical considerations are given, which can not be considered evidence.

To date, it is not known about any real applications of the SPECTRE protocol, which, in particular, may indicate that the authors themselves understand its lack of validity. Shortly after SPECTRE, they proposed their new protocol.

In 2018, the authors of the SPECTRE protocol proposed a new, completely different protocol—the so-called blockchain PHANTOM [4]. According to the authors, this protocol solves many of the above problems of the blockgraph. It is not at all similar to the GHOST and SPECTRE protocols. Instead of working with individual blocks or their pairs, the new protocol works with so-called clusters—sets of blocks in which the blocks are "strongly connected". Other blocks, on the other hand, have little "connection" with the blocks that belong to the cluster. It is believed that blocks created by honest miners will be "strongly" related to other blocks that are also created by honest miners. Therefore, it is assumed that the largest cluster, with a predominant probability, will contain only blocks created by honest miners.

There is also a protocol for constructing a linear order, which significantly depends on the selected main cluster.

The consensus protocol is to find the largest (or almost largest) cluster. The protocol is quite complex and contradictory. Since it is also not clearly described and allows for ambiguous interpretation, it is possible to simulate a situation in which the implementation of this protocol will lead to looping. Or in which the protocol does not include in the cluster a block that has "strong" links with other blocks in the cluster. It should be noted that the problem of finding the "correct" cluster is NP-complete, so in the implementation of the protocol it is proposed to use only some simplified solution of the problem. This leads to incorrect operation of the protocol. The authors of PHANTOM declare the high bandwidth of the protocol and the presence of a linear order, but the time of confirmation of the transaction, as well as the time of stabilization of the blockgraph is very long. A block is considered stable when it is referenced by a special block called an hourglass. This block must be a reference to all blocks in the primary cluster. The authors tried to estimate the probability of such a block and, accordingly, the average time to wait. To do this, they used the apparatus of probability theory, but with huge mathematical errors. However, they still failed to obtain any specific estimates of the waiting time of such a block, except that it is finite, because the probability of such a block is non-zero.

To reduce consensus latency, the authors also propose to use some symbiosis of these two protocols—first to build a basic cluster using the PHANTOM protocol, and then to use the SPECTRE protocol to establish a linear order on the blocks of this cluster. This proposal is not yet sufficiently developed and is currently only being studied. By the way, a new attack on the liveness property of the PHANTOM protocol was recently published.

The Graphchain protocol [5] promises not only to address the issue of scaling, but also to reduce the oligopoly of large mining pools. Unlike SPECTRE and PHANTOM, Graphchain works with individual transactions at once, without combining them into blocks. Each transaction refers to selected "parent" transactions, i.e. those that are valid and have been created and processed before. Some work is required to process each transaction, i.e. this protocol is PoW-like. The algorithm for resolving the conflict situation between transactions is similar to that implemented in Bitcoin. For each of the conflicting transactions, its "height" is calculated—a value equal to the total work (PoW) of this transaction and all others to which this transaction refers (directly or indirectly), i.e. transactions that are its "ancestors". The transaction with the highest "height" is considered valid, and the corresponding conflicting transaction is "destroyed". The authors of the protocol claim that this way of confirming transactions encourages miners "not to conflict, but to cooperate." In addition, the relatively small amount of PoW reduces the oligopoly of mining pools. Unlike Bitcoin, the processing of each individual transaction in Graphchain is rewarded: transaction verification is considered a useful job that must be paid for.

The following three protocols look promising in practice, although they do not solve all network problems. In a sense, they are based on similar ideas.

The Prism protocol [10] divides blocks into several classes, according to their functions: proposers, vouters (i.e. those who vote) and blocks that process transactions. Suggestion blocks refer to transaction blocks, thereby confirming these blocks. The vouter blocks decide which of the proposition blocks will be selected to form the main chain. This main chain will determine the state of the transaction log—as a chain of blocks in Bitcoin. Vouter blocks are combined into several different chains (depending on the value of the hashes in this block), which operate in parallel. This organization of blocks allows you to scale the various functions of the blockchain (transaction processing, confirmation, etc.) in parallel. The authors argue that the bandwidth of this protocol and its latency can be scaled up to physical limitations on the network, while maintaining resistance to major attacks.

Fruit Chain[7] protocol also divides blocks by functions. Some blocks, with a large value of PoW, form the main chain, as in Bitcoin. In addition to these blocks, there are also blocks with significantly lower PoW values, which are used only for transaction processing. These blocks are called "fruits". Blocks from the main chain confirm the fruit blocks.

The stability of such a protocol can be proved by the same principles as for Bitcoin—it is enough to prove the relevant statements for the main chain. One of the most significant advantages of Fruit Chain is the short waiting time for transaction processing, due to the large number of "small" fruit blocks. However, this protocol does not reduce the waiting time to confirm the transaction, i.e. after its first confirmation by a block from the main chain still need to wait until a certain number of blocks with a "big" PoW.

The Parallel Chains protocol recently proposed in [9] has characteristics similar to those of the Fruit Chain and Prism protocols. One of its main advantages is that it can be used for both PoW and PoS. According to the protocol, the blocks are organized into several (from several tens to several hundreds) practically independent chains. The first chain is synchronizing. Each of its blocks refers only to the previous block of the same chain. Blocks of another chain refer to the previous block of the same chain and to the last block of the first chain, which they see at the time of creation. The linear order of the blocks in this protocol is ensured precisely by the presence of a synchronization chain. Moreover, due to some additional details of the transaction protocol, the blocks of different chains do not intersect, as a result of which the speed of transaction processing increases in direct proportion to the number of chains. Such a protocol does provide significantly higher bandwidth, and the rationale for its resilience to major attacks is the same as in a traditional blockchain—for each chain separately. However, again, this protocol does not make it possible to reduce the waiting period for stabilization of the graph, in particular the waiting period for full confirmation of the transaction.

In recent studies, we have been able to propose a new consensus protocol on the blockgraph, which in the case of conflicting transactions is a complete analogue of the longer chain rule. Resistance to a double spend attack is fully justified for such a

---

[7] J. Dilley, A. Poelstra, and J. Wilkins. Unfreezable blockchain. bitcoin forum., 2016. URL: https://bitcointalk.org/index.php?topic=57647.msg686497#msg68649.

protocol, but resistance to other attacks has not yet been studied. Such research may be the subject of further work.

3 Improvement of the SPECTRE Consensus Protocol and Substantiation of the Resistance of the Obtained Modification to a Double Spend Attack

3.1 Description of the SPECTRE Consensus Protocol

SPECTRE [3] is Proof-of-Work protocol, in which the blocks are ordered in the form of a directed acyclic graph (hereinafter—DAG, direct acyclic graph) $G = (C, E)$, where C is a set of blocks (graph vertices), and E is the set of corresponding hash references (graph edges). The rib-arrow is placed from the block z_1 to the block z_2 if and only if the block z_1 contains the hash value of the block in its header z_2.

The rules of mining are very simple [3]:

(1)   as soon as a miner has created a new block or received it from another miner, he must distribute it throughout the network;
(2)   when creating a block, embed in its header a list containing the hash value of all the "sheets" that it "sees".

To prevent conflicting transactions, a procedure that can be called "voting" is used. In this case, each block $z \in C$ has exactly one vote, which he gives for one of these transactions. To describe the voting process, we need to enter some notation with [3]:

- $past(z, G) \subset C$ means the set of all blocks (vertices of the graph) to which there is a path from the block z;
- $future(z, G) \subset C$ means the set of all blocks (graph vertices) from which there is a path to the block z;
- $cone(z, G) \subset C$ means the set of all blocks (vertices of the graph) that belong to or to $past(z, G) \subset C$, or to $future(z, G) \subset C$;
- $virtual(G)$ means some hypothetical block of this graph, such that $past(virtual(G)) = G$.

If in some two blocks $x, y \in G$ if there are conflicting transactions, then you need to use the voting procedure to determine which of the blocks (a, consequently, which of the transactions) has priority. If a blockvotes for a block x, then it is denoted as $x \prec y$, and the corresponding voice of the block is equal to -1. If a blockvotes for the block y, then it is denoted as $y \prec x$, and the corresponding voice of the block is equal to $+1$. Some blocks may make an uncertain decision, then the vote is equal to 0. The final decision is made by a majority vote (in case of uncertainty, an arbitrary decision is chosen).

The voting rules, according to [3], are as follows:

(1) if executed for the block $z \in G$: $z \in future(x)$ $z \notin future(y)$, then this block votes for $x \prec y$, that is, his voice is equal -1;

(2) if $z \in future(x) \cap future(y)$, then the vote of this block is determined recursively, namely according to the majority of votes in the subgraph formed by its past (in other words, such a blockvotes as $virtual(past(z))$;

(3) if $z \notin future(x) \cup future(y)$, then this block votes like the majority in $future(z)$;

(4) if $z = virtual(G)$, then he votes like most blocks in his past (i.e. like most blocks G);

(5) finally, if $z = x$ or $z = y$, then such a block votes for itself, imitating each block from its past and setting a precedent for blocks that did not fall into its past.

In the original work [3], the authors of the SPECTRE protocol considered two types of attacks: the double spend attack and the censorship attack. When performing a double spend attack, an attacker violates the first rule of mining: after creating a block, he does not distribute it for some time, and then distributes two subgraphs simultaneously, which contain one of the conflicting transactions. When performing a censorship attack, the attacker violates the second rule: he does not confirm all the sheets, but only some of them selectively, namely those blocks that are related to the conflicting transaction that he wants to impose.

We will consider further some hybrid attack at which performance the malefactor breaks both rules of mining at once.

3.2 Description of a Hybrid Attack on the SPECTRE Protocol

For convenience, we will break this hybrid attack into several stages, which differ in certain actions of the attacker. Figure 1 shows all the stages of this attack.

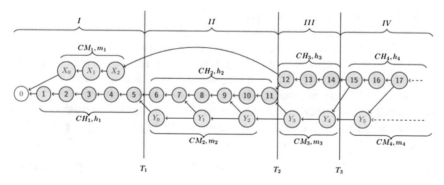

Fig. 1 Scheme of a hybrid attack on the SPECTRE protocol

The stages of the attack are separated by moments of time T_1, T_2 and T_3. The physical meaning of these points is explained below.

To simplify the presentation, we have shown in this figure only a small number of blocks combined into several chains. Nevertheless, when constructing an attack and calculating its probability, we will assume that any number of blocks can be created at each stage of the attack. It should also be noted that if we replace any of the chains with the appropriate "tree", then neither in the attack scheme itself, nor in the calculations and proofs, nothing will change.

Thus, the attack consists of such stages.

Stage 1. It starts from the moment when the attacker creates one of the two conflicting transactions (which sends the coin to the vendor and which then plans to cancel):

- the attacker creates one of the two conflicting transactions and includes it in the block X_0, then confirms this block with other blocks, creating a subgraph (in the figure—a chain) CM_1 with the length m_1; he does not distribute this subgraph, and does not send it to anyone at all;
- honest miners at this time create a subgraph (in the figure—a chain) CH_1 with the length of h_1.

Stage 2. It starts from the moment T_1, when an attacker decides that he has created a sufficient number of confirmation blocks for the block X_0:

- the attacker creates the second of the conflicting transactions (which sends the same coin to his wallet) and includes it in the block Y_0, then confirms this block with other blocks, creating a subgraph (in the figure—a chain) CM_2 with the length m_2; he also does not distribute this subgraph, and does not send it to anyone at all;
- honest miners at this time create a subgraph (in the picture—a chain) CH_2 with the length of h_2, which is a continuation of the subgraph CH_1.

Stage 3. It starts from the moment T_2, when an attacker decides that he has created a sufficient number of confirmation blocks for the block Y_0:

- the attacker distributes a subgraph CM_1;
- after the vendor saw the subgraph CM_1 with "his" transaction, he waits for some (possibly zero) time Δ and then sends the goods to the attacker;
- while waiting, honest miners create a subgraph (in the picture—a chain) CH_3 with the length of $h_3 \geq 0$ (in our modification we will assume that $h_3 \geq z$ for some acceptable z, the meaning of which we will explain below);
- the attacker during this time creates a subgraph (in the figure—a chain) CM_3 with the length of m_3, which continues to confirm only the block Y_0; however, it does not yet distribute subgraphs CM_2 and CM_3.

Stage 4. It starts from the moment T_3, when the vendor sends the goods and the attacker finds out about it:

- the vendor sends the goods at the time T_3;
- at the same time the attacker distributes subgraphs CM_2 and CM_3;

- honest miners confirm both subgraphs, CH_3 and CM_3, creating a subgraph CH_4 from h_4 blocks;
- the attacker continues to confirm only the subgraph CM_3 by a subgraph CM_4 of m_4 the blocks (combining a double spend attack with a censorship attack).

The attack will be successful if the block Y_0 will vote for more blocks than for X_0.

According to the attack scheme and voting protocol defined in SPECTRE, the votes of the blocks will be distributed as follows:

- all blocks from CM_1 vote for X_0 (m_1 blocks);
- all blocks from CH_3 vote for X_0 (h_3 blocks, and in our modification—$h_3 \geq z$ blocks);
- all blocks from CM_2, CM_3 and CM_4 vote for Y_0 ($m_2 + m_3 + m_4$ blocks);
- all blocks from CH_1 and CH_2 vote as a majority in their future ($h_1 + h_2$ blocks);
- all blocks from CH_4 vote like the majority in their past (h_4 blocks).

Now let's explain what is the vulnerability of the source protocol with the conflict resolution algorithm. [3] does not provide a clearly described protocol according to which the vendor should behave in order to reduce the risks of disappearance of the transaction with payment for the goods. It is only said that it must wait for a certain number of confirmation blocks that refer to the block with the corresponding transaction. It does not specify how to determine a sufficient number of confirmation blocks and what it depends on. The authors (without any proof) claim that the probability of an attack decreases exponentially with the number of confirmation blocks.

Now consider the stages of our attack. It is no coincidence that an attacker distributes a subgraph CM_1 only after the block X_0 received a sufficient number of confirmation blocks. In this case, the vendor will see almost simultaneously all the blocks of this subgraph and, according to the protocol, can immediately send the goods. If the attacker is lucky and the vendor sends the goods almost instantly (in this case $h_3 = m_3 = 0$), inequality is sufficient for the success of the attack $m_2 > m_1$, and to ensure its implementation in full within the capabilities of the attacker, regardless of its computing power.

Our proposed modification is that the vendor must wait for a certain number of confirmation blocks to be created from the moment he first sees the block with the corresponding transaction, even if at that moment the block already had some, even very large, number of confirmations. This requirement guarantees inequality $h_3 \geq z$, where z—set number of confirmation blocks (how to determine it we will explain below). In this case, as shown in the next paragraph, the probability of an attack will depend on the network parameters (in particular, the hashrate of the attacker) and will really quickly (according to experimental data—exponentially) decrease with increasing number of confirmation blocks.

3.3   Construction of the Upper Estimate of the Probability of a Hybrid Attack and Determination of the Required Number of Confirmation Blocks

At this point we will prove the theorem on the upper estimate of the probability of a hybrid attack. For further presentation we will need the following notations and supporting statements, most of which are given in [7, 8] and the work of Grunspan.[8]

Let $\alpha_H, \alpha_M > 0$—be the values that characterize the intensity of block generation by honest miners and the attacker, respectively, and $\alpha = \alpha_H + \alpha_M$—be the total intensity of the output blocks. Then these values will be the parameters of the distribution functions of random variables T_H and T_M, which determine, respectively, the time until the release of the next block for honest miners and for the attacker:

$$F_{T_H} = P(T_H < t) = 1 - e^{-\alpha_H t}, \quad F_{T_M} = P(T_M < t) = 1 - e^{-\alpha_M t}. \qquad (1)$$

Then the probabilities that the next block will be created, respectively, by honest miners and an attacker, are equal to

$$p_H = \frac{\alpha_H}{\alpha} \quad \text{and} \quad p_M = \frac{\alpha_M}{\alpha} \qquad (2)$$

Let's mark D_H—as the maximum time it takes for an honest miner to send an already created block to all (at least all honest) miners (network synchronization time).

Next, we will make assumptions in favor of the attacker and will assume that the appropriate time $D_M = 0$, and that an attacker has the ability to detain blocks distributed by honest miners at any time not exceeding D_H. It can also delay its blocks at any time, and can distribute them instantly, and in general can violate all mining rules.

We also determine the probabilities of subsequent events:

- p'_H—the likelihood that honest miners created and distributed the next block faster than the attacker did;
- p'_M—the probability of an alternative event.

Then, according to Lemma 1 in [7], we obtain the following equations:

$$p'_H = e^{-\alpha_M D_H} p_H \quad \text{and} \quad p'_M = 1 - p'_H = 1 - e^{-\alpha_M D_H} p_H \qquad (3)$$

Let's mark $P_z(k)$ the probability that the attacker will create exactly z blocks. Then, according to Lemma 4 in [7],

---

[8] Cyril Grunspan and Ricardo Pérez-Marco. 2017. Double spend races. CoRRabs/1702.02867 (2017). URL: http://arxiv.org/abs/1702.02867.

$$P_z(k) = \sum_{i=0}^{k} \left[C_{z+i-1}^i p_H^z p_M^k e^{-\alpha_M z D_H} \cdot \frac{(\alpha_M z D_H p_H)^{k-i}}{(k-i)!} \right]. \tag{4}$$

Note that in the case of $D_H = 0$ formula (4) looks much simpler:

$$P_z(k) = C_{z+k-1}^k p_H^z p_M^k, \tag{5}$$

since in this case in the sum (4) there will be only one term at $k = i$, and all others disappear due to the presence of a multiplier $(\alpha_M z D_H p_H)^{k-i}$.

Now we are ready to formulate and prove the main result.

Theorem 1 *Let the vendor wait for receipt according to the rules of transaction acceptance z confirmation blocks that appeared after he first saw the block with this transaction. Then, in the notation (1)–(4), the upper probability limit $P_z(\alpha_M, \alpha_H, D_H)$ the success of the hybrid attack equals to:*

$$P_z(\alpha_M, \alpha_H, D_H) \leq 1 - \sum_{k=0}^{z-1} P_z(k) \cdot \left(1 - \left(\frac{p_M'}{p_H'} \right)^{z-k} \left((p_H')^{z-k-1} + 1 \right) \right), \tag{6}$$

in particular in the case when $D_H = 0$:

$$P_z(\alpha_M, \alpha_H, 0) \leq 1 - \sum_{k=0}^{z-1} p_H^z p_M^k \cdot C_{z+k-1}^k \cdot \left(1 - \left(\frac{p_M}{p_H} \right)^{z-k} \left((p_H)^{z-k-1} + 1 \right) \right). \tag{7}$$

In this case, if $p_M' \geq p_H'$, then $P_z(\alpha_M, \alpha_H, D_H) = 1$ at any value $z \in N$.

Proof. Let B_1 is the first block that contains both blocks X_0 and Y_0 in their past (in Fig. 1 it is block 15). Such a block, as well as all blocks in its future, will vote like most blocks in their past. Thus, the voices of the blocks from the subgraph CH_4 significantly depend on how the blocks belonging to $past(B_1, G)$. Let's analyze how these blocks can vote:

- all the blocks with CH_3 will vote for X_0;
- all the blocks with CM_3 will vote for Y_0;
- the last block in CH_2, let's mark it B_2 (in Fig. 1 it is block 11), will vote for Y_0, if $h_3 \leq m_3$, and in this case all blocks with will vote in the same way CH_1 and CH_2.

Thus, one of the conditions for a successful attack is a condition $h_3 \leq m_3$, which hereinafter we will denote the condition (C1).

If (C1) is not executed, the attacker can still carry out a successful attack if the blocks in $past(B_2, G)$ vote for Y_0. Let's analyze when it is possible. If (C1) is not executed, then B_2 votes for X_0. Let $h_3 = m_3 + l$ for some $l \in N$. Then the block placed directly in front B_2, let's call it B_3 (in Fig. 1 it is block 10), will vote for Y_0 only if during the time between the appearance of blocks B_3 and B_2 the attacker created at least $l + 1$ block. Similarly, if B_3 and B_2 vote for X_0, a necessary condition for the block B_4, placed directly in front of B_3, to vote for Y_0, is a condition, when during the time between the appearance of blocks B_4 and B_3 the attacker created at least $l + 2$ blocks, etc.

So if the condition (C1) is not executed, a necessary condition for some block with a subgraph CH_2 to vote for Y_0, there is such a condition:

- in a subgraph CH_2 there is such a block (call it B_5), that between the appearance of this block and the next one, the attacker created at least $l + 1 + u$ blocks, where $u = \#\{CH_2 \cap future(B_5, G)\}$.

Let us denote this condition (C2).

If none of these conditions (C1) and (C2) is executed, the attacker still has a chance to carry out his attack. A necessary condition for this is the following condition:
at some point in time $T > T_3$ inequality will be satisfied $m_4 = h_4 + l$.
Let us denote this condition (C3).

Therefore, for the successful implementation of the attack, it is necessary that it is carried out at least from the conditions (C1), (C2) or (C3).

Note that, since the time before the release of the blocks has an exponential distribution, the number of blocks created over a period of time is described by the Poisson distribution. Therefore, the processes of generating blocks at different time intervals and on different subgraphs are independent. Hence the independence of events, which consist in the fulfillment of conditions (C1), (C2) and (C3).

By the condition of the theorem, $h_3 \geq z$. Then, using Lemma 4 from [7], we obtain:

$$P(C1) = 1 - \sum_{k=0}^{z-1} P_z(k). \tag{8}$$

Then,

$$P(\neg(C1) \cap (C2)) = P(\neg(C1)) \cdot P((C2))$$

$$= \sum_{k=0}^{z-1} P_z(k) \left(\left(p'_M\right)^{z-k} + \left(p'_M\right)^{z-k+1} + \cdots + \left(p'_M\right)^{z-k+(k-1)} \right)$$

$$\leq \sum_{k=0}^{z-1} P_z(k) \left(\frac{\left(p'_M\right)^{z-k}}{1 - p'_M} \right) = \sum_{k=0}^{z-1} P_z(k) \left(\frac{\left(p'_M\right)^{z-k}}{p'_H} \right). \tag{9}$$

In addition, equality is fair

$$P(\neg(C1) \cap \neg(C2) \cap (C3)) = (1 - P((C1))) \cdot (1 - P((C2))) \cdot P((C3))$$
$$\leq (1 - P((C1))) \cdot P((C3)).$$

Now note that fulfilling condition (C3) means that the attacker was able to "overtake" honest miners in the number of created blocks after falling behind them by $z - k$ blocks. Therefore

$$P((C3)) \leq \left(\frac{p'_M}{p'_H}\right)^{z-k}.$$

Then

$$P(\neg(C1) \cap \neg(C2) \cap (C3)) \leq \sum_{k=0}^{z-1} P_z(k)\left(\frac{p'_M}{p'_H}\right)^{z-k}. \tag{10}$$

Therefore, applying (8)–(10), we obtain:

$$P_z(\alpha_M, \alpha_H, D_H) \leq 1 - \sum_{k=0}^{z-1} P_z(k) + \sum_{k=0}^{z-1} P_z(k)\frac{(p'_M)^{z-k}}{p'_H} + \sum_{k=0}^{z-1} P_z(k)\left(\frac{p'_M}{p'_H}\right)^{z-k}$$

$$= 1 - \sum_{k=0}^{z-1} P_z(k)\left(1 - \frac{(p'_M)^{z-k}}{p'_H} - \left(\frac{p'_M}{p'_H}\right)^{z-k}\right)$$

$$= 1 - \sum_{k=0}^{z-1} P_z(k) \cdot \left(1 - \left(\frac{p'_M}{p'_H}\right)^{z-k}\left((p'_H)^{z-k-1} + 1\right)\right),$$

which completes the proof of the theorem. □

Note that formula (6) is also convenient to use to calculate the minimum number of confirmation blocks required to ensure that the probability of a hybrid attack was less than some predetermined value.

3.4 Numerical Results for the Probability of Success of a Hybrid Attack

Here we present tables with numerical results obtained using formula (6). Tables 1 and 2 show the smallest value of the number of confirmation blocks for which the probability of a hybrid attack is not greater than 10^{-3}, at different values of network parameters. These values are calculated for two different values of the intensity of the output blocks: for $\alpha = \frac{1}{600} = 0.00167$, both for BTC (Table 1) and for $\alpha = \frac{1}{60} = 0.0167$, i.e. 10 times greater intensity (Table 2).

Note that these Tables differ significantly from Tables 4 and 5 in [7], which also give the lowest values for the number of confirmation blocks. The difference is that in [7] we investigated the "classic" blockchain, while in this section we consider a blockgraph with a much more complex consensus protocol.

Table 1 The smallest value of the number of confirmation blocks, for which the probability of a hybrid attack is not more than 10^{-3}, at different values of the share of the hashrate of the attacker and at $\alpha = \frac{1}{600} = 0.00167$

p_M	D_H			
	0 c	15 c	30 c	60 c
0.1	6	6	7	7
0.15	9	9	10	10
0.20	14	14	14	15
0.25	21	21	22	24
0.3	33	35	36	40
0.35	60	65	69	81
0.4	137	154	174	228

Table 2 The smallest value of the number of confirmation blocks, for which the probability of a hybrid attack is not more than 10^{-3}, at different values of the share of the hashrate of the attacker and at $\alpha = \frac{1}{60} = 0.0167$

p_M	D_H			
	0 c	15 c	30 c	60 c
0.1	6	7	8	11
0.15	9	11	13	20
0.20	14	17	23	43
0.25	21	29	44	172
0.3	33	55	114	–
0.35	60	139	–	–
0.4	137	203	–	–

To calculate the number of confirmation blocks, we used formulas (6) and (7) and by iteration over z calculated $\min\{z \geq 1 : P_z(\alpha_M, \alpha_H, D_H) \leq 10^{-3}\}$.

In Table 2, the blank cells mean that the probability of an attack is 1 for any number of confirmation blocks.

We have chosen synchronization time values D_H to ensure that the interval that often occurs in real networks is guaranteed. For example, for the Bitcoin network, the synchronization time often varies from 0 to 20 s.

You should also note that provided $D_H = 0$ the first column in Tables 1 and 2 is the same, while all the others are significantly different. The explanation for this fact is as follows: the longer the synchronization time of honest miners, the greater the advantage of the attacker. Moreover, under the condition of non-zero synchronization time, the advantage of the attacker significantly depends on the intensity of block generation: the greater the intensity, the greater the advantage. This fact, in particular, was analytically proved in [7] for the blockchain, here we can also see it in formula

(3): the larger the value $\alpha_M D_H$, the greater the probability p'_M; and the higher this probability, the lower the so-called safety margin, i.e. the critical fate of the attacker (when $D_H = 0$ it is equal to 50%). For example, as we see in the last column of Table 2, at $D_H = 60$ s and $\alpha = \frac{1}{60} = 0.0167$, the security limit is less than 30%.

4 Summary

In this section, we have considered and analyzed the use of monitoring systems in a decentralized environment, which has significant conveniences and advantages over a centralized one. It is shown what characteristics such an environment should have and how you can try to provide them. As an example of a protocol that can be used in decentralized monitoring environments, a modification of the SPECTRE protocol is considered. Estimates of the stability of this modified protocol are constructed and some numerical results are given.

In recent years, the possibility of moving from a blockchain to a more general structure, which has additional capabilities, but remains resistant to major attacks, has been actively explored. Basically, the new structure is a blockgraph (DAG), which has certain features. Works [2–4, 6] offer different ways of such generalizations and partial analysis of their properties. At the same time, the results of such studies raise a large number of questions. Next we will highlight the main ones.

1. Conflict resolution algorithms are quite questionable, in terms of practicality and resistance to attacks. In many works, they are not even described correctly.
2. Impossibility (or only partial possibility) to establish a linear order on a set of blocks. Some protocols establish only a partial order, some bypass this issue altogether.
3. Long waiting time for full confirmation of the transaction, i.e. the time from the moment the transaction is included in the block before the transaction becomes irreversible with a high probability
4. Insufficient mathematical substantiation of the main statements, especially regarding the resistance of protocols to the main attacks. Thus, mathematical calculations in many works contain significant errors, and in others instead of mathematical calculations heuristic considerations are given.

Therefore, the field for research in this direction remains extremely large, and the task of scaling the blockchain is very important.

Over the last 5–7 years, many consensus protocols have been proposed on block-graphs. It should be noted that the use of such protocols, although it leads to a significant increase in the speed of transaction processing, but carries certain risks. At present, none of these protocols have rigorous evidence of resistance to major attacks, such as double spend attack, network splitting attack, and so on. The construction of such proofs requires overcoming great analytical difficulties, even if you use a very simplified mathematical model of the network.

In this section, we took the first step toward constructing consensus protocol stability estimates on blockgraphs by constructing upper estimates for the probability of a double spend attack on the SPECTRE consensus protocol. To increase its security, we have proposed a somewhat non-standard algorithm for accepting a transaction by a vendor: it must wait for a certain number of confirmation blocks to be created, so to speak, under its supervision.

An important property of the results obtained is that, in addition to the probability estimate itself, we can also calculate the required number of confirmation blocks sufficient to guarantee the irreversibility of the transaction with a probability as close to 1 as desired.

It should also be noted that the method of constructing such estimates proposed here can only be applied to the SPECTRE protocol, and cannot be easily transferred to other protocols on the blockgraphs. But the very idea of the method may be useful for other protocols.

References

1. Satoshi Nakamoto. Bitcoin: A peer-to-peer electronic cash system (2008)
2. Sompolinsky Y, Zohar A (2013) Accelerating bitcoin's transaction processing: fast money grows on trees, not chains. IACR Cryptology ePrint Archive 2013:881
3. Sompolinsky Y, Lewenberg Y, Zohar A (2016) SPECTRE: a fast and scalable cryptocurrency protocol. IACR Cryptology ePrint Archive 2016:1159
4. Sompolinsky Y, Zohar A (2018) Phantom. IACR Cryptology ePrint Archive, Report 2018/104
5. Boyen X, Carr C, Haines T (2018) Graphchain: a blockchain-free scalable decentralised ledger. In Proceedings of the 2nd ACM workshop on blockchains, cryptocurrencies, and contracts, pp 21–33
6. Popov S (2016) The tangle. cit. on (2016), p 131
7. Kovalchuk L, Kaidalov D, Nastenko A, Rodinko M, Shevtsov O, Oliynykov R (2020) Decreasing security threshold against double spend attack in networks with slow synchronization. Comput Commun 154:75–81
8. Kudin A, Kovalchuk L, Kovalenko B (2019) Theoretical foundations and application of blockchain: implementation of new protocols of consensus and crowdsourcing computing. Mathematical and computer modelling. Technical sciences, vol 19, pp 62–68. https://doi.org/10.32626/2308-5916.2019-19 (in Ukrainian)
9. Fitzi M, Gaži P, Kiayias A, Russell A (2018) Parallel chains: improving throughput and latency of blockchain protocols via parallel composition. Cryptology ePrint Archive, Report 2018/1119. https://eprint.iacr.org/2018/1119
10. Bagaria V, Kannan S, Tse D, Fanti G, Viswanath P (2018) De-constructing the blockchain to approach physical limits. arXiv preprint arXiv:1810.08092

Models and Methods of Secure Routing and Load Balancing in Infocommunication Networks

Oleksandr Lemeshko⬥, Oleksandra Yeremenko⬥,
and Maryna Yevdokymenko⬥

Abstract An analysis of existing network security solutions shows that a promising direction is to improve traffic management and routing process that should consider network performance parameters and network security parameters that characterize the efficiency of network intrusion detection systems and analyze vulnerabilities and risks. A flow-based model of secure routing with load balancing has been developed with modified load balancing conditions and considering the probability of its compromise. It has been found that in the process of load balancing, links with a lower compromise probability are loaded more intensively, unloading more vulnerable from the point of view of links compromise.

Keywords Network security · Secure routing · Security metric · Packet loss · Self-similar traffic

1 Introduction

Traditionally, the main requirements for modern infocommunication systems and networks are ensuring a given Quality of Service and network security level. These tasks should be solved comprehensively, interconnectedly, and based on the optimal use of available network resources. However, modern network technologies and protocols do not meet these requirements, focusing on improving the performance Quality of Service (QoS) or network security (NS) separately [1–4].

Providing multiservice in the infocommunication network (ICN), the quite different information traffic circulates with its requirements for QoS and NS, the characteristics of which can be described by the theory of fractals and self-similar processes [5, 6]. Therefore, an important scientific and practical task is to develop mathematical models, methods, and protocols that would provide the specified values of the main QoS indicators and focus on improving network security. Recently, a central role in the set of tasks to ensure QoS and NS is given to the Network Layer,

O. Lemeshko · O. Yeremenko (✉) · M. Yevdokymenko
Kharkiv National University of Radio Electronics, 14 Nauky Ave, Kharkiv 61000, Ukraine
e-mail: oleksandra.yeremenko.ua@ieee.org

© The Author(s), under exclusive license to Springer Nature Switzerland AG 2022
R. Oliynykov et al. (eds.), *Information Security Technologies in the Decentralized Distributed Networks*, Lecture Notes on Data Engineering and Communications Technologies 115, https://doi.org/10.1007/978-3-030-95161-0_10

namely models, methods, and protocols of the routing process [7–17]. Consequently, such technological tools should receive their theoretical and protocol development to provide QoS and NS in routing self-similar traffic.

2 Basic Flow-Based Models of Traffic Routing in Infocommunication Networks

In the general case, the structure of the flow-based mathematical model of routing in ICN contains the following conditions [7–20]:

- flow conservation on network routers;
- preventing overload of communication links;
- implementation of a single path or multipath routing;
- optimal use of available network resources (link, buffer, path, etc.).

Namely, these conditions must consider the peculiarities of ensuring the guaranteed Quality of Service and network security. Moreover, they will be the basis of the corresponding optimization problems: determining the basis of constraints imposed on routing variables or criteria for optimality routing decisions.

2.1 Basic Flow-Based Model of Traffic Routing in Infocommunication Networks Without Packet Loss

The basic model is the mathematical routing model in ICN described in detail in [12, 15, 17]. Then, oriented graph $\Gamma = (U, W)$ describes the network structure where.

- $U = \{u_i, i = \overline{1, m}\}$—set of vertices that models routers;
- $W = \{w_{i,j}, i, j = \overline{1, m}; \ i \neq j\}$—set of edges of graph Γ when $w_{i,j}$ models network link between ith and jth routers;
- $\varphi_{i,j}$—link $w_{i,j}$ bandwidth (packets per second, pps).

Let us define K as the set of flows transmitting in the network ($k \in K$). Then, as a result of solving the routing problem in ICN, it is necessary to calculate a set of routing variables. Each of these variables $x_{i,j}^k$ determines the fraction of the intensity of the kth flow, which is sent in the absence of packet loss to the communication link $w_{i,j}$. Depending on the type of routing supported in the network, the following conditions are imposed on routing variables:

$$0 \leq x_{i,j}^k \leq 1. \tag{1}$$

It is necessary to ensure compliance with the flow conservation conditions [12, 15, 17] to ensure the connectivity of the calculated unicast routes in the absence of

ICN routers overload:

$$\begin{cases} \sum_{j:w_{i,j}\in W} x_{i,j}^k - \sum_{j:w_{j,i}\in W} x_{j,i}^k = 1, \ k \in K, \ u_i = s_k; \\ \sum_{j:w_{i,j}\in W} x_{i,j}^k - \sum_{j:w_{j,i}\in W} x_{j,i}^k = 0, \ k \in K, \ u_i \neq s_k, d_k; \\ \sum_{j:w_{i,j}\in W} x_{i,j}^k - \sum_{j:w_{i,j}\in W} x_{j,i}^k = -1, \ k \in K, \ u_i = d_k, \end{cases} \quad (2)$$

where s_k and d_k are the source and destination nodes for the kth flow.

Additional restrictions are imposed on routing variables $x_{i,j}^k$ to prevent network links' overload:

$$\sum_{k\in K} \lambda_{req}^k x_{i,j}^k \leq \varphi_{i,j}, \quad (i,j) \in V \quad (3)$$

where λ_{req}^k is the kth packet flow average intensity (in packets per second, pps) arriving at the network. The value λ_{req}^k directly determines the bandwidth requirements for this flow.

In turn, the network links utilization can be calculated using the following expression:

$$\rho_{i,j} = \frac{\sum_{k\in K} \lambda_{req}^k x_{i,j}^k}{\varphi_{i,j}} \quad (4)$$

Accordingly, the intensity of successfully transmitted kth flow packets in the link is determined as follows:

$$\lambda_{i,j}^k = \lambda_{req}^k x_{i,j}^k \quad (5)$$

The advantage of the routing model (1)–(5) is its linear nature, which greatly simplifies the required routing variables calculation in real-time.

3 Quality of Service Ensuring Conditions in the Infocommunication Network

In terms of previously described basic model (1)–(5), the conditions for ensuring the Quality of Service for the average end-to-end packet delay, in the general case, are as follows [12]:

$$\tau_{MP}^k \leq \tau_{req}^k, \quad (6)$$

where τ_{req}^k requirements for the allowable value of the average end-to-end packet delay of the kth flow; τ_{MP}^k is the actual value of the average end-to-end packet delay of the kth flow, which is calculated by the following formula [12]:

$$\tau_{MP}^k = \sum_{p=1}^{|P^k|} x_p^k \tau_p^k, \tag{7}$$

where x_p^k is the proportion of the kth packet flow that was successfully delivered to the receiving router via the pth path; τ_p^k is the average end-to-end packet delay of the kth flow transmitted via the pth path; $|P^k|$ is the size of the set P^k, the value of which determines the total number of paths available for kth flow routing.

In the general case, the following expression can be used for the x_p^k calculation:

$$x_p^k = \frac{\lambda_p^k}{\lambda_{req}^k}, \tag{8}$$

where λ_p^k is the kth packet flow intensity that was delivered to the receiving router via the pth path.

In turn, the average end-to-end packet delay of the kth flow transmitted via the pth network path is calculated according to the additive scheme:

$$\tau_p^k = \sum_{w_{i,j} \in p} \tau_{i,j}^k, \tag{9}$$

where $\tau_{i,j}^k$ is the average end-to-end packet delay of the kth flow in the link $w_{i,j}$, the value of which is entirely determined by the selected service model of packet flows in the network.

3.1 Service Models of Self-Similar Packet Flows in Infocommunication Networks

Numerical values of the main QoS indicators directly depend on the following basic functional parameters of the network and its elements:

- characteristics of network traffic: number of flows, their intensities (packet transmission rate), packet lengths, etc.;
- router interface parameters: bandwidth, utilization level, etc.

The variant was considered when the routers' interfaces were modeled by queuing systems (QS) with a self-similar input flow and a deterministic service time fBM/D/1. Then the average packet delay in the link $w_{i,j}$ can be calculated using the formula [18]:

$$\tau_{i,j} = \frac{\rho_{i,j}^{(H-1/2)/(1-H)}}{(1 - \rho_{i,j})^{H/(1-H)} \varphi_{i,j}} \tag{10}$$

Table 1 The value of the Hurst parameter when modeling different types of traffic (according to the recommendations of ITU-T Q.3925)

Traffic types		Flow types	H values range
WWW traffic		Self-similar	$H = 0.7...0.9$
File transfer			$H = 0.85...0.95$
E-mail traffic			$H = 0.75$
Peer-to-peer traffic			$H = 0.55...0.6$
IPTV traffic	Unicast		$H = 0.75...0.8$
	Multicast		$H = 0.55...0.6$
Telemetry, stationary and mixed sensors			$H = 0.67...0.69$

In the general case, the self-similarity parameter (Hurst parameter) varies in the range $0.5 \leq H \leq 1$, and its value depends on the type of traffic transmitted in the network (Table 1).

As the analysis showed [12, 15, 16], the formulation in the analytical form of the conditions for QoS ensuring on a set of indicators (3) and (6) by routing means can be implemented using a tensor methodology.

3.2 Tensor Generalization of the Infocommunication Network Mathematical Model

The tensor methodology of ICN modeling and research is based on the following main stages:

- geometrization of the network structure with a reasonable introduction of geometric space and bases (coordinate systems);
- determination of rules of co- and contravariant coordinates transformation of the introduced tensor—ICN model;
- metrization of the ICN tensor model with a determination of the functional dependence between co- and contravariant components of the introduced tensors;
- obtaining the required dependencies for the calculation of the main QoS indicators.

Within the study conducted in this work, the ICN structure was modeled by a one-dimensional network $S = (U, V)$ [7, 12, 19]. As before, ICN routers have been described by a set of network nodes $U = \{u_i, i = \overline{1, m}\}$, where m is the total number of nodes in the network S. While a set of edges $V = \{v_z, z = \overline{1, n}\}$ simulate ICN links, where n is the total number of edges in the network S. That is, their continuous numbering is used to describe network elements (Fig. 1).

The poles of the network S are nodes that simulate the routers of the packets' source and destination in implementing unicast routing in ICN. The following structural characteristics will be defined and used for the network:

Fig.1 Network S example
that models the ICN structure

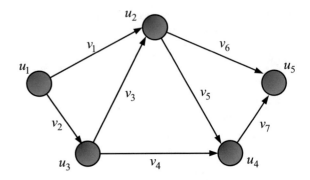

- $\mu(S)$ is the cyclomatic number, which determines the number of linearly independent (basic) circuits in the network;
- $\phi(S)$ is the network rank, which sets the number of basic node pairs in S;
- $\kappa(S)$ is the number of linearly independent interpolar paths in the network;
- $\vartheta(S)$ is the number of basic internal node pairs in the network, where the set of internal node pairs contains all node pairs except the pole.

The following relationships relate to these structural characteristics:

$$\phi(S) = m - 1, \mu(S) = n - m + 1, \kappa(S) = \mu + 1 = n - m + 2,$$

$$\vartheta(S) = \phi - 1 = m - 2 \qquad (11)$$

Therefore, a discrete n-dimensional geometric space is introduced on the ICN structure, and depending on the aspect of network consideration in the introduced discrete space, several coordinate systems (CS) can be determined [12–14]:

- coordinate system of network edges $\{v_z, z = \overline{1, n}\}$, projections of tensors in which will be denoted by an index v;
- coordinate system of interpolar paths $\{\gamma_i, i = \overline{1, \kappa}\}$ and internal node pairs $\{\varepsilon_j, j = \overline{1, \vartheta}\}$ of the network S, the tensor projections in which will be denoted by the index $\gamma\varepsilon$.

The orthogonality of these coordinate systems is justified by the fact that under expressions (11) the condition $n = \kappa(S) + \vartheta(S)$ is met.

In the introduced n-dimensional space, a mixed bivalent tensor can describe the network concerning each individually selected packet flow for which QoS conditions need to be obtained [12]:

$$Q = T \otimes \Lambda, \qquad (12)$$

where $\otimes$ is tensor multiplication operator; T is the univalent covariant tensor of average packet delays; Λ is the univalent contravariant tensor of average flow intensities in the coordinate paths of the network.

In the general case, the components of the mixed bivalent tensor Q (12) are interconnected by a metric tensor [12]:

$$T = E\Lambda, \tag{13}$$

where E is a twice covariant metric tensor.

Tensor Eq. (13) in one or another coordinate system takes the appropriate vector–matrix form. For example, in the introduced coordinate systems, the tensor Eq. (13) will look like this:

$$T_v = E_v \Lambda_v \text{ and } T_{\gamma\varepsilon} = E_{\gamma\varepsilon} \Lambda_{\gamma\varepsilon}, \tag{14}$$

where Λ_v and T_v, $\Lambda_{\gamma\varepsilon}$ and $T_{\gamma\varepsilon}$ are projections of tensors Λ and T in the CS of edges and CS of interpolar paths and internal nodal pairs, respectively, which are represented by n-dimensional vectors.

$E_v = \left\| e_{ij}^v \right\|$ and $E_{\gamma\varepsilon} = \left\| e_{ij}^{\gamma\varepsilon} \right\|$ are projections of the twice covariant metric tensor in the selected CS, represented by $n \times n$-matrices.

As is known [7, 12–16], the tensors projections in different coordinate systems are interconnected by linear transformation laws. According to [12], the law of covariant transformation can be described by a nonsingular $n \times n$-matrix $A_{\gamma\varepsilon}^v$, which determines the rules of transition from interpolar paths and internal node pairs CS to edges CS of the network. Then the projections of the covariant tensor under conditions of CS change are transformed as follows:

$$T_v = A_{\gamma\varepsilon}^v T_{\gamma\varepsilon} \tag{15}$$

The law of contravariant coordinate transformation under the condition of change of the considered CS can be described by a nonsingular $n \times n$-matrix $C_{\gamma\varepsilon}^v$ [12, 13]:

$$\Lambda_v = C_{\gamma\varepsilon}^v \Lambda_{\gamma\varepsilon}, \tag{16}$$

where n-dimensional vector $\Lambda_{\gamma\varepsilon}$, which is a tensor Λ projection in the CS of the interpolar paths and internal node pairs, has the following structure:

$$\Lambda_{\gamma\varepsilon} = \begin{bmatrix} \Lambda_\gamma \\ -- \\ \Lambda_\varepsilon \end{bmatrix}; \quad \Lambda_\gamma = \begin{bmatrix} \lambda_\gamma^1 \\ \vdots \\ \lambda_\gamma^j \\ \vdots \\ \lambda_\gamma^\kappa \end{bmatrix}; \quad \Lambda_\varepsilon = \begin{bmatrix} \lambda_\varepsilon^1 \\ \vdots \\ \lambda_\varepsilon^p \\ \vdots \\ \lambda_\varepsilon^\vartheta \end{bmatrix}, \tag{17}$$

where Λ_γ is the κ-dimensional vector of flow intensities along the basic interpolar paths of the network; Λ_ε is the ϑ-dimensional vector of flow intensities between nodes that form the basic internal node pairs; λ_γ^j is the flow intensity along the jth basic interpolar path (γ_j); λ_ε^p is the intensity of the flow arriving at the network and outgoing the network through nodes that create the pth basic internal node pair (ε_p).

The matrices of co- and contravariant coordinate transformation in the case of change of these bases are interconnected by orthogonality conditions $C_{\gamma\varepsilon}^v\left(A_{\gamma\varepsilon}^v\right)^t = I$, where I is a unit matrix of $n \times n$ size; $[\cdot]^t$ is the matrix transposition operation.

The rules of matrices formation of co- and contravariant coordinates transformation of the introduced tensors under the condition of change of the described coordinate systems are considered in detail in works [12–14].

Therefore, if expression (10) is generalized to the whole set of communication links and leads to the vector–matrix form (14), then the projection coordinates of the twice covariant metric tensor E in the coordinate system of network edges can be represented by the values of diagonal matrix E_v elements [12]:

$$e_{ii}^v = \frac{\rho_i^{(H-1/2)/(1-H)}}{\lambda_v^i(1 - \rho_i)^{H/(1-H)}\varphi_i}, \tag{18}$$

where, as mentioned above, λ_v^i is the intensity of the packet flow, which is considered from the point of view of constructing a tensor model in the ith network link (i.e., the continuous numbering of links is used).

The law of projections transformation of a twice covariant tensor E under the condition of coordinate systems change from the basis of edges to the basis of interpolar paths and internal nodal pairs has the following form [12]:

$$E_{\gamma\varepsilon} = (C_{\gamma\varepsilon}^v)^t E_v C_{\gamma\varepsilon}^v \tag{19}$$

Let us decompose the projection of the metric tensor in the basis of interpolar paths and internal nodal pairs:

$$E_{\gamma\varepsilon} = \left\|
\begin{array}{c|c}
E_{\gamma\varepsilon}^{\langle 1\rangle} & E_{\gamma\varepsilon}^{\langle 2\rangle} \\
\hline
E_{\gamma\varepsilon}^{\langle 3\rangle} & E_{\gamma\varepsilon}^{\langle 4\rangle}
\end{array}
\right\|, \tag{20}$$

where $E_{\gamma\varepsilon}^{\langle 1\rangle}$ is the square $\kappa \times \kappa$ submatrix; $E_{\gamma\varepsilon}^{\langle 4\rangle}$ is the square $\vartheta \times \vartheta$ submatrix; $E_{\gamma\varepsilon}^{\langle 2\rangle}$ is the $\kappa \times \vartheta$ submatrix; $E_{\gamma\varepsilon}^{\langle 3\rangle}$ is the $\vartheta \times \kappa$ submatrix.

Then, as shown in [12]:

$$\tau_{MP} = \frac{\Lambda_\gamma^t E_{\gamma\varepsilon}^{\langle 1\rangle} \Lambda_\gamma}{\lambda^{req}} \tag{21}$$

Taking into account inequality (6), the condition of QoS ensuring for the average end-to-end packet delay and bandwidth will take the form

$$\tau_{req} \lambda^{req} \geq \Lambda_\gamma^t E_{\gamma\varepsilon}^{\langle 1 \rangle} \Lambda_\gamma \tag{22}$$

Thus, inequalities (3) and (22) are the required conditions for ensuring the Quality of Service in terms of bandwidth and average end-to-end packet delay in ICN.

In this study, it is proposed to consider the aspects related to improving the level of network security in ICN by forming the optimal routing solutions criterion appropriately. Then conditions (1)–(3) and (22) will act as restrictions imposed on routing variables.

4 Classification of Secure Routing Metrics in Infocommunication Network

In the general case, the following linear optimality criterion will be used to calculate the optimal paths in ICN [21]:

$$\sum_{k \in K} \sum_{w_{i,j} \in W} c_{i,j} x_{i,j}^k \Rightarrow \min, \tag{23}$$

where $c_{i,j}$ are routing metrics, which should consider the essential security characteristics of network links and routers.

Initially, it is proposed to use the links compromise probabilities as secure routing metrics:

$$c_{i,j} = p_{i,j}^{comp}, \tag{24}$$

where $p_{i,j}^{comp}$ is the link $w_{i,j} \in W$ compromise probability.

The link compromise probability belongs to the class of multiplicative metrics (metrics that multiply along the path), i.e., the probability of the lth path compromise in the ICN is calculated as:

$$p_l^{comp} = 1 - \prod_{w_{i,j} \in l} (1 - p_{i,j}^{comp}) \tag{25}$$

Since the optimality criterion (23) is additive, the metrics of secure routing should be written as follows:

$$c_{i,j} = -\log_{10}\left(1 - p_{i,j}^{comp}\right) \tag{26}$$

Then, it is proposed to calculate secure routing metrics based on information security risk (ISR) assessment of network links and routers [22–26]. ISR can be computed using the vulnerability criticality metrics specified in NIST CVSS v3 [24]: basic, time, and user environmental metrics. Based on the results obtained in [22–26], in the calculation of weights $c_{i,j}$, the basic metrics selected, which, in contrast to time metrics and metrics of the user environment, characterize the time-insensitive vulnerabilities of network elements and allow assessment of the information security risk of the communication network in general, and not for individual cases of compromising network elements.

We introduce the following notation [22–24]:

- $\mathbf{U} = \{U_i^q; q = \overline{1, Q}, i = \overline{1, m}\}$ is the set of vulnerabilities that are detected on the network nodes (routers), where U_i^q is the qth vulnerability on the ith network node;
- $U_i^* \subset \mathbf{U}$ is the set of vulnerabilities on the ith network node;
- BS_i^q criticality index of the qth vulnerability on the ith network node, which is calculated using the basic metrics of the vulnerability assessment system, which are presented in the recommendation NIST CVSS v3 [24], and characterizes the conditional losses from the exploitation of U_i^q vulnerability by the attacker;
- P_i^q is the probability of using the qth vulnerability by an attacker on the ith network node, which is the compromise probability in a physical sense.

According to [22, 24], to calculate the risk of information security from the exploitation of existing vulnerabilities on the ith network node, the following expression is used:

$$R^i = \sum_{U_i^q \in U_i^*} BS_i^q \cdot P_i^q \qquad (27)$$

According to the NIST recommendation, losses relative to basic vulnerability metrics at network nodes [22, 24] are calculated as:

$$BS_i^q = (0.6 \cdot \mathrm{Imp_i^q} + 0.4 \cdot \mathrm{Ex_i^q} - 1.5) \cdot f(\mathrm{Imp_i^q}), \qquad (28)$$

where $\mathrm{Imp_i^q}$ is the potential loss from the use of qth vulnerability by an attacker on the ith network node; $\mathrm{Ex_i^q}$ is the difficulty of exploiting the qth vulnerability by an attacker on the ith network node; $f(\mathrm{Imp_i^q})$ is the function from potential loss in case of exploitation of qth vulnerability by the attacker on ith network element.

The potential impact from the vulnerability exploitation is calculated as [22, 24]:

$$\mathrm{Imp_i^q} = 10, 41\big[1 - (1 - Conf_i^q) \cdot (1 - Int_i^q) \cdot (1 - Av_i^q)\big], \qquad (29)$$

where $Conf_i^q$ are losses from breach of information confidentiality transmitted over the network and cannot be obtained by an unauthorized, for example, the external user (attacker); Int_i^q are losses from the violation of the integrity, characterized by modification, alteration, and destruction of information by an unauthorized network user (attacker); Av_i^q are losses from violation of network resource availability in case of using the qth vulnerability on the ith network node.

These are three metrics of the base group $Conf_i^q$, Int_i^q, and Av_i^q determine the possible consequences of the attacker's exploitation of the qth vulnerability on the ith network node.

In each of these metrics, the damage (impact) from the exploitation of the vulnerability can be [22, 24]:

- *absent* with a value of 0;
- *partial* with a value of 0.275;
- *complete* with a value of 0.66.

The values of metrics $Conf_i^q$, Int_i^q, and Av_i^q are presented in Table 2 [22, 24].

The complexity of exploiting the vulnerability is calculated using the following expression:

$$\text{Ex}_i^q = 20 \cdot Ac_i^q \cdot Au_i^q \cdot AcV_i^q, \tag{30}$$

where Ac_i^q is the indicator of the vulnerability assessment system, which characterizes the difficulty of obtaining access (access vector); Au_i^q is the indicator of the vulnerability assessment system that is responsible for authentication requirements; AcV_i^q is the indicator of the vulnerability assessment system, which reflects the method of using the qth vulnerability on the ith network node, which is physically characterized by "remoteness" of the attacker, i.e., the number of devices and/or access restrictions through which the attacker can reach the ith network node to attack.

Table. 2 Values of indicators for calculation of basic metrics of network elements vulnerabilities

Designation	Description	Metric value
Confidentiality impact $Conf_i^q$		
None (N)	There is no impact on the confidentiality of network information	0.0
Partial (P)	There is considerable network information disclosure. Access to some system files is possible but limited	0.275
Complete (C)	There is a total information disclosure	0.66
Integrity impact Int_i^q		
None (N)	There is no impact on the integrity of network information	0.0
Partial (P)	It is possible to modify network information or system files partially	0.275
Complete (C)	It is possible to modify any network information of the network nodes	0.66
Availability impact Av_i^q		
None (N)	There is no impact on the availability of the network	0.0
Partial (P)	There is a possibility of reduced performance or failure of some functions of the network nodes	0.275
Complete (C)	There is a possibility of total failure of the network nodes	0.66

These indicators are the basic metrics [22–26], which characterize the overall complexity of the attack when exploiting a vulnerability on the ith network node (Table 3).

According to [22–26], the potential loss function $f(\text{Imp}_i^q)$ takes a zero value in the absence of damage, i.e., $f(\text{Imp}_i^q) = 0$. Here we will consider the case when the potential damage is present when $\text{Imp}_i^q \neq 0$. Therefore, in further calculations, we use $f(\text{Imp}_i^q) = 1.176$ [22–26]. Then, to quantify the worst-case information security risk when compromising the communication link $w_{i,j} \in W$ outgoing from the ith node, we use the following expression of the exponential form [22–24]:

Table 3 Values of indicators of the vulnerability assessment system, which characterize the complexity of the vulnerabilities exploitation

Designation	Description	Metric value
Access vector Ac_i^q		
Local (L)	To exploit the vulnerability, the attacker needs physical access to the local network	0.395
Adjacent network (A)	To exploit the vulnerability, an attacker needs access to either the broadcast or collision domain of the vulnerable software	0.646
Network (N)	An attacker could exploit the vulnerability remotely from any part of the network, including the Internet	1.0
Authentication Au_i^q		
Multiple (M)	An attacker needs to perform more than one authentication procedure to exploit a node vulnerability	0.45
Single (S)	It is sufficient for an attacker to authenticate once to exploit a node vulnerability	0.56
None (N)	An attacker does not need authentication to access and exploit the vulnerability	0.704
Access complexity AcV_i^q		
High (H)	There are several strict restrictions on access to a network node. For example, exploitation of a node vulnerability is possible only in a short time or requires the use of social engineering, in which the attacker can be identified	0.35
Medium (M)	There are some restrictions on access to the network node. For example, a connection to a vulnerable node is only possible from specific nodes, or the affected configuration of the network node is non-default	0.61
Low (L)	There are no special conditions for access to the vulnerability of the network node. For example, when the network is available to many users at the same time or when a vulnerable configuration is running on multiple network nodes	0.71

$$R_{i,j} = c_{i,j} \cdot \ln \sum_{U_i^q \in U_i^*} e^{BS_i^q}, \tag{31}$$

where $c_{i,j}$ are the weighting coefficients (weight of compromise) used to assess the risk posed by exploiting vulnerabilities on the ith network node. In fact, the coefficients $c_{i,j}$ quantitatively characterize the compromise probability and the potential damage from the attacker's exploitation of existing vulnerabilities on the ith network node.

Note that in the ase when the link $w_{i,j} \in W$ compromise occurs only due to the vulnerabilities exploitation on the ith node, then the information security risks of the node and communication link are equal, i.e.,

$$\sum_{U_i^q \in U_i^*} BS_i^q \cdot P_i^q = c_{i,j} \cdot \ln \sum_{U_i^q \in U_i^*} e^{BS_i^q} \tag{32}$$

The calculation of weights $c_{i,j}$ is based on the assumption that the communication link $w_{i,j} \in W$ will be compromised due to the compromise of the ith network node, i.e., due to the exploitation of existing vulnerabilities on this node. In this case, the node compromise probability depends on the vulnerabilities' presence and exploitation and is calculated as an information security risk. Based on (27)–(31), the value of each of the weights $c_{i,j}$ (routing metrics) in expression (32) can be calculated using the following expression:

$$c_{i,j} = \frac{\sum_{U_i^q \in U_i^*} BS_i^q \cdot P_i^q}{\ln \sum_{U_i^q \in U_i^*} e^{BS_i^q}} \tag{33}$$

Therefore, based on the proposed routing model (1)–(33), several methods can be offered to solve the problem of secure routing in the infocommunication network depending on the traffic type and selected routing metrics. That is, the methods of secure routing are based on the solution of the optimization problem with the optimality criterion (23). Given the low heterogeneity of ICN and the variety of network equipment, it is advisable to use routing metrics (24) and (26), otherwise—the metric (33). In the absence of overload, or when routing flows that are not sensitive to packet loss, routing variables should be subject to restrictions (3) and (22), which are associated with ensuring the QoS level in terms of performance and average packet delay. In the conditions of probable ICN overload in the process of secure routing of flows sensitive to the packet loss level, it is necessary to use the routing method with QoS conditions (3), (6), and (22) to provide additional quality assurance and reliability – probability packet loss.

5 Research of Secure Routing Processes in Infocommunication Network Under Proposed Models and Methods

A variant of the infocommunication network structure within the research framework was selected, presented in Fig. 2. The ICN contained 12 routers and 17 communication links. Figure 2 in the breaks of communication links shows their bandwidth (packets per second, pps). The source of the packets was the first router, and the destination of the packets was the twelfth router. The operation of each of the router interfaces, as an example, was modeled by the fBM/D/1 queuing system (10).

The network structure (Fig. 2) defined a 17-dimensional space. The matrix that determined the law of coordinate transformation in expression (15) has the form:

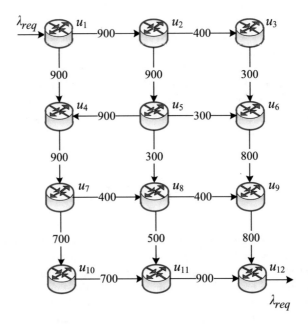

Fig. 2 Example of the network structure

$$A_{\gamma\varepsilon}^{v} = \begin{Vmatrix} 0 & 0 & 0 & 0 & 0 & 0 & 0 & 0 & 0 & 0 & 0 & 0 & 0 & 0 & 0 & 0 & 1 \\ 0 & 0 & 0 & 0 & 0 & 0 & 0 & 0 & 0 & 0 & 0 & 0 & 0 & 0 & 1 & 0 & -1 \\ 1 & 0 & 0 & 0 & 0 & 0 & -1 & 0 & 0 & 0 & 0 & 0 & 0 & 0 & 0 & 1 & 0 \\ 0 & 0 & 0 & 0 & 0 & 0 & 0 & 0 & 0 & 0 & 0 & 0 & 0 & 1 & 0 & 0 & -1 \\ 0 & 1 & 0 & 0 & 0 & 0 & -1 & 0 & 0 & 0 & 0 & 1 & 0 & 0 & -1 & 0 & 0 \\ 0 & 0 & 0 & 0 & 0 & 0 & 0 & 0 & 0 & 0 & 0 & 0 & 0 & -1 & 0 & 1 & 0 \\ 0 & 0 & 0 & 0 & 0 & 0 & 0 & 0 & 0 & 0 & 0 & 1 & 0 & -1 & 0 & 0 & 0 \\ 0 & 0 & 1 & 0 & 0 & 0 & -1 & 0 & 0 & 0 & 0 & 0 & 1 & 0 & 0 & -1 & 0 \\ 0 & 0 & 0 & 0 & 0 & 0 & 0 & 0 & 0 & 0 & 1 & 0 & 0 & -1 & 0 & 0 & 0 \\ 0 & 0 & 0 & 1 & 0 & 0 & -1 & 0 & 1 & 0 & 0 & -1 & 0 & 0 & 0 & 0 & 0 \\ 0 & 0 & 0 & 0 & 0 & 0 & 0 & 0 & 0 & 0 & 1 & 0 & -1 & 0 & 0 & 0 & 0 \\ 0 & 0 & 0 & 0 & 0 & 0 & 0 & 0 & 1 & 0 & -1 & 0 & 0 & 0 & 0 & 0 & 0 \\ 0 & 0 & 0 & 0 & 1 & 0 & -1 & 0 & 0 & 1 & 0 & 0 & -1 & 0 & 0 & 0 & 0 \\ 0 & 0 & 0 & 0 & 0 & 0 & 0 & 1 & 0 & 0 & -1 & 0 & 0 & 0 & 0 & 0 & 0 \\ 0 & 0 & 0 & 0 & 0 & 1 & 0 & 0 & -1 & 0 & 0 & 0 & 0 & 0 & 0 & 0 & 0 \\ 0 & 0 & 0 & 0 & 0 & 0 & 0 & 1 & 0 & -1 & 0 & 0 & 0 & 0 & 0 & 0 & 0 \\ 0 & 0 & 0 & 0 & 0 & 0 & 1 & -1 & 0 & 0 & 0 & 0 & 0 & 0 & 0 & 0 & 0 \end{Vmatrix}.$$

The following data were used as input:

$$\lambda_{req} = 150\,\text{pps}; \quad \tau_{req} = 40\,\text{ms}. \tag{34}$$

As part of the study, the problem of secure routing in ICN, for example, was solved for two probable scenarios of links compromise (Table 4).

According to the first scenario of compromising network communication links (Table 4), the optimal path from the point of view of applying metrics (24) is

Table 4 Compromise probabilities of network links for two scenarios

Scenario	Link compromise probability								
1	$p_{1,2}^{comp}$	$p_{2,3}^{comp}$	$p_{1,4}^{comp}$	$p_{2,5}^{comp}$	$p_{3,6}^{comp}$	$p_{4,5}^{comp}$	$p_{5,6}^{comp}$	$p_{4,7}^{comp}$	$p_{5,8}^{comp}$
	0.1	0.2	0.3	0.15	0.21	0.18	0.16	0.22	0.11
	$p_{6,9}^{comp}$	$p_{7,8}^{comp}$	$p_{8,9}^{comp}$	$p_{7,10}^{comp}$	$p_{8,11}^{comp}$	$p_{9,12}^{comp}$	$p_{10,11}^{comp}$	$p_{11,12}^{comp}$	
	0.23	0.31	0.17	0.25	0.13	0.19	0.21	0.11	
2	$p_{1,2}^{comp}$	$p_{2,3}^{comp}$	$p_{1,4}^{comp}$	$p_{2,5}^{comp}$	$p_{3,6}^{comp}$	$p_{4,5}^{comp}$	$p_{5,6}^{comp}$	$p_{4,7}^{comp}$	$p_{5,8}^{comp}$
	0.2	0.1	0.0	0.25	0.11	0.28	0.01	0.12	0.01
	$p_{6,9}^{comp}$	$p_{7,8}^{comp}$	$p_{8,9}^{comp}$	$p_{7,10}^{comp}$	$p_{8,11}^{comp}$	$p_{9,12}^{comp}$	$p_{10,11}^{comp}$	$p_{11,12}^{comp}$	
	0.3	0.21	0.2	0.11	0.17	0.09	0.27	0.31	

$$u_1 \rightarrow u_2 \rightarrow u_5 \rightarrow u_8 \rightarrow u_{11} \rightarrow u_{12}, \qquad (35)$$

which met the bandwidth requirements and the average end-to-end packet delay (3), (6). Thus, the average packet delay along this path was 34.6 ms, and the probability of its compromise (25) was 0.4728.

According to the second scenario of network links compromise (Table 4), the following path is optimal from the point of view of the metrics (24) use

$$u_1 \rightarrow u_2 \rightarrow u_5 \rightarrow u_8 \rightarrow u_9 \rightarrow u_{12}, \qquad (36)$$

which met the requirements for bandwidth and average packet delay (3), (6). Thus, the average packet delay along this path was 37.4 ms, and the probability of its compromise (25) was 0.5676. For the second compromise scenario (Table 4), the optimally secure path in the first case (35) had a slightly higher than 0.5676 compromise probability—0.6598. That is, the use of the optimality criterion (23) with routing metrics (24) provided the calculation of the optimal in terms of the compromise probability path, which met the QoS requirements (3), (6), (22) that is very important for secure routing of multimedia traffic.

6 Conclusion

It is established that the main requirements for modern infocommunication systems and networks are traditionally ensuring a given level of Quality of Service and network security. In addition, the multiservice providing means that information traffic with different requirements for QoS and NS circulates in the infocommunication network that can be described by the theory of fractals and self-similar processes. Therefore, an important scientific and practical task is developing mathematical models, methods, and routing protocols that would provide the set values of the main QoS indicators of self-similar traffic and focus on improving network security.

Basic flow-based mathematical routing models are selected for research that adequately describe the network process without overload and the probable network elements (links and queue buffer) overload. A tensor analysis apparatus was used to formalize the conditions for ensuring the QoS in terms of bandwidth and average end-to-end delay. The resulting QoS conditions can act as additional restrictions imposed on routing variables.

In order to optimize the processes of secure routing, the optimality criterion and the system of metrics for secure routing in the infocommunication network are proposed, which should be used in different scenarios of network equipment (routers and communication links) compromise. The most straightforward routing metrics can be based on the network routers and links compromise probabilities. More complex secure routing metrics are obtained by assessing the information security risk of network elements.

The study results confirmed the adequacy and efficiency of the proposed models and secure routing methods in calculating the most secure routes, further meeting the QoS level requirements of self-similar packet flows.

References

1. Gupta S (2018) Security and QoS in wireless sensor networks, 1st edn. eBooks2go Inc.
2. Kiser Q (2020) Computer networking and cybersecurity: a guide to understanding communications systems, internet connections, and network security along with protection from hacking and cyber security threats. Kindle Edition
3. Barreiros M, Lundqvist P (2015) QoS-enabled networks: tools and foundations. Wiley, New York
4. Dodd, A.Z.: The Essential Guide to Telecommunications (Essential Guide Series). Pearson. (2019)
5. Sheluhin O, Smolskiy S, Osin A (2007) Self-similar processes in telecommunications. Wiley, New York
6. Park K, Willinger W (eds) Self-similar network traffic and performance evaluation. Wiley, New York
7. Lemeshko OV, Yevseyeva OY, Garkusha SV (2013) A tensor model of multipath routing based on multiple QoS metrics. In: 2013 International Siberian conference on control and communications (SIBCON) proceedings. IEEE, pp 1–4. https://doi.org/10.1109/SIBCON.2013.6693645
8. Yeremenko O, Lemeshko O, Persikov A (2018) Secure routing in reliable networks: proactive and reactive approach. In: Shakhovska N, Stepashko V (eds) CSIT 2017. AISC, vol 689. Springer, Cham, pp 631–655. https://doi.org/10.1007/978-3-319-70581-1_44
9. Lemeshko O, Yeremenko O, Yevdokymenko M, Shapovalova A, Hailan AM, Mersni A (2019) Cyber resilience approach based on traffic engineering fast reroute with policing. In: 2019 10th IEEE international conference on intelligent data acquisition and advanced computing systems: technology and applications (IDAACS) Proceedings, vol 1. IEEE, pp 117–122 (2019). https://doi.org/10.1109/IDAACS.2019.8924294
10. Medhi D, Ramasamy K (2017) Network routing: algorithms, protocols, and architectures. Morgan Kaufmann
11. Rak J, Hutchison D (eds) Guide to disaster-resilient communication networks (computer communications and networks), 1st edn. 2020 Edition. Springer, Berlin
12. Lemeshko O, Papan J, Yeremenko O, Yevdokymenko M, Segec P (2021) Research and development of delay-sensitive routing tensor model in IoT core networks. Sensors 21(11):1–23. https://doi.org/10.3390/s21113934
13. Lemeshko OV, Yeremenko OS, Hailan AM (2016) QoS solution of traffic management based on the dynamic tensor model in the coordinate system of interpolar paths and internal node pairs. In: 2016 International conference radio electronics & info communications (UkrMiCo) proceedings. IEEE, pp 1–6. https://doi.org/10.1109/UkrMiCo.2016.7739625
14. Lemeshko O, Yeremenko O (2016) Routing tensor model presented in the basis of interpolar paths and internal node pairs. In: 2016 third international scientific-practical conference problems of infocommunications science and technology (PIC S&T) Proceedings. IEEE, pp 201–204. https://doi.org/10.1109/INFOCOMMST.2016.7905381
15. Lemeshko O, Yeremenko O, Yevdokymenko M, Radivilova T (2021) Research of the QoE fast ReRouting processes with differentiated R-factor maximization for VoIP-flows using the tensor model of the corporate telecommunication network. In: Vorobiyenko P, Ilchenko M, Strelkovska I (eds) Current trends in communication and information technologies. IPF 2020. Lecture notes in networks and systems, vol 212. Springer, Cham. https://doi.org/10.1007/978-3-030-76343-5_6

16. Lemeshko O, Yeremenko O, Yevdokymenko M, Hailan AM (2021) Tensor multiflow routing model to ensure the guaranteed quality of service based on load balancing in network. In: Hu Z, Petoukhov S, Dychka I, He M (eds) Advances in computer science for engineering and education III. ICCSEEA 2020. Advances in intelligent systems and computing, vol 1247. Springer, Cham. https://doi.org/10.1007/978-3-030-55506-1_11
17. Lee Y, Seok Y, Choi Y, Kim C (2002) A constrained multipath traffic engineering scheme for MPLS networks. In: 2002 IEEE international conference on communications. ICC 2002 (Cat. No.02CH37333) Proceedings, vol 4. IEEE, pp 2431–2436. https://doi.org/10.1109/ICC.2002.997280
18. Norros I (1994) A storage model with self-similar input. Queueing Syst 16:387–396. https://doi.org/10.1007/BF01158964
19. Shah SA, Issac B (2018) Performance comparison of intrusion detection systems and application of machine learning to snort system. Futur Gener Comput Syst 80:157–170. https://doi.org/10.1016/j.future.2017.10.016
20. Kron G (1965) Tensor analysis of networks. Wiley, New York
21. Akimaru H, Kawashima K (2012) Teletraffic: theory and applications. Springer Science & Business Media, Berlin
22. Yevdokymenko M, Yeremenko O, Shapovalova A, Shapoval M, Porokhniak V, Rogovaya N (2021) Investigation of the secure paths set calculation approach based on vulnerability assessment. In: Modern machine learning technologies and data science workshop. Proceedings of 3rd international workshop (MoMLeT&DS 2021), 2917, pp 207–217
23. Scarfone K, Mell P (2007) Guide to intrusion detection and prevention systems (IDPS). NIST special publication 800-94
24. Common Vulnerability Scoring System v3.0: Examples, Forum of Incident Response and Security Teams. https://www.first.org/cvss/examples
25. Abedin M, Nessa S, Al-Shaer E, Khan L (2006) Vulnerability analysis for evaluating quality of protection of security policies. QoP '06. https://doi.org/10.1145/1179494.1179505
26. Peltier TR (2005) Information security risk analysis. CRC Press, Boca Raton

The Methods of Data Comparison in Residue Numeral System

Victor Krasnobayev⑩, Sergey Koshman⑩, Alexandr Kuznetsov⑩, and Tetiana Kuznetsova⑩

Abstract The object of research is information processing in the systems of transmission and processing of digital data, which are built based on a non-positional number system in a residue number system (RNS). The purpose of the research is to develop a methods of data comparison, presented in RNS. To make this goal achieved, the existing methods of data comparison in RNS were analyzed and due to this a method of data comparison was developed, which allows performing the operation of number comparison in positive and negative numerical ranges. The following methods were used: the methods of analysis and synthesis, as well as the methods of numbers theory. It is recommended to use the presented algorithm, which is based on the developed method, in the practical implementation of the operation of comparing two numbers in RNS. The work presents the implementation of examples of the comparison operation based on the usage of the developed methods.

Keywords Residue numeral system · Method and algorithm of comparison integers · Artificial form of representing numbers · Single-row code

1 Introduction

One of the main advantages of the residue number system (RNS), comparing with the positional number system, is the possibility of organizing the process of the quick realization of the purpose of arithmetic addition, subtraction, and multiplication [1–7]. We achieve this by using such properties of RNS such as low-bit and independence of residues by effectively applying the tabular arithmetic.

V. Krasnobayev (✉) · S. Koshman · A. Kuznetsov · T. Kuznetsova
V. N. Karazin, Kharkiv National University, 4 Svobody Sq., Kharkiv 61022, Ukraine

© The Author(s), under exclusive license to Springer Nature Switzerland AG 2022
R. Oliynykov et al. (eds.), *Information Security Technologies in the Decentralized Distributed Networks*, Lecture Notes on Data Engineering and Communications Technologies 115, https://doi.org/10.1007/978-3-030-95161-0_11

At the same time, the need to perform comparisons of two numbers in positive and negative numerical ranges, by solving the tasks and algorithms for various purposes, reduces the overall efficiency of RNS [8–12]. This is due to the considerable time of the operation of comparing two numbers in RNS (in comparison with the execution time of the above-mentioned arithmetic operations) [1, 3–5]. Therefore, the research and development of mathematic models, methods, and algorithms of number comparison in RNS is an important and urgent task.

Nowadays, there are three groups of methods of comparing the numbers in RNS [13–20]. The first one includes methods of direct comparison, based on the conversion of numbers A_{RNS} and B_{RNS} from the code of RNS in the positional system code (PSC) $A_{PSC} = \overline{\alpha_1, \alpha_2 \ldots \alpha_\rho}$ and $B_{PSC} = \overline{\beta_1, \beta_2, \ldots \beta_\rho}$ (ρ- digit of numbers A_{PSC} and B_{PSC}) and their further comparison. The second group is based on the principle of zeroing. The process of zeroing is the transition of the original number represented RNS $A_{RNS} = (a_1, a_2, \ldots, a_{i-1}, a_i, a_{i+1}, \ldots a_n)$ to the number of type $A_{RNS}^{(N)} = (0, 0, \ldots, 0, \gamma_n^{(A)})$. Then, by the value $\gamma_n^{(A)}$ the interval $[jm_i, (j+1)m_i)$ of inclusion of the number A_{RNS} is determined. The process of zeroing the number $B_{RNS} = (b_1, b_2, \ldots, b_{i-1}, b_i, b_{i+1}, \ldots, b_n)$ is carried similarly, where the values are obtained from $\gamma_n^{(B)}$. Positional comparison of obtained values $\gamma_n^{(A)}$ and $\gamma_n^{(B)}$ determines the result of comparing the numbers A_{RNS} and B_{RNS}. To the third group of methods, we assign the methods which are based on the formation of special features, the so-called positional attributes of the non-positional code (PANC). Through the formation of PANC single-row code (SRC) is created, with the hip of which you are able to compare numbers.

2   The Main Part

Let the SRC be set by ordered ($m_i < m_{i+1}$). Let the SRC be set by ordered ($m_i < m_i + 1$) mutually pairwise prime natural numbers (bases) $m_1, m_2, \ldots, m_n$, and let the compared operands be represented as:

$$A = (a_1, a_2, \ldots, a_n), B = (b_1, b_2, \ldots, b_n)$$

It is assumed that the source operands lie in the corresponding intervals:

$$\left[\frac{j_1 M}{m_n}, \frac{(j_1 + 1)M}{m_n}\right) \text{ and } \left[\frac{j_2 M}{m_n}, \frac{(j_2 + 1)M}{m_n}\right)$$

where $M = \prod_{i=1}^{n} m_1$, and the number $j_k + 1$ of the interval is determined by the well-known expression $j_k = \gamma_n \overline{m}_n (\mathrm{mod}\, m_n)$, where the value of m_n is determined from the solution of the comparison $\overline{m}_n M / m_n \equiv l (\mathrm{mod}\, m_n)$. For $j_1 \neq j_2$ the operation of arithmetic comparison can be implemented by comparing the number of intervals, namely: if $j_1 < j_2$, then $A < B$, if $j_1 > j_2$, then $A > B$. When $j_1 = j_2$ the number $j_2 +$

1 of interval $\left[\frac{j_3 M}{m_n}, \frac{(j_3+1)M}{m_n}\right)$ is determined, in which the number $A-B$ is located. If $0 \le j_3 < (m_n + 1)/2$, then $A < B$, and if $\frac{m_n+1}{2} \le j_3 < m$, then $A > B$.

The method of arithmetic comparison of numbers in the SRC involves the conversion of numbers to the form $A^{(H)} = (0, 0, ..., \gamma_n)$, which requires $n-1$ cycles of the operation of zerofication. In addition, it is necessary to make a positional comparison of numbers of the $(j_1 + 1)$ и $(j_2 + 1)$ intervals of the source operands A and B. Il this complicates the comparison algorithm and increases the time of comparison of numbers, which leads to the need to develop effective methods of comparison of operands in the SRC, not requiring determination of positional characteristics.

Consider arithmetic parallel subtraction method. If we consider the compared operands in arbitrary intervals:

$$[jm_i, (j+1)m_i),$$

where

$$j = 1, N-1 \left(N = \prod_{\substack{k=l \\ k \neq i}}^{n} m_k\right)$$

The source operands A, B are reduced to multiples to m_i by modular subtraction of the following form:

$$A_{m_i} = A - a_i = (a_1^{(i)}, a_2^{(i)}, \ldots, a_{i-1}^{(i)}, 0, a_{i+1}^{(i)}, \ldots, a_n^{(i)},$$
$$B_{m_i} = B - b_i = (b_1^{(i)}, b_2^{(i)}, \ldots, b_{i-1}^{(i)}, 0, b_{i+1}^{(i)}, \ldots, b_n^{(i)},$$

where

$$a_i = (a_1', a_2', \ldots, a_i, a_n') \quad b_i = (b_1', b_2', \ldots, b_i, b_n').$$

Further, by means of a set of constants $0, m_i, 2m_i, \ldots, (N-1)m_i$, represented on the $(n-1)$th basis of the SRC $m_1, m_2, \ldots, m_{i-1}, m_{i+1}, \ldots, m_n$, a SRC is constructed, respectively

$$K_N^{(n_A)} = \{z_N z_{N-1} \ldots z_2 z_1\},$$
$$z_{n_A} = 0 \left(z_1 = 1; 1 = \overline{1, N}, 1 \neq n_A\right),$$
$$K_N^{(n_B)} = \{z_N' z_{N-1}' \ldots z_2' z_1'\},$$
$$z_{n_B}' = 0 \left(z_1' = 1, 1 = \overline{1, N}, 1 \neq n_B\right).$$

Algorithm for constructing a SRC in the SRC can be represented as follows:

$$
\begin{cases}
A_{m_i} - 0 = z_1, \\
A_{m_i} - m_i = z_2, \\
A_{m_i} - 2m_i = z_3, \\
\cdots \\
A_{m_i} - \left(\prod_{\substack{\rho=1 \\ \rho \neq 1}}^{n-1} m_\rho - 1 \right) m_i = z_N
\end{cases}
\qquad
\begin{cases}
B_{m_i} - 0 = z'_1, \\
B_{m_i} - m_i = z'_2, \\
B_{m_i} - 2m_i = z'_3, \\
\cdots \\
B_{m_i} - \left(\prod_{\substack{\rho=1 \\ \rho \neq 1}}^{n-1} m_\rho - 1 \right) m_i = z_N
\end{cases}
\tag{1}
$$

In this case, we have:

$$z_{n_A} = 0, \text{ while } A_{n_2} - n_A \cdot m_i = 0;$$

$$z_{n_A} = 1, \text{ while } A_{m_t} \neq n_A \cdot m_i;$$

$$z'_{n_B} = 0, \text{ while } B_{m_i} - n_B \cdot m_i = 0;$$

$$z'_{n_s} = 1, \text{ while } B_{m_z} \neq n_B \cdot m_i.$$

Geometrically, this comparison method can be explained as follows. The interval $[0, \prod_{i=1}^n m_i)$ is divided into segments. The source operands A and B, by subtracting the type $\prod_{\substack{\rho=1 \\ \rho \neq i}}^{n-1} m_\rho$ constants, to the left edge of their hit interval. This is equivalent to converting the compared operands to numbers multiple to the base m_i of the SRC. On this basis, the accuracy W of the comparison of the operands depends on the value of the base m_i, i.e. $W = W(m_i)$. However, with maximum accuracy of comparison $W_{max} = W(m_{min})$ (for an ordered SRC $W_{max} = W(m_1)$) the number of equipment for technical devices that implement the comparison operation in the SRC is dramatically increased. Indeed, the number of equipment N_0 of the comparison device in the SRC is significantly dependent on the number of adders N_1, performing the parallel subtraction operation.

When $W_{max} = W(m_1)$ we have $N_1 = N_{max} = \prod_{i=2}^n m_i$, and when $W_{min} = W(m_n)$,—$N_1 = N_{min} = \prod_{i=1}^{n-1} m_i$.

Thus, the need to ensure a high degree of accuracy of comparison requires a significant amount of equipment, which reduces the effectiveness of the use of comparison devices in the SRC. This circumstance determines the relevance and importance of finding more effective methods of comparing numbers in the SRC, providing high accuracy W comparison with an acceptable amount of N_0 equipment of comparing devices.

In the general form, the task is formulated as follows. It is necessary to find $N_0 = \min$ at W_{max}, i.e. $N_0(W_{max}) = \min$. As shown above $W_{max} = W(m_1)$. The change in the size of the base m_i $(i = \overline{1,n})$ affects only the number N_1 of the equipment of the group of adders.

In this case, it is legitimate to formulate the task in the form of a definition

$$N_1(W_{max}) = \min \tag{2}$$

Obviously, with maximum accuracy equal to the unit length of the interval, $W_{max} = W(m_i = 2)$. However, in this case $N_1(W_{max}) = \max$ and this result does not satisfy the condition (2). On the other hand—$N_1(W_{min}) = \min$. Thus, it is necessary to develop a method of comparison that would satisfy the condition (2) to the maximum extent.

Parallel subtraction method with residual comparison. We introduce the operation of comparing directly the residues a_n, b_n, on the base of m_n. In this case, the result of this comparison, simultaneously with the result of the comparison of a SRC $K_N^{(n_A)}$ and $K_N^{(n_B)}$, will be determined with maximum precision W_{max} and with a minimum amount of equipment N_{min}. The algorithm for determining the result of an arithmetic comparison operation can be represented as follows:

$$\begin{cases} \text{if } n_A > n_B, \text{ then } A > B; \\ \text{if } n_A < n_B, \text{ then } A < B; \\ \text{if } n_A = n_B, \text{ and then} \\ \quad \begin{cases} a_n = b_n, \text{ TO } A = B, \\ a_n > b_n, \text{ TO } A > B, \\ a_n < b_n, \text{ To } A < B. \end{cases} \end{cases} \tag{3}$$

The set of relations (3) represents the general algorithm for the implementation of the operation of arithmetic comparison of operands in the SRC. It is advisable to consider examples of specific performance of the operation of arithmetic comparison of numbers in the SRC. Let the SRC be given by bases $m_1 = 2$, $m_2 = 3$ and $m_3 = 5$. Code words are given in Table 1. In Table 2, the constants a_n (b_n) are defined in the given SRC, and in Table 3 by $n - 1$th basis of the SRC m_i $(i = \overline{1, n-1})$ the constants of a SRC $\Delta \cdot m_n$ $\left(\Delta = \overline{0, \prod_{i=1}^{n-1} m_i}\right)$ are given.

Let the compared operands be represented as $A_{23} = (1, 10, 011)$ and $B_{21} = (1, 00, 001)$. In this case, the values of the constants (Table 2) determine the values $A_{m_n} = A_{23} - a_n = (0, 10, 000)$, $B_{m_n} = B_{21} - b_n = (01, 00, 000)$, which corresponds to the shift of the operands A and B to the left edge of the interval $[20, 25)$. Next, using the constants of the SRC (Table 3), we determine the SRC for the input operands in the form:

$$K_N^{(n_A)} = K_6^{(4)} = \{110111\}; \quad K_N^{(n_B)} = K_6^{(n_B)} = K_6^{(4)} = \{110111\}$$

where

Table. 1 Table of code words

A	A in SRC			A	A in SRC		
	$m_1 = 2$	$m_2 = 3$	$m_3 = 5$		$m_1 = 2$	$m_2 = 3$	$m_3 = 5$
0	0	00	000	15	1	00	000
1	1	01	001	16	0	01	001
2	0	10	010	17	1	10	010
3	1	00	011	18	0	00	011
4	0	01	100	19	1	01	100
5	1	10	001	20	0	10	000
6	0	00	001	21	1	00	001
7	1	01	010	22	0	01	010
8	0	10	011	23	1	10	011
9	1	00	100	24	0	00	100
10	0	01	000	25	1	01	000
11	1	10	001	26	0	10	001
12	0	00	010	27	1	00	010
13	1	01	011	28	0	01	011
14	0	10	100	29	1	10	100

Table 2 Table of constants for γ_3

γ_3	Constants
000	$(0, 00, 000)$
001	$(1, 01, 001)$
010	$(0, 10, 010)$
011	$(1, 00, 011)$
100	$(0, 01, 100)$

Table 3 Table of constants for $m_n = m_3$

$(0 \div N - 1) \cdot m_n$	SRC		Position number of zero
	$m_1 = 2$	$m_2 = 3$	
$0 \cdot m_3 = 0$	0	00	1
$1 \cdot m_3 = 5$	1	10	2
$2 \cdot m_3 = 10$	0	01	3
$3 \cdot m_3 = 15$	1	00	4
$4 \cdot m_3 = 20$	0	10	5
$5 \cdot m_3 = 25$	1	01	6

$$N_1 = \prod_{i=1}^{n-1} m_i = 6$$

At the same time, the comparison result $a_n = 011 > b_n = 001$ is determined in parallel with the $(n = [\log_2(m_n - 1)] + 1)$-h bit comparison circuit in time. Since

$n_A = n_B = 4$, then, in accordance with the above algorithm, we determine that $A_{23} > B_{21}$.

Let the compared operands be represented as $A_{23} = (1, 10, 011)$, $B_3 = (1, 00, 011)$. In this case, the following differences are determined from the values of the constants (Table 2):

$$A_{m_n} = A_{23} - a_n = (0, 10, 00); B_{m_n} = B_3 - b_n = (1, 00, 000),$$

which corresponds to the shift of the operand A_{23} o the left edge of the interval $[20, 25)$, and the operand B_3 to the left edge of the interval $[0, 5)$. Next, using the constants of the SRC (Table 3), we determine the SRC for the considered input operands A_{23}, and B_3:

$$K_N^{(n_A)} = K_6^{(5)} = \{101111\}; K_N^{(n_B)} = K_6^{(1)} = \{111110\}$$

Since $n_A = 5 > n_B = 1$, then, in accordance with the above algorithm, we determine that $A_{23} > B_3$.

Method and algorithm for comparing with constant. Consider an effective method and algorithm for implementing the operation of arithmetic comparison of operands in the SRC. The essence of this algorithm is that not the operands A and B, are directly compared, but the values of $\chi = (A - B) \bmod M = (\gamma_1, \gamma_2, ..., \gamma_n)$ and m_1.

In this case, the value is determined:

$$\chi_{m_1} = \chi - \gamma_1 = (0, \gamma_2', \gamma_3', ..., \gamma_n'),$$

where constants $\gamma_1 = (\gamma_1, \gamma_2', \gamma_3', ..., \gamma_n')$ and $\gamma_1 = (a_1 - b_1) \bmod m_1$ presented in a given SRC.

Then the general algorithm for comparison of the operands is presented in the form.

$$\begin{cases} A > B, & \text{while } \chi_{m_1} \leq m_1; \\ A < B, & \text{while } \chi_{m_1} > m_1; \\ A = B, & \text{while } \chi_{m_1} = 0. \end{cases} \quad (4)$$

By means of a set of constants $0, m_1, 2m_1, ..., (N - 1) \cdot m_1 \left(N = \prod_{i=1}^{n} m_i\right)$, represented in the SRC with bases $m_2, m_3, ..., m_n$, the construction of a SRC in the following form is carried out:

$$\begin{cases} A > B, & \text{while } \chi_{m_1} \leq m_1; \\ A < B, & \text{while } \chi_{m_1} > m_1; \\ A = B, & \text{while } \chi_{m_1} = 0. \end{cases} \quad (5)$$

$$K_n^{(n_\chi)} = \{z_N z_{N-1} \ldots z_2 z_1\}$$

,

where

$$\begin{cases} \chi_{m_1} - 0 = z_1, \\ \chi_{m_1} - m_1 = z_2, \\ \chi_{m_1} - 2m_1 = z_3, \\ \cdots \cdots \cdots \\ \chi_{m_1} - (N-1)m_1 = z_n. \end{cases}$$

while

$$z_{n_z} = 0, \ \text{for} \chi_{m_1} - n_z m_1 = 0,$$

$$z_{n_2} = 1, \ \text{for} \ \chi_{m_1} - n_\chi m_1 \neq 0.$$

The first module m_1 COK of SRC is also represented by a SRC of length N binary digits, in which the second place on the right will be zero ($m_1 - 1 \cdot m_1 = 0$), and on the rest—ones, i.e. single-row unitary code will be presented in the form:

$$K_n^{(2)} = \{11 \ldots 101\}$$

Further, by known methods, in accordance with algorithm (4), operands A and B, represented by a SRC, are compared.

For the above SSRC, we consider an example implementation of this method. Let $A = (0, 01, 010)$ and $B = (1, 10, 011)$. We define the value:

$$\chi = (A - B) \bmod M = [(0 - 1) \bmod 2, \ (01 - 10) \bmod 3, \ (010 - 011) \bmod 5]$$
$$= (1, 10, 100).$$

By value $\gamma_1 = 1$ we define a constant in the form $\gamma_1 = (1, 01, 001)$ (Table 4). Next we perform the operation:

$$\chi_{m_1} = \chi - \gamma = (00, 01, 011)$$

Table 4 Constants for γ_1

γ_1	Constants
00	$(0, 00, 000)$
01	$(1, 01, 001)$

Table 5 Table of constants for $m_n = m_1$

$(0 \div N - 1) \cdot m_1$ $\Delta \cdot 2$	SRC			Position number of zero
	$m_2 = 3$	$m_3 = 5$		
$0\ m_1 = 0$	00	000		1
$1\ m_1 = 2$	10	010		2
$2\ m_1 = 4$	01	100		3
$3\ m_1 = 6$	00	001		4
$4\ m_1 = 8$	10	011		5
$5\ m_1 = 10$	01	000		6
$6\ m_1 = 12$	00	010		7
$7\ m_1 = 14$	10	100		8
$8\ m_1 = 16$	01	001		9
$9\ m_1 = 18$	00	011		10
$10\ m_1 = 20$	10	000		11
$11\ m_1 = 22$	01	010		12
$12\ m_1 = 24$	00	100		13
$13\ m_1 = 26$	10	001		14
$14\ m_1 = 28$	01	011		15

The operand χ_{m_1}, which is a multiple of the module value $m_1 = 2$, goes to the first inputs of the corresponding adders, the second inputs of which receive the corresponding constants (Table 5).

Since $\chi_{m_1} - 14_{m_i} = 0.$, then the SRC will look like:

$$K_n^{(n_\chi)} = K_{15}^{(14)} = \{011111111111111\}.$$

In accordance with algorithm (4), we determine that $A < B$.

The advantage of the considered algorithm is to ensure maximum accuracy of comparison with an acceptable amount of equipment for its implementation.

Consider the implementation of the algebraic comparison operation of two numbers represented in RNS. In this case, we describe two fundamentally possible options of organizing the algebraic comparison procedure: introducing the sign in an explicit and implicit (using the artificial form (AF) representation of the compared numbers) forms.

Consider the first option. In this case, the compared operands $A_{RNS} = (a_1, a_2, ..., a_{i-1}, a_i, a_{i+1}, ..., a_n)$ and $B_{RNS} = (b_1, b_2, ..., b_{i-1}, b_i, b_{i+1}, ..., b_n)$ additionally have two significant digits $\Omega_{+A}(\Omega_{+B})$ and $\Omega_{-A}(\Omega_{-B})$, where

$$\Omega_{+A}(\Omega_{+B}) = \begin{cases} 1, & if\ A_{RNS}(B_{RNS}) > 0, \\ 0, & if\ A_{RNS}(B_{RNS}) < 0; \end{cases} \quad and \quad \Omega_{-A}(\Omega_{-B}) = \begin{cases} 0, & if\ A_{RNS}(B_{RNS}) > 0, \\ 1, & if\ A_{RNS}(B_{RNS}) < 0. \end{cases}$$

$$(6)$$

Thus, the compared operands are represented as following

$$A_{RNS}^{(*)} = \{\Omega_{+A}, \Omega_{-A}; A_{cox}\} = \{\Omega_{+A}, \Omega_{-A};$$
$$(a_1, a_2, \ldots, a_{i-1}, a_i, a_{i+1}, \ldots, a_n)\};$$
$$B_{RNS}^{(*)} = \{\Omega_{+B}, \Omega_{-B}; B_{RNS}\} = \{\Omega_{+B}, \Omega_{-B};$$
$$(b_1, b_2, \ldots, b_{i-1}, b_i, b_{i+1}, \cdots, b_n)\},$$
(7)

where $\Omega_{+A}(\Omega_{+B})$—positive and Ω_{-A}, Ω_{-B}—negative signs of algebraic numbers $A_{RNS}^{(*)}$ and $B_{RNS}^{(*)}$ in RNS respectively.

Equality $A_{RNS}^{(*)} = B_{RNS}^{(*)}$ of two algebraic numbers $A_{RNS}^{(*)}$ and $B_{RNS}^{(*)}$ is determined by a logical equation. In this case, $A_{RNS}^{(*)} = B_{RNS}^{(*)}$, if

$$[(A_{RNS} = B_{RNS}) \wedge (\Omega_{+A} \wedge \Omega_{+B})] \vee$$
$$\vee [(A_{RNS} = B_{RNS}) \wedge (\Omega_{-A} \wedge \Omega_{-B})]$$
(8)

Taking into account the above-mentioned relationship to determine the conditions of the equality $A_{RNS} = B_{RNS}$, the expressions (7) are defined in the following way

$$\{[(n_A = n_B) \wedge (a_n = b_n)] \wedge (\Omega_{+A} \wedge \Omega_{+B})\} \vee$$
$$\vee \{[(n_A = n_B) \wedge (a_n = b_n)] \wedge (\Omega_{-A} \wedge \Omega_{-B})\}.$$
(9)

The inequality $A_{RNS}^{(*)} > B_{RNS}^{(*)}$ is based on the following logical conditions

$$[(A_{RNS} = B_{RNS}) \wedge (\Omega_{+A} \wedge \Omega_{-B})] \vee [(A_{RNS} > B_{RNS}) \wedge$$
$$\wedge (\Omega_{+A} \wedge \Omega_{+B})] \vee [(A_{RNS} > B_{RNS}) \wedge$$
$$\wedge (\Omega_{+A} \wedge \Omega_{-B})] \vee [(A_{RNS} < B_{RNS}) \wedge (\Omega_{+A} \wedge \Omega_{-B})] \vee$$
$$\vee [(A_{RNS} < B_{RNS}) \wedge (\Omega_{-A} \wedge \Omega_{-B})].$$
(10)

Considering the above-mentioned relations, the analytical expressions (9) can be represent as following

$$\{[(n_A = n_B) \wedge (a_n = b_n)] \wedge (\Omega_{+A} \wedge \Omega_{-B})\} \vee$$
$$\vee \{(n_A > n_B) \vee [(n_A = n_B) \wedge (a_n > b_n)]\} \wedge$$
$$\wedge \{(\Omega_{+A} \wedge \Omega_{+B})\} \vee (n_A > n_B) \vee$$
$$\vee [(n_A = n_B) \wedge (a_n > b_n)]\} \wedge \{(\Omega_{+A} \wedge \Omega_{-B})\} \vee$$
$$d \vee \{(n_A < n_B) \vee [(n_A = n_B) \wedge (a_n < b_n)]\} \wedge$$
$$\vee \{(\Omega_{+A} \wedge \Omega_{-B})\} \vee \{(n_A < n_B) \vee$$
$$\vee [(n_A = n_B) \wedge (a_n < b_n)]\} \wedge \{(\Omega_{-A} \wedge \Omega_{-B})\}.$$
(11)

The inequality $A_{RNS}^{(*)} < B_{RNS}^{(*)}$ is true if the following logical conditions are true

$$\left[(A_{RNS} = B_{RNS}) \wedge (\Omega_{-A} \wedge \Omega_{+B})\right] \vee \left[(A_{cox} > B_{cox}) \wedge \right.$$
$$\wedge (\Omega_{-A} \wedge \Omega_{+B})\right] \vee \left[(A_{RVS} > B_{RNS}) \wedge \right.$$
$$\wedge (\Omega_{-A} \wedge \Omega_{-B})\right] \vee \left[(A_{RNS} < B_{RVS}) \wedge (\Omega_{+A} \wedge \Omega_{+B})\right] \vee$$
$$\vee \left[(A_{RNS} < B_{RNS}) \wedge (\Omega_{-A} \wedge \Omega_{+B})\right] \tag{12}$$

Considering the expressions above, the collective equality (11) of logical equalities can be written as

$$\{[(n_A = n_B) \wedge (a_n = b_n)] \wedge (\Omega_{-A} \wedge \Omega_{+B})\} \vee$$
$$\vee \{(n_A > n_B) \vee [(n_A = n_B) \wedge (a_n > b_n)]\} \wedge$$
$$\wedge \{(\Omega_{-A} \wedge \Omega_{+B})\} \vee \{(n_A > n_B) \vee$$
$$\vee [(n_A = n_B) \wedge (a_n > b_n)]\} \wedge \{(\Omega_{-A} \wedge \Omega_{-B})\} \vee$$
$$\vee \{(n_A < n_B) \vee [(n_A = n_B) \wedge$$
$$\wedge (a_n < b_n)]\} \wedge \{(\Omega_{+A} \wedge \Omega_{+B})\} \vee$$
$$\vee \{(n_A < n_B) \vee [(n_A = n_B) \wedge$$
$$\wedge (a_n < b_n)]\} \wedge \{(\Omega_{-A} \wedge \Omega_{+B})\}. \tag{13}$$

The set of mathematical relations $(7) \div (12)$, without taking into account the values of a_n, b_n underlies the existing method of the algebraic comparison of numbers in RNS, which is presented below.

- Representation of the compared numbers in RNS
  $$A_{RNS}^{(*)} = \{\Omega_{+A}, \Omega_{-A} A_{x\varepsilon}\}$$
  $$= \{\Omega_{+A}, \Omega_{-A}; (a_1, a_2, \ldots, a_{i-1}, a_i, a_{i+1}, \ldots, a_n)\}$$
  and
  $$B_{RNS}^{(*)} = \{\Omega_{+B}, \Omega_{-B}; B_{ke}\}$$
  $$= \{\Omega_{+B}, \Omega_{-B}; (b_1, b_2, \ldots, b_{i-1}, \ldots, b_i, b_{i+1}, \ldots, b_n)\}$$
- Selection of the constant of the process of zeroing by the values a_n and b_n of the numbers A and B $KN_{m_n}^{(A_{RNS})} = (a'_1, a'_2, \ldots, a'_{i-1}, a'_i, a'_{i+1}, \ldots, a_n)$, and $KN_{m_n}^{(B_{RNS})} = (b'_1, b'_2, \ldots, b'_{i-1}, b'_i, b'_{i+1}, \ldots, b_n)$.
- Determination of the difference between numbers $A_{m_n} = A_{RNS} - KN_{m_n}^{(A)}$
  $$KN_{m_n}^{(A)} = (a_1, a_2, \ldots, a_{i-1}, a_i, a_{i+1}, \ldots, a_n) - (a'_1, a'_2, \ldots, a'_{i-1}, a'_i, a'_{i+1}, \ldots, a_n)$$
  and
  $$= (a_1^{(1)}, a_2^{(1)}, \ldots, a_{i-1}^{(1)}, a_i^{(1)}, a_{i+1}^{(1)}, \ldots, 0)$$
  $$B_{m_n} = B_{RVS} - KN_{m_*}^{(B)} = (b_1, b_2, \ldots, b_{i-1}, b_i, b_{i+1}, \ldots, b_n)$$
  $$- (b'_1, b'_2, \ldots, b'_{i-1}, b'_i, b'_{i+1}, \ldots, b_n)$$
  $$= (b_1^{(1)}, b_2^{(1)}, \ldots, b_{i-1}^{(1)}, b_i^{(1)}, b_{i+1}^{(1)}, \ldots, 0)$$

- Component definition $z_i^{(A)}$ и $z_j^{(B)}$ $(0, m_n, \ldots, (N-1) \cdot m_n)$ $A_{m_n} - K_A \cdot m_n = Z_{K_A}^{(A)}$ and $B_{m_n} - K_B \cdot m_n = Z_{K_B}^{(B)}$; $K_{N_{m_n}}^{(n_A)} = \{Z_{N_{m_n}-1}^{(A)} \ Z_{N_{m_n}-2}^{(A)} \ \ldots Z_2^{(A)} Z_1^{(A)} \ Z_0^{(A)}\}$, and $K_{N_{m_n}}^{(n_B)} = \left\{ Z_{N_{m_n}-1}^{(B)} \ Z_{N_{m_n}-2}^{(B)} \ \ldots Z_2^{(B)} \ Z_1^{(B)} \ Z_0^{(B)} \right\}$.

- The formation of PANK $K_{N_{m_n}}^{(n_A)} = \left\{ Z_{N_{m_n}-1}^{(A)} \ Z_{N_{m_n}-2}^{(A)} \ \ldots Z_2^{(A)} Z_1^{(A)} \ Z_0^{(A)} \right\}$, $K_{N_{m_n}}^{(n_B)} = \left\{ Z_{N_{m_n}-1}^{(B)} \ Z_{N_{m_n}-2}^{(B)} \ \ldots Z_2^{(B)} Z_1^{(B)} \ Z_0^{(B)} \right\}$

- Algorithm of implementing the algebraic comparison operation:

$$A_{RNS}^{(*)} = B_{RNS}^{(*)}, \ if \ \{(n_A = n_B) \wedge (\Omega_{+A} \wedge \Omega_{+B})\} \vee \{(n_A = n_B) \wedge (\Omega_{-A} \wedge \Omega_{-B})\}$$

$$A_{RNS}^{(*)} > B_{RNS}^{(*)}, \ if \ \{(n_A = n_B) \wedge (\Omega_{+A} \wedge \Omega_{-B})\} \vee \{(n_A > n_B) \wedge (\Omega_{+A} \wedge \Omega_{+B})\} \vee$$
$$\vee \{(n_A > n_B) \wedge (\Omega_{+A} \wedge \Omega_{-B})\} \vee \{(n_A < n_B)\} \wedge \{(\Omega_{+A} \wedge \Omega_{-B})\} \vee \{(n_A < n_B)\} \wedge$$
$$\wedge \{(\Omega_{-A} \wedge \Omega_{-B})\}.$$

$$A_{RNS}^{(*)} < B_{RNS}^{(*)}, \ if \ \{(n_A = n_B) \wedge (\Omega_{-A} \wedge \Omega_{+B})\} \vee \{(n_A > n_B) \vee (\Omega_{-A} \wedge \Omega_{+B})\} \vee$$
$$\vee \{(n_A > n_B) \wedge (\Omega_{-A} \wedge \Omega_{-B})\} \vee \{(n_A < n_B) \wedge (\Omega_{+A} \wedge \Omega_{+B})\} \vee \{(n_A < n_B) \wedge$$
$$\wedge (\Omega_{-A} \wedge \Omega_{+B})\}.$$

Based on the described method, we will develop an algorithm for algebraic comparison of numbers in RNS (Table 6).

Table. 6 Algebraic number comparison algorithm $A_{RNS}^{(*)}$ and $B_{RNS}^{(*)}$

№ i.o	Result comparisons numbers $A_{RNS}^{(*)}$ and $B_{RNS}^{(*)}$	The condition for performing comparison operations
1	$A_{RNS}^{(*)} = B_{RNS}^{(*)}$	$\{(n_A = n_B) \wedge (\Omega_{+A} \wedge \Omega_{+B})\} \vee \{(n_A = n_B) \wedge (\Omega_{-A} \wedge \Omega_{-B})\}.$
2	$A_{RNS}^{(*)} > B_{RNS}^{(*)}$	$\{(n_A = n_B) \wedge (\Omega_{+A} \wedge \Omega_{-B})\} \vee \{(n_A > n_B) \wedge (\Omega_{+A} \wedge \Omega_{+B})\} \vee \{(n_A > n_B) \wedge (\Omega_{+A} \wedge \Omega_{-B})\} \vee \vee \{(n_A < n_B)\} \wedge \{(\Omega_{+A} \wedge \Omega_{-B})\} \vee \vee \{(n_A < n_B)\} \wedge \{(\Omega_{-A} \wedge \Omega_{-B})\}.$
3	$A_{RNS}^{(*)} < B_{RNS}^{(*)}$	$\{(n_A = n_B) \wedge (\Omega_{-A} \wedge \Omega_{+B})\} \vee \{(n_A > n_B) \vee \vee (\Omega_{-A} \wedge \Omega_{+B})\} \vee \{(n_A > n_B) \wedge (\Omega_{-A} \wedge \Omega_{-B})\} \vee \vee \{(n_A < n_B) \wedge (\Omega_{+A} \wedge \Omega_{+B})\} \vee \{(n_A < n_B) \wedge \wedge (\Omega_{-A} \wedge \Omega_{+B})\}.$

It is recommended to use the algebraic comparison algorithm in the practical realization of the comparing operation of two numbers in RNS [4–7].

Consider an example of the algebraic comparison of two numbers $A_{RNS}^{(*)}$ and $B_{RNS}^{(*)}$.

Example 1 Compare two numbers $A_{RNS}^{(*)} = 21$ and $B_{RNS}^{(*)} = -24$. Considering the expressions (5), (6) we represent the comparing operands as $A_{21}^{(*)} = \{1, 0; \ (1, 00, 001)\}$ and $B_{-24}^{(*)} = \{0, 1; \ (0, 00, 100)\}$. According to the values of residues $a_n = a_3 = 001$ and $b_n = b_3 = 100$ we will choose zeroing constants that are represented in the following way $KN_{m_n}^{(A)} = (1, \ 00, \ 001)$ and $KN_{m_k}^{(B)} = (1, 00, 001)$. Next, we define the numbers $A_{m_1} = A_{RNS} - KN_{m_t}^{(A)} = (1, 10, 001) - (1, 00, 001) = (0, 10, 000)$ and $B_{m_n} = B_{RNS} - KN_{m_n}^{(B)} = (1, 00, 001) - (1, 01, 001) = (0, 10, 000)$, that corresponds to the shift of the operands $A_{21} = \left|A_{21}^{(*)}\right|$ and $B_{24} = \left|B_{24}^{(*)}\right|$ on the left edge of the interval $[20, 25)$.

Next, SRC for the input operands is formed $\left|A_{21}^{(*)}\right|$ and $\left|B_{24}^{(*)}\right|$ as $K_{N_{m_n}}^{(n_A)} = K_6^{(4)} = \{101111\}$ $K_{N_{m_n}}^{(n_B)} = K_6^{(4)} = \{101111\}$. At the same time, through the ($n = [\log_2(m_n - 1) + 1]$) bit comparison circuit, the result of the residual comparison is determined in parallel in time $a_n = 001 < b_n = 100$. Because of $n_A = n_B = 4$, $\Omega_{+A} = 1$, and $\Omega_{-B} = 1$, then in accordance with the algorithm of the algorithmic comparison (item 2, Table 1) we determine that $A_{21}^{(*)} > B_{-24}^{(*)}$. Check: $21 > -24$.

Consider the second variant of the method of the algebraic comparison of numbers in RNS. That is the representation of the comparing numbers $A_{RSN} = (a_1, a_2, ..., a_n)$ and $B_{RSN} = (b_1, b_2, ..., b_n)$ in AF, i.e. $A'_{RSN} = (a'_1, a'_2, ..., a'_n)$ and $B'_{RSN} = (b'_1, b'_2, ..., b'_n)$. In this case, the following algorithm compares two numbers $A'_{RSN} = (a'_1, a'_2, ..., a'_n)$ and $B'_{RSN} = (b'_1, b'_2, ..., b'_n)$ in RNS

$$
\begin{cases}
A'_{RNS} = B'_{RNS}, \\
\quad if \left\{ (n_{A'} - n_{B'}) \wedge \left[(a'_1 + b'_1) = 0(\text{mod}2) \right] \right\}; \\
A'_{RNS} > B'_{RNS}, \\
\quad if \left\{ (n_{A'} > n_{B'}) \vee \left\{ (n_{A'} = n_{B'}) \wedge \left[(a'_1 = 1) \wedge (b'_1 = 0) \right] \right\} \right\}; \\
A'_{RNS} < B'_{RNS}, \\
\quad if \left\{ (n_{A'} < n_{B'}) \vee \left\{ (n_{A'} = n_{B'}) \wedge \left[(b'_1 = 1) \wedge (a'_1 = 0) \right] \right\} \right\}.
\end{cases}
\tag{14}
$$

Initial numbers A and B are submitted to AF

$$
\begin{cases}
A'_{RNS}(B'_{RNS}) = \dfrac{M}{2} + |A_{RNS}|(|B_{RNS}|), \\
\quad if \ A_{RNS}(B_{RNS}) \geq 0, \\
A'_{RNS}(B'_{RNS}) = \dfrac{M}{2} - |A_{RNS}|(|B_{RNS}|), \\
\quad if \ A_{RNS}(B_{RNS}) < 0,
\end{cases}
\tag{15}
$$

i.e. for positive numbers, we have that $A'_{RNS} = \frac{M}{2} + |A_{RNS}|$, and for negative – $A'_{RNS} = \frac{M}{2} - |A_{RNS}|$. To determine the result of the comparison operation between two numbers, the following obvious relationships are used in RNS:

$$\begin{aligned} &\text{if } A'_{RNS} = B'_{RNS}, \quad \text{then } A_{RNS} = B_{RNS}, \\ &\text{if } A'_{RNS} > B'_{RNS}, \quad \text{then } A_{RNS} > B_{RNS}, \\ &\text{if } A'_{RNS} < B'_{RNS}, \quad \text{then } A_{RNS} < B_{RNS}. \end{aligned} \tag{16}$$

The simplified process of determining the result of the operation of comparing two numbers $A_{RNS} = (a_1, a_2, ..., a_n)$ and $B_{RNS} = (b_1, b_2, ..., b_n)$ in RNS can be represented as follows. At the beginning, in accordance with the relation (14), AF are formed $A'_{RNS} = (a'_1, a'_2, ..., a'_n)$ and $B'_{RNS} = (b'_1, b'_2, ..., b'_n)$ of the initial numbers $A_{RNS} = (a_1, a_2, ..., a_n)$ and $B_{RNS} = (b_1, b_2, ..., b_n)$. To ensure the maximum accuracy of two numbers, we will perform the process of zeroing on the basis of $m_1 = 2$. In this case, by the values a'_1 and b'_1 the data is defined $[a'_1, (a'_2)', ..., (a'_n)']$ and $[b'_1, (b'_2)', ..., (b'_n)']$. Further values are determined, respectively, $A'_{m_1} = A' - KN^{(A')} = = (a'_1, a'_2, ..., a'_n) - [a'_1, (a'_2)', ..., (a'_n)'] = [0, a_2^{(1)}, ..., a_n^{(1)}]$ and $B'_{m_1} = B' - KN^{(B')} = = (b'_1, b'_2, ..., b'_n) - [b'_1, (b'_2)', ..., (b'_n)'] = [0, b_2^{(1)}, ..., b_n^{(1)}]$. Next, SRC is determined as following $K_N^{(n_{A'})} = \{z_{N-1}, z_{N-2}, ..., z_2, z_1, z_0\}$ and $K_N^{(n_B)} = \{z_{N-1}^*, z_{N-2}^*, ..., z_2^*, z_1^*, z_0^*\}$. Wherein $z_{n_A} = 0$, if $A_{m_1} - n_A \cdot m_1 = 0$ and $z_{n_A} = 1$, if $A_{m_1} - n_A \cdot m_1 \neq 0$ and $z_{n_B} = 0$, if $B_{m_1} - n_B \cdot m_1 = 0$ and $z_{n_B} = 1$, if $B_{m_1} - n_B \cdot m_1 \neq 0$.

When implementing the algebraic comparison operation of two numbers, three options are possible. The first option $n_{A'} = n_{B'}$, the second option $n_{A'} > n_{B'}$ and the third one $n_{A'} < n_{B'}$.

Consider the examples of specific performance of the operation of comparing two numbers for RNS, with the given bases $m_1 = 2$, $m_2 = 3$ and $m_3 = 5$ wherein $\frac{M}{2} = 15 = (1, 00, 000)$. Table 7 presents the volume of code words in RNS, Table 8 presents the zeroing constants for $m_1 = 2$, in Table 9, there are constants for groups of matrices and values of SRC, Table 10 presents the algorithm of forming the result of comparing two numbers A' and B' in RNS for the option $n_{A'} = n_{B'}$.

Example 2 . Let be $A = 3$ ($A = (1, 00, 011)$) and $B = -4$ ($B = (0, 01, 100)$). $A' = \frac{M}{2} + A = (1, 00, 000) + (1, 00, 011) = (0, 00, 011)$ and $B' = \frac{M}{2} - B = (1, 00, 000) - -(0, 01, 100) = (1, 10, 001)$.

The procedure of comparing two numbers $A' = (0, 00, 011)$ and $B' = (1, 10, 001)$ is carried out in accordance with the algorithm (13). By value $a'_1 = 0$ numbers $A' = (0, 00, 011)$ (Table 8) choose a constant $KN^{(A')} = (0, 00, 000)$. The adder performs the operation $A'_{m_1} = A' - KN^{(A')} = (0, 00, 011) - (0, 00, 000) =$

Table. 7 The number of code words in RNS

$A(B)$ In PNS	$A'(B')$ In PNS	$A'(B')$ in RNS		
		$m_1 = 2$	$m_2 = 3$	$m_3 = 5$
-15	0	0	00	000
-14	1	1	01	001
-13	2	0	10	010
-12	3	1	00	011
-11	4	0	01	100
-10	5	1	10	000
-9	6	0	00	001
-8	7	1	01	010
-7	8	0	10	011
-6	9	1	00	100
-5	10	0	01	000
-4	11	1	10	001
-3	12	0	00	010
-2	13	1	01	011
-1	14	0	10	100
0	15	1	00	000
1	16	0	01	001
2	17	1	10	010
3	18	0	00	011
4	19	1	01	100
5	20	0	10	000
6	21	1	00	001
7	22	0	01	010
8	23	1	10	011
9	24	0	00	100
10	25	1	01	000
11	26	0	10	001
12	27	1	00	010
13	28	0	01	011
14	29	1	10	100

Table. 8 Constant zeroing for $m_1 = 2$

$a_1(b_1)$	Constant $KN^{(A)} \ KN^{(B)}$ in RNS		
	$m_1 = 2$	$m_2 = 3$	$m_3 = 5$
0	0	00	000
1	1	01	001

Table. 9 Constants for groups of adders and values SRC $K_{15}^{(n_{A'})}\left(K_{15}^{(n_{B'})}\right)(m_1 = 2)$

Value $n_{A'}(n_{B'}) = \overline{0,\ 14}$	$K_{15}^{(n_{A'})},\left(K_{15}^{(n_{B'})}\right)$	Value $j \cdot m_1\ (j = \overline{0,\ 14})$			
		PNS	RNS		
			$m_1 = 2$	$m_2 = 3$	$m_3 = 5$
0	{111111111111110}	0	0	00	000
1	{111111111111101}	2	0	10	010
2	{111111111111011}	4	0	01	100
3	{111111111110111}	6	0	00	001
4	{111111111101111}	8	0	10	011
5	{111111111011111}	10	0	01	000
6	{111111110111111}	12	0	00	010
7	{111111101111111}	14	0	10	100
8	{111111011111111}	16	0	01	001
9	{111110111111111}	18	0	00	011
10	{111101111111111}	20	0	10	000
11	{111011111111111}	22	0	01	010
12	{110111111111111}	24	0	00	100
13	{101111111111111}	26	0	10	001
14	{011111111111111}	28	0	01	011

Table 10 Algorithm of the result formation of comparing two numbers A' AND B' in RNS for option $n_{A'} = n_{B'}$

The first ($n_{A'} = n_{B'}$) output of the comparison circuit	The output of the comparison circuit	Output
a'_1	b'_1	
0	0	The first ($A' = B'$)
0	1	Third ($A' < B'$)
1	0	Second ($A' > B'$)
1	1	The first ($A' = B'$)

$(0, 00, 011)$. Similarly, by number value $b'_1 = 1$ of number $B' = (1, 10, 001)$ choose $KN^{(B')} = (1, 01, 001)$ and the adder performs the operation $B_{m_1} = B' - KN^{(B')} = (1, 10, 001) - (1, 01, 001) = (0, 01, 000)$. Because of $A'_{m_1} - n_{A'} \cdot m_1 = 18 - 9 \cdot 2 = 0$, then SRC is formed as $K_{15}^{(9)} = \{111110111111111\}$ (Table 9).

Because of $B'_{m_1} - n_{B'} \cdot m_1 = 10 - 5 \cdot 2 = 0$, then SRC is formed as $B'_{m_1} - n_{B'} \cdot m_1 = 10 - 5 \cdot 2 = 0$. If $n_{A'} = 9 > n_{B'} = 5$ and $a'_1 + b'_1 = (0 + 1) = 1 (\mathrm{mod}\,2)$ (Table 10), then we have that $A' > B'$. In accordance with relation (9), we have the result of a comparison operation, i.e. $A > B$. Check: $A = 3 > B = -4$.

Because of $B'_{m_1} - n_{B'} \cdot m_1 = 10 - 5 \cdot 2 = 0$, then SRC is formed as $B'_{m_1} - n_{B'} \cdot m_1 = 10 - 5 \cdot 2 = 0$. If $n_{A'} = 9 > n_{B'} = 5$ and $a'_1 + b'_1 = (0 + 1) = 1 (\mod 2)$ (Table 10), then we have that $A' > B'$. In accordance with relation (9), we have the result of a comparison operation, i.e. $A > B$. Check: $A = 3 > B = -4$.

Example 3 Let $A = -2$ ($A = (0, 10, 010)$) and $B = -3$ ($B = (1, 00, 011)$). $A' = \frac{M}{2} - A = (1, 00, 000) - (0, 10, 010) = (1, 01, 011)$ and $B' = \frac{M}{2} - B = (1, 00, 000) - -(1, 00, 011) = (0, 00, 010)$.

By value $a'_1 = 1$ of the number $A' = (1, 01, 011)$ choose a constant $KN^{(A')} = (1, 01, 001)$ (Table 8). The adder implements the $A'_{m_1} = A' - KN^{(A')} = (1, 01, 011) - (1, 01, 001) = (0, 00, 010)$. By value $b'_1 = 0$ of the number $B' = (0, 00, 010)$ choose $KN^{(B')} = (0, 00, 000)$. The adder implements the operation $B'_{m_1} = B' - KN^{(B')} = (0, 00, 010) - (0, 00, 000) = (0, 00, 010)$. Because of $A'_{m_1} - n_{A'} \cdot m_1 = = 12 - 6 \cdot 2 = 0$, and $B'_{m_1} - n_{B'} \cdot m_1 = 12 - 6 \cdot 2 = 0$, then A' and B' have identical SRC, equal to $K_{15}^{(6)} = \{111111110111111\}$. i.e. we have that $n_{A'} = n_{B'} = 6$. The inequality $A' > B'$ holds for $(n_{A'} = n_{B'})$ (Table 9). According to the relation (15), we have the result of the comparison operation $A > B$. Check: $A = -2 > B = -3$.

3 Conclusions of Research

Thus, it is advisable to use the methods and algorithms developed in the article for the practical implementation of high-speed devices, where it is needed to perform an operation of data comparison that is presented in RNS. The developed methods are based on the usage of the positional feature of the non-positional code structure by which a SRC is formed. The obtained theoretical results are confirmed by the instances presented in the article.

References

1. Akushskii, I.Y., Yuditskii, D.I.: Machine arithmetic in residual classes [in Russian]. Sov Radio Moscow (1968)
2. Ananda Mohan, P.V.: Residue Number Systems: Theory and Applications: Springer International Publishing Switzerland. Birkhäuser Basel (2016)
3. Yanko, A., Koshman, S., Krasnobayev, V.: Algorithms of data processing in the residual classes system. In: 2017 4th International Scientific-Practical Conference Problems of Info Communications. Science and Technology (PIC S&T), Kharkov, pp. 117–121 (2017)

4. Krasnobayev, V., Koshman, S., Yanko, A., Martynenko, A.: Method of error control of the information presented in the modular number system. In: 2018 International Scientific-Practical Conference on Problems of Info Communications Science and Technology (PIC S&T). Kharkov, 2018, pp. 39–42

5. Krasnobayev V, Koshman S, Mavrina M (2014) A method for increasing the reliability of verification of data represented in a residue number system. Cybern Syst Anal 50(6):969–976

6. Krasnobayev, V., Kuznetsov, A., Koshman, S., Moroz, S.: Improved method of determining the alternative set of numbers in residue number system. In: Chertov, O., Mylovanov, T., Kondratenko, Y., Kacprzyk, J., Kreinovich, V., Stefanuk, V. (eds.) Recent Developments in Data Science and Intelligent Analysis of Information. ICDSIAI 2018. Advances in Intelligent Systems and Computing, vol. 836, pp. 319–328. Springer, Cham (2019), 05 Aug 2018. https://doi.org/10.1007/978-3-319-97885-7_31

7. Tiwari, A., Tomko, K.: Enhanced reliability of finite state machines in FPGA through efficient fault detection and correction. IEEE Trans. Reliab. **54**(3), 459–467

8. Reddy, C.M., Nalini, N.: FT2R2Cloud: Fault tolerance using time-out and retransmission of requests for cloud applications, In: 2014 International Conference on Advances in Electronics Computers and Communications, Bangalore, pp. 1–4 (2014)

9. Braun, C., Wunderlich, H.: Algorithm-based fault tolerance for many-core architectures. In: 2010 15th IEEE European Test Symposium, Praha, pp. 253–253 (2010)

10. Krasnobayev, V., Kuznetsov, A., Kononchenko, A., Kuznetsova, T.: Method of data control in the residue classes. In: Proceedings of the Second International Workshop on Computer Modelling and Intelligent Systems (CMIS-2019). Zaporizhzhia, Ukraine, April 15–19, 2019, pp. 241–252 (2019)

11. Popov, D.I., Gapochkin, A.V.: Development of algorithm for control and correction of errors of digital signals, represented in system of residual classes. In: 2018 International Russian Automation Conference (RusAutoCon) Sochi, pp. 1–3 (2018)

12. Kocherov, Y.N., Samoylenko, D.V., Koldaev, A.I.: Development of an antinoise method of data sharing based on the application of a two-step-up system of residual classes. In: 2018 International Multi-Conference on Industrial Engineering and Modern Technologies (FarEastCon), Vladivostok, pp. 1–5 (2018)

13. Krasnobayev, V., Kuznetsov, A., Zub, M., Kuznetsova, K.: Methods for comparing numbers in non-positional notation of residual classes. In: Proceedings of the Second International Workshop on Computer Modelling and Intelligent Systems (CMIS-2019). Zaporizhzhia, Ukraine, 15–19 April 2019, pp. 581–595 (2019)

14. Kasianchuk, M., Yakymenko, I., Pazdriy, I., Melnyk, A., Ivasiev, S.: Rabin's modified method of encryption using various forms of system of residual classes. In: 2017 14th International Conference the Experience of Designing and Application of CAD Systems in Microelectronics (CADSM), Lviv, pp. 222–224 (2017)

15. Krasnobayev, V., Kuznetsov, A., Lokotkova, I., Dyachenko, A.: The method of single errors correction in the residue class. In: 2019 3rd International Conference on Advanced Information and Communications Technologies (AICT). Lviv, Ukraine, pp. 125–128 (2019). https://doi.org/10.1109/AIACT.2019.8847845

16. Dubrova, E.: Fault-tolerant design: an introduction course notes. Royal Institute of Technology, Stockholm, Sweden, 2013, p. 147. https://pld.ttu.ee/IAF0530/draft.pdf

17. Krasnobaev, V., Kuznetsov, A., Babenko, V., Denysenko, M., Zub, M., Hryhorenko, V.: The method of raising numbers, represented in the system of residual classes to an arbitrary power of a natural number. In: 2019 IEEE 2nd Ukraine Conference on Electrical and Computer Engineering (UKRCON), Lviv, Ukraine, pp. 1133–1138 (2019). https://doi.org/10.1109/UKRCON.2019.8879793

18. Radu, M.: Reliability and fault tolerance analysis of FPGA platforms. In: IEEE Long Island Systems, Applications and Technology (LISAT) Conference 2014, Farmingdale, NY, pp. 1–4 (2014)

19. Krasnobayev VA, Yanko AS, Koshman SA (2016) A method for arithmetic comparison of data represented in a residue number system. Cybern Syst Anal 52(1):145–150. https://doi.org/10.1007/s10559-016-9809-2

20. Kasianchuk, M., Yakymenko, I., Pazdriy, I., Zastavnyy, O.: Algorithms of findings of perfect shape modules of remaining classes system. In: The Experience of Designing and Application of CAD Systems in Microelectronics, Lviv, pp. 316–318 (2015)

The Data Control in the System of Residual Classes

Victor Krasnobayev⦿, Sergey Koshman⦿, and Alexandr Kuznetsov⦿

Abstract This work discusses issues related to the development and application of methods for improving the efficiency of operation based on the use of the codes of the system of residual classes (SRC). The research is aimed at presents the methods of data operational control in the SRC. The main focus is given to the control methods based on the principle of nullification of numbers in the SRC. It is shown that the resulting redundancy with the introduction of one additional (control) base provides control and correction of errors in the process of performing operations. The algorithms developed and presented in the work allow modeling the data control unit for the practical implementation of the proposed methods. The analysis is performed and the calculated data of the implementation time of the control operation for the considered methods of nullification is obtained, and the conditional amount of the control system equipment is also calculated.

Keywords System of residual classes · Operational data control · Number nulling · Non-modular operations

1 Introduction

The efficiency of operation of information processing systems in the SRC largely depends on the methods used to control data. One of the methods for determining the correctness of a number is the nulling (nullification) method, which consists in the transition from the initial number

$$A = (a_1||a_2|| \dots ||a_n||a_{n+1})$$

to the number

V. Krasnobayev · S. Koshman (✉) · A. Kuznetsov
V. N. Karazin Kharkiv National University, 4 Svobody Sq., Kharkiv 61022, Ukraine
e-mail: s.koshman@karazin.ua

A. Kuznetsov
e-mail: kuznetsov@karazin.ua

R. Oliynykov et al. (eds.), *Information Security Technologies in the Decentralized Distributed Networks*, Lecture Notes on Data Engineering and Communications Technologies 115, https://doi.org/10.1007/978-3-030-95161-0_12

$$A^{(Z)} = (0||0||\dots||0||\gamma_{n+1})$$

with the help of such a sequence of transformations in which there is no output outside the working range of the SRC.

The process of nulling a number consists in successively subtracting from this number the nulling constants of the form

$$(m_{1.1}, m_{1.2}, \dots, m_{n+1}), \quad \text{where } m_{1.1} = (1, 2, \dots, m_1 - 1);$$
$$(0, m_{2.2}, \dots, m_{n+1,2}), \quad \text{where } m_{2.2} = (1, 2, \dots, m_2 - 1).$$

In this case, the number $A = (a_1||a_2||\dots||a_n||a_{n+1})$ is sequentially converted to a form $A' = (0||a_2'||\dots||a_n'||a_{n+1}')$, then to a number $A'' = (0||0||\dots||a_n''||a_{n+1}'')$ etc.

By repeating the process n times, we get

$$\Lambda^{(Z)} = (0||0||\dots||0||\gamma_{n+1}).$$

If $\gamma_{n+1} = 0$, then the original number is correct and lies in the range $[0, \ M)$, if $\gamma_{n+1} \neq 0$, then the number is incorrect and lies in the range $[jM, (j+1)M)$, for $j = (1, 2, \dots, m_{n+1} - 1)$, where $(j+1)$ is the value of the number of the interval in which the operand $\widetilde{A}$ falls.

We show that an error ΔA can transfer the correct number A, lying in the interval $[0, \ M)$, only in one of the two intervals.

We write $\widetilde{A} = A + \Delta A$, i.e.

$$\widetilde{A} = (a_1||a_2||\dots||a_n||a_{n+1}) + (0||0||\dots||\Delta a_i||\dots||1).$$

Obviously, ΔA is not in the first (working) interval $[0, \ M)$, since the first interval contains the correct number

$$A_0 = (0||0||\dots||0||0).$$

Let ΔA be in the k-th interval

$$(k-1)M \leq \Delta A \leq kM.$$

We write the system of two inequalities

$$\begin{cases} 0 \leq A < M, \\ (k-1)M \leq \Delta A < kM. \end{cases}$$

Add up the two inequalities

$$(k-1)M \leq A + \Delta A \leq (k+1)M.$$

Let $j = k - 1$, then we can write

$$jM \leq \tilde{A} < (j + 2)M,$$

i.e. an error may cause the correct operand to be incorrect, lying only in one of two intervals $[jM, (j + 1)M)$ or $[(j + 1)M, (j + 2)M)$.

Consider another important property of the SRC, which consists in the ability to change the ratio between the number of information and control bases in the process of solving the problem and flexibly use the reserves of accuracy and reliability [1–3].

A known method of variable scaling, which allows to reduce the number of digits in the presentation of numerical information in the PNS. Due to this, it is possible to introduce additional discharges for the organization of hardware operative control in the presence of restrictions on the increase in weight, size and cost of computer systems and components of fast processing of integer data (CSCPID). At the same time, you can flexibly control the accuracy, speed and reliability of calculations. The specificity of the PNS imposes the following restrictions on the variable scaling method:

- before each step of the program execution, it is necessary to perform additional shift operations that reduce the actual CSCPID speed by $\approx 10\%$;
- the use of variable scaling implies (before creating a program) the implementation of a large amount of theoretical work on the definition of rational scale coefficients;
- variable scaling makes sense to apply only a certain class of tasks; This method is hardly advisable for real-time CSCPID.

Completely different results can be obtained in the SRC. The time required to perform a non-modular operation in the SRC is proportional to the number of information bases, i.e. the number of bases determining the accuracy of calculations. Turning to calculations with less accuracy, you can increase the speed of CSCPID. If some ordered SRC is expanded by adding bases, each of which is larger than the base of the original SRC, the minimum code distance d_{min} is automatically increased by an amount l. The same can be achieved by reducing the number of information bases, i.e. turning to calculations with less accuracy. Consequently, there is an inversely proportional relationship between the corrective capabilities of the codes in the SRC and the accuracy of the calculations. At the same CSCPID, you can perform the same calculations with high accuracy, but with a lower value of reliability, and others—with less accuracy, but with a higher value of reliability and speed [4–6].

Thus, CSCPID in SRC has the property of adaptation to the problems to be solved depending on the requirements for accuracy, speed and reliability of calculations.

Figure 1a, b schematically depicts the dependence of the characteristics of CSCPID on the ratio between the number of information and control bases

Provided that $y_1 + y_2 = const$, where:

y_1—number of information bases n;
y_2—number of control bases k;
y_3—CSCPID performance;

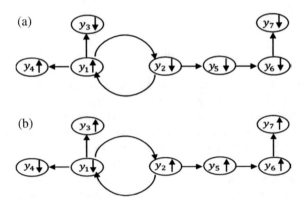

Fig. 1 The dependence of the main characteristics of CSCPID in the SRC with $y_1 + y_2 = const$: **a** with an increase in the number of information bases; **b** with a decrease in the number of information bases

y_4—CSCPID accuracy;
y_5—corrective ability of the code in the SRC;
y_6—reliability of calculations of CSCPID;
y_7—reliability of CSCPID.

2 Method of Successive Subtractions

Consider the nulling procedure in terms of the time of its implementation [7–9]. The essence of the first ($Z1$), the basic in the theory of SRC, nullification procedure [the procedure of sequential nullification (SN)] consists of a sequence of operations for subtracting in form

$$A^{(i+1)} = A^{(i)} - NC^{(i)}, \tag{1}$$

by means of a set of NC of the form (2)

$$NC^{(0)} = [t_1^{(0)} || t_2^{(0)} || t_3^{(0)} || \cdots || t_{i-1}^{(0)} || t_i^{(0)} || t_{i+1}^{(0)} || \cdots || t_{n-3}^{(0)} || t_{n-2}^{(0)} || t_{n-1}^{(0)} || t_n^{(0)} || t_{n+1}^{(0)}],$$
$$t_1^{(0)} = a_1^{(0)}, \quad t_1^{(0)} = \overline{0, m_1 - 1};$$
$$NC^{(1)} = [0 || t_2^{(1)} || t_3^{(1)} || \cdots || t_{i-1}^{(1)} || t_i^{(1)} || t_{i+1}^{(1)} || \cdots || t_{n-3}^{(1)} || t_{n-2}^{(1)} || t_{n-1}^{(1)} || t_n^{(1)} || t_{n+1}^{(1)}],$$
$$t_2^{(1)} = a_2^{(1)}, \quad t_2^{(1)} = \overline{0, m_2 - 1};$$
$$NC^{(2)} = [0 || 0 || t_3^{(2)} || \cdots || t_{i-1}^{(2)} || t_i^{(2)} || t_{i+1}^{(2)} || \cdots || t_{n-3}^{(2)} || t_{n-2}^{(2)} || t_{n-1}^{(2)} || t_n^{(2)} || t_{n+1}^{(2)}],$$
$$t_3^{(2)} = a_3^{(2)}, \quad t_3^{(2)} = \overline{0, m_3 - 1};$$
$$NC^{(i-1)} = [0 || 0 || 0 || \cdots || 0 || t_i^{(i-1)} || t_{i+1}^{(i-1)} || \cdots || t_{n-3}^{(i-1)} || t_{n-2}^{(i-1)} || t_{n-1}^{(i-1)} || t_n^{(i-1)} || t_{n+1}^{(i-1)}],$$
$$t_i^{(i-1)} = a_i^{(i-1)}, \quad t_i^{(i-1)} = \overline{0, m_i - 1};$$

$$NC^{(n-2)} = [0||0||0|| \ldots ||0||0||0|| \ldots ||0||0||t_{n-1}^{(n-2)}||t_n^{(n-2)}||t_{n+1}^{(n-2)}],$$
$$t_{n-1}^{(n-2)} = a_{n-1}^{(n-2)}, \quad t_{n-1}^{(n-2)} = \overline{0, m_{n-1} - 1};$$
$$NC^{(n-1)} = [0||0||0|| \ldots ||0||0||0|| \ldots ||0||0||0||t_n^{(n-1)}||t_{n+1}^{(n-1)}],$$
$$t_n^{(n-1)} = a_n^{(n-1)}, \quad t_n^{(n-1)} = \overline{0, m_n - 1}, \tag{2}$$

from the corresponding numbers

$$A^{(0)} = [a_1^{(0)}||a_2^{(0)}||a_3^{(0)}|| \ldots ||a_{i-1}^{(0)}||a_i^{(0)}||a_{i+1}^{(0)}|| \ldots ||a_{n-3}^{(0)}||a_{n-2}^{(0)}||a_{n-1}^{(0)}||a_n^{(0)}||a_{n+1}^{(0)}],$$
$$A^{(1)} = [0||a_2^{(1)}||a_3^{(1)}|| \ldots ||a_{i-1}^{(1)}||a_i^{(1)}||a_{i+1}^{(1)}|| \ldots ||a_{n-3}^{(1)}||a_{n-2}^{(1)}||a_{n-1}^{(1)}||a_n^{(1)}||a_{n+1}^{(1)}],$$
$$A^{(i-1)} = [0||0||0|| \ldots ||0||a_i^{(i-1)}||a_{i+1}^{(i-1)}|| \ldots ||a_{n-3}^{(i-1)}||a_{n-2}^{(i-1)}||a_{n-1}^{(i-1)}||a_n^{(i-1)}||a_{n+1}^{(i-1)}$$

and so on.

For example: performing the first subtraction operation

$$A^{(1)} = A^{(0)} - NC^{(0)} = [a_1^{(0)}||a_2^{(0)}||a_3^{(0)}|| \ldots ||a_{i-1}^{(0)}||a_i^{(0)}||a_{i+1}^{(0)}|| \ldots$$
$$\ldots ||a_{n-3}^{(0)}||a_{n-2}^{(0)}||a_{n-1}^{(0)}||a_n^{(0)}||a_{n+1}^{(0)}] - [t_1^{(0)}||t_2^{(0)}||t_3^{(0)}|| \ldots ||t_{i-1}^{(0)}||t_i^{(0)}||t_{i+1}^{(0)}|| \ldots$$
$$\ldots ||t_{n-3}^{(0)}||t_{n-2}^{(0)}||t_{n-1}^{(0)}||t_n^{(0)}||t_{n+1}^{(0)}] = \left\{ [a_1^{(0)} - t_1^{(0)}] \bmod m_1 ||[a_2^{(0)} - t_2^{(0)}] \bmod m_2 || \right.$$
$$||[a_3^{(0)} - t_3^{(0)}] \bmod m_3 || \ldots ||[a_{i-1}^{(0)} - t_{i-1}^{(0)}] \bmod m_{i-1} ||[a_i^{(0)} - t_i^{(0)}] \bmod m_i ||$$
$$||[a_{i+1}^{(0)} - t_{i+1}^{(0)}] \bmod m_{i+1} || \ldots ||[a_{n-3}^{(0)} - t_{n-3}^{(0)}] \bmod m_{n-3} ||[a_{n-2}^{(0)} - t_{n-2}^{(0)}] \bmod m_{n-2} ||$$
$$\left. ||[a_{n-1}^{(0)} - t_{n-1}^{(0)}] \bmod m_{n-1} ||[a_n^{(0)} - t_n^{(0)}] \bmod m_n || [a_{n+1}^{(0)} - t_{n+1}^{(0)}] \bmod m_{n+1} \right\} =$$
$$= [0||a_2^{(1)}||a_3^{(1)}|| \ldots ||a_{i-1}^{(1)}||a_i^{(1)}||a_{i+1}^{(1)}|| \ldots ||a_{n-3}^{(1)}||a_{n-2}^{(1)}||a_{n-1}^{(1)}||a_n^{(1)}||a_{n+1}^{(1)}];$$

performing the second subtraction

$$A^{(2)} = A^{(1)} - NC^{(1)} =$$
$$= [0||a_2^{(1)}||a_3^{(1)}|| \ldots ||a_{i-1}^{(1)}||a_i^{(1)}||a_{i+1}^{(1)}|| \ldots ||a_{n-3}^{(1)}||a_{n-2}^{(1)}||a_{n-1}^{(1)}||a_n^{(1)}||a_{n+1}^{(1)}] -$$
$$- [0||t_2^{(1)}||t_3^{(1)}|| \ldots ||t_{i-1}^{(1)}||t_i^{(1)}||t_{i+1}^{(1)}|| \ldots ||t_{n-3}^{(1)}||t_{n-2}^{(1)}||t_{n-1}^{(1)}||t_n^{(1)}||t_{n+1}^{(1)}] =$$
$$= [a_i^{(1)} - t_i^{(1)}] \bmod m_i ||[a_{i+1}^{(1)} - t_{i+1}^{(1)}] \bmod m_{i+1} || \ldots ||[a_{n-3}^{(1)} - t_{n-3}^{(1)}] \bmod m_{n-3} ||$$
$$||[a_{n-2}^{(1)} - t_{n-2}^{(1)}] \bmod m_{n-2} ||[a_{n-1}^{(1)} - t_{n-1}^{(1)}] \bmod m_{n-1} ||[a_n^{(1)} - t_n^{(1)}] \bmod m_n ||$$
$$\left. ||[a_{n+1}^{(1)} - t_{n+1}^{(1)}] \bmod m_{n+1} \right\} =$$
$$= [0||0||a_3^{(2)}|| \ldots ||a_{i-1}^{(2)}||a_i^{(2)}||a_{i+1}^{(2)}|| \ldots ||a_{n-3}^{(2)}||a_{n-2}^{(2)}||a_{n-1}^{(2)}||a_n^{(2)}||a_{n+1}^{(2)}];$$

performing the third subtraction

$$A^{(3)} = A^{(2)} - NC^{(2)} =$$
$$= [0||0||a_3^{(2)}|| \ldots ||a_{i-1}^{(2)}||a_i^{(2)}||a_{i+1}^{(2)}|| \ldots ||a_{n-3}^{(2)}||a_{n-2}^{(2)}||a_{n-1}^{(2)}||a_n^{(2)}||a_{n+1}^{(2)}] -$$

$$- [0||0||t_3^{(2)}|| \dots ||t_{i-1}^{(2)}||t_i^{(2)}||t_{i+1}^{(2)}|| \dots ||t_{n-3}^{(2)}||t_{n-2}^{(2)}||t_{n-1}^{(2)}||t_n^{(2)}||t_{n+1}^{(2)}] =$$

$$= \left\{ 0||0||[a_3^{(2)} - t_3^{(2)}] \bmod m_3|| \dots ||[a_{i-1}^{(2)} - t_{i-1}^{(2)}] \bmod m_{i-1}||[a_i^{(2)} - t_i^{(2)}] \bmod m_i|| \right.$$

$$||[a_{i+1}^{(2)} - t_{i+1}^{(2)}] \bmod m_{i+1}|| \dots ||[a_{n-3}^{(2)} - t_{n-3}^{(2)}] \bmod m_{n-3}||[a_{n-2}^{(2)} - t_{n-2}^{(2)}] \bmod m_{n-2}||$$

$$\left. ||[a_{n-1}^{(2)} - t_{n-1}^{(2)}] \bmod m_{n-1}||[a_n^{(2)} - t_n^{(2)}] \bmod m_n|| [a_{n+1}^{(2)} - t_{n+1}^{(2)}] \bmod m_{n+1} \right\} =$$

$$= [0||0||0||a_4^{(3)}||a_5^{(3)}|| \dots ||a_{i-1}^{(3)}||a_i^{(3)}||a_{i+1}^{(3)}|| \dots ||a_{n-3}^{(3)}||a_{n-2}^{(3)}||a_{n-1}^{(3)}||a_n^{(3)}||a_{n+1}^{(3)}],$$

etc. The algorithm for performing the SN procedure is presented in Table 1. In accordance with this algorithm, the initial number $A = A^{(0)} = (a_1^{(0)}||a_2^{(0)}|| \dots ||a_i^{(0)}|| ||a_{i+1}^{(0)}|| \dots ||a_n^{(0)}||a_{n+1}^{(0)})$ according to the formula (1) is sequentially converted to the form $A^{(Z)} = (0||0|| \dots ||0||\gamma_{n+1})$ with the help of a sequence of operations that does not result in the output of the numerical value of the number $A^{(0)}$ over the working range $[0, M)$ of the SRC. In this case, the initial number $A = A^{(0)} = (a_1^{(0)}||a_2^{(0)}|| \dots ||a_i^{(0)}||a_{i+1}^{(0)}|| \dots ||a_n^{(0)}||a_{n+1}^{(0)})$ is sequentially reduced to the form $A^{(H)}$, i.e.

$$A = A^{(0)} = (a_1^{(0)}||a_2^{(0)}|| \dots ||a_i^{(0)}||a_{i+1}^{(0)}|| \dots ||a_n^{(0)}||a_{n+1}^{(0)}),$$
$$A^{(1)} = (0||a_2^{(1)}||a_3^{(1)}|| \dots a_n^{(1)}||a_{n+1}^{(1)}),$$
$$A^{(2)} = (0||0||a_3^{(2)}|| \dots ||a_n^{(2)}||a_{n+1}^{(2)}),$$
$$A^{(3)} = (0||0||0||a_4^{(3)}|| \dots ||a_n^{(3)}||a_{n+1}^{(3)})$$

Table 1 SN algorithm

Operation number	Contents of operation
1	Appeal by value $a_1^{(0)}$ and number $A^{(0)}$ in BNC_0 for the $NC^{(0)}$
2	Performing a subtraction operation $A^{(1)} = A^{(0)} - NC^{(0)}$
3	Appeal by value $a_2^{(1)}$ and number $A^{(1)}$ in BNC_1 for the $NC^{(1)}$
4	Performing a subtraction operation $A^{(2)} = A^{(1)} - NC^{(1)}$
5	Appeal by value $a_3^{(2)}$ and number $A^{(2)}$ in BNC_2 for the $NC^{(2)}$
6	Performing a subtraction operation $A^{(3)} = A^{(2)} - NC^{(2)}$
7	Appeal by value $a_4^{(3)}$ and number $A^{(3)}$ in BNC_3 for the $NC^{(3)}$
8	Performing a subtraction operation $A^{(4)} = A^{(3)} - NC^{(3)}$
$\vdots$	$\dots$
$2n - 3$	Appeal by value $a_{n-1}^{(n-2)}$ and number $A^{(n-2)}$ in BNC_{n-2} for the $NC^{(n-2)}$
$2n - 2$	Performing a subtraction operation $A^{(n-1)} = A^{(n-2)} - NC^{(n-2)}$
$2n - 1$	Appeal by value $a_n^{(n-1)}$ and number $A^{(n-1)}$ in BNC_{n-1} for the $NC^{(n-1)}$
$2n$	Performing a subtraction operation $A^{(n)} = A^{(n-1)} - NC^{(n-1)}$. Getting a nullified number $A^{(N)} = A^{(n)} = [0||0|| \dots ||0|| \dots ||0||0||\gamma_{n+1} = a_{n+1}^{(n)}]$

and so on.

By repeating the subtraction n times, we get the value $A^{(Z)} = (0||0|| \ldots ||0||a_{n+1}^{(n)})$, or $A^{(Z)} = (0||0|| \ldots ||0||\gamma_{n+1})$, where $\gamma_{n+1} = a_{n+1}^{(n)}$.

Denoting the sampling time of the NC from the corresponding nullification block (NB) of the CS functioning in the SRC as t_1, and the time of subtracting from the number $A^{(i-1)}$ the constant $NC^{(i-1)}$, i.e. performing the operation $A^{(i)} = A^{(i-1)} - NC^{(i-1)}$—as t_2, we get the total time T_{Z1} of the nullification procedure for the first $Z1$ method

$$T_{Z1} = n(t_1 + t_2). \tag{3}$$

3 Parallel Subtraction Method

Consider the following method (Z3) of operational control of data in the SRC [parallel nullification method (PNM)]. The essence of the proposed control method is that the nullification procedure is carried out parallel in time for two reasons. For n-even numbers, we have $a_i^{(i-1)}$, $a_{n-i+1}^{(i-1)}$ $(i = \overline{1, n/2})$, namely $a_1^{(0)}, a_n^{(0)}; a_2^{(1)}, a_{n-1}^{(1)}; a_3^{(2)}, a_{n-2}^{(2)}; \ldots a_{n/2}^{(n/2)}, a_{n/2+1}^{(n/2)}$ (Fig. 2). For n-odd numbers, we have $a_1^{(0)}, a_n^{(0)}, a_2^{(1)}, a_{n-1}^{(1)}; a_3^{(2)}, a_{n-2}^{(2)}; \ldots a_{(n+1)/2}^{((n+1)/2-1)}$ (Fig. 3). In this case, for an arbitrary value i of NC for the corresponding number, have the following form

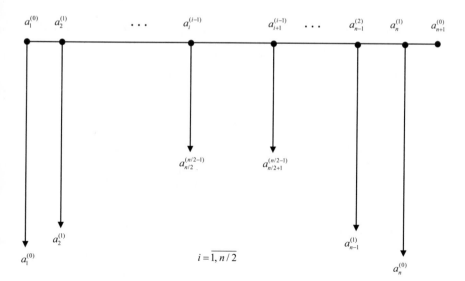

Fig. 2 Sampling scheme for nullification constants for PNM method (n-even number)

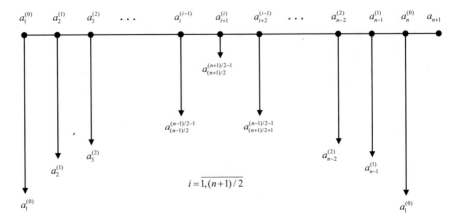

Fig. 3 Sampling scheme for nullification constants for PNM method (*n*-odd number)

$$A^{(i)} = [\overbrace{0\,||\,0\,||\,\dots\,||\,0\,||}^{i-zerores} a_{i+1}^{(i)}\,||a_{i+2}^{(i)}\,||\dots\,||\,a_{n-i-1}^{(i)}\,||\,a_{n-i}^{(i)}\,||\,\overbrace{0\,||\dots\,||\,0\,||\,0\,||}^{i-zeroes} a_{n+1}^{(i)}],$$

$$NC^{(i)} = [0\,||\,0\,||\,\dots\,||\,0\,||t_{i+1}^{(i)}\,||\,t_{i+2}^{(i)}\,||\dots\,||\,t_{n-i-1}^{(i)}\,||\,t_{n-i}^{(i)}\,||\,0\,||\dots\,||\,0\,||\,0\,||\,t_{n+1}^{(i)}];$$

$$t_{i+1}^{(i)} = \overline{0,\,m_{i+1}}, \quad t_{n-i}^{(i)} = \overline{0,\,m_{n-i}}; \quad t_{i+1}^{(i)} = a_{i+1}^{(i)}, \quad t_{n-i}^{(i)} = a_{n-i}^{(i)}.$$

For an arbitrary value i we have that

$$A^{(i+1)} = A^{(i)} - NC^{(i)}$$

$$= [0||0||\dots||0||a_{i+1}^{(i)}||a_{i+2}^{(i)}||a_{i+3}^{(i)}||\dots||a_{n-i-2}^{(i)}||a_{n-i-1}^{(i)}||a_{n-i}^{(i)}||0||\dots||0||a_{n+1}^{(i)}] -$$

$$- [0||0||\dots||0||t_{i+1}^{(i)}||t_{i+2}^{(i)}||t_{i+3}^{(i)}||\dots||t_{n-i-2}^{(i)}||t_{n-i-1}^{(i)}||t_{n-1}^{(i)}||0||\dots||0||t_{n+1}^{(i)}] =$$

$$= \Big\{0||0||\dots||0||[a_{i+1}^{(i)} - t_{i+1}^{(i)}] \bmod m_{i+1}||[a_{i+2}^{(i)} - t_{i+2}^{(i)}]$$

$$\bmod m_{i+2}||[a_{i+3}^{(i)} - t_{i+3}^{(i)}] \bmod m_{i+3}||\dots$$

$$\dots||[a_{n-i-2}^{(i)} - t_{n-i-2}^{(i)}] \bmod m_{n-i-2}||[a_{n-i-1}^{(i)} - t_{n-i-1}^{(i)}] \bmod m_{n-i-1}||$$

$$||[a_{n-i}^{(i)} - t_{n-i}^{(i)}] \bmod m_{n-i}||0||\dots||0||[a_{n+1}^{(i)} - t_{n+1}^{(i)}] \bmod m_{n+1}\Big\} =$$

$$= [0||0||\dots||0||0||a_{i+2}^{(i+1)}||a_{i+3}^{(i+1)}||\dots||a_{n-i-2}^{(i+1)}||a_{n-i-1}^{(i+1)}||0||0||\dots||0||0||a_{n+1}^{(i+1)}].$$

The algorithm for performing the PNM procedure is presented in Table 2. Before getting the value $\gamma_{n+1} = a_{n+1}^{(n/2)}$ for *n*-even number, we have that

$$A^{(n/2-1)} = [\overbrace{0||0||\dots||0||}^{n/2-1\ zeroes} a_{n/2}^{(n/2-1)}||a_{n/2+1}^{(n/2-1)}||\overbrace{0||\dots||0||0||}^{n/2-1\ zeroes} a_{n+1}^{(n/2-1)}].$$

Table 2 PNM algorithm

Operation number	Contents of operation
1	Appeal by the value of the remainders $a_1^{(0)}$ and $a_n^{(0)}$ of a number $A^{(0)}$ in BNC_0 for the $NC^{(0)}$
2	Performing a subtraction operation $A^{(1)} = A^{(0)} - NC^{(0)}$
3	Appeal by the value of the remainders $a_2^{(1)}$ and $a_{n-1}^{(1)}$ of a number $A^{(1)}$ in BNC_1 for the $NC^{(1)}$
4	Performing a subtraction operation $A^{(2)} = A^{(1)} - NC^{(1)}$
5	Appeal by the value of the remainders $a_2^{(2)}$ and $a_{n-2}^{(2)}$ of a number $A^{(2)}$ in BNC_2 for the $NC^{(2)}$
6	Performing a subtraction operation $A^{(3)} = A^{(2)} - NC^{(2)}$
...	...
i	Performing a subtraction operation $A^{(i)} = A^{(i-1)} - NC^{(i-1)}$
$i+1$	Appeal by the value of the remainders $a_{i+1}^{(i)}$ and $a_{n-i}^{(i)}$ of a number $A^{(i)}$ in BNC_i for the $NC^{(i)}$
$i+2$	Performing a subtraction operation $A^{(i+1)} = A^{(i)} - NC^{(i)}$
...	...
$n-3$	Appeal by the value of the remainders $a_{n/2-1}^{(n/2-2)}$ and $a_{n/2+2}^{(n/2-2)}$ of a number $A^{(n/2-2)}$ in $BNC_{n/2-2}$ for the $NC^{(n/2-2)}$
$n-2$	Performing a subtraction operation $A^{(n/2-1)} = A^{(n/2-2)} - NC^{(n/2-2)}$
$n-1$	Appeal by the value of the remainders $a_{n/2}^{(n/2-1)}$ and $a_{n/2+1}^{(n/2-1)}$ of a number $A^{(n/2-1)}$ in $BNC_{n/2-1}$ for the $NC^{(n/2-1)}$
n	Performing a subtraction operation $A^{(n/2)} = A^{(n/2-1)} - NC^{(n/2-1)}$ Getting nullified number $A^{(Z)}$ $A^{(Z)} = A^{(n/2)} = [0\|\|0\|\| \ldots \|\|0\|\|, \ldots, \|\|0\|\|0\|\|\gamma_{n+1} = a_{n+1}^{(n/2)}]$

$$NC^{(n/2-1)} = [\overbrace{0\|\|0\|\| \ldots \|\|0}^{n/2-1\ zeroes}\| \, |t_{n/2}^{(n/2-1)}\|\|t_{n/2+1}^{(n/2-1)}\| \overbrace{\|0\|\| \ldots \|\|0\|\|0\|\|}^{n/2-1\ zeroes} t_{n+1}^{(n/2-1)}],$$

$$t_{n/2}^{(n/2-1)} = \overline{0, m_{n/2}}, \quad t_{n/2+1}^{(n/2-1)} = \overline{0, m_{n/2+1}}, \quad t_{n/2}^{(n/2-1)} = a_{n/2}^{(n/2-1)},$$

$$t_{n/2+1}^{(n/2-1)} = a_{n/2+1}^{(n/2-1)}. \quad A^{(z)} = A^{(n/2)} = A^{(n/2-1)} - NC^{(n/2-1)} =$$

$$\left\{ 0\|\|0\|\| \ldots \|\|0\|\| \left[a_{n/2}^{(n/2-1)} - t_{n/2}^{(n/2-1)}\right] \bmod m_{n/2}\|\| \left[a_{n/2+1}^{(n/2-1)} - t_{n/2+1}^{(n/2-1)}\right] \bmod \right.$$

$$\left. m_{n/2+1}\|\|0\|\|0\|\| \ldots \|\|0\|\|0\|\| \left[a_{n+1}^{(n/2-1)} - t_{n+1}^{(n/2-1)}\right] \bmod m_{n+1} \right\} =$$

$$\left[0\|\|0\|\| \ldots \|\|0\|\|0 \|\| 0 \|\| \ldots \|\| 0 \|\| 0 \|\| a_{n+1}^{(n/2)} \right], \text{ where } \gamma_{n+1} = a_{n+1}^{(n/2)}.$$

Before getting the value $\gamma_{n+1} = a_{n+1}^{(n/2)}$ for n-odd number, we have that

$$A^{((n+1)/2-1)} = [\overbrace{0\|\|0\|\| \ldots \|\|0}^{\frac{n+1}{2}-1\ zeroes}\| \, a_{(n+1)/2}^{((n+1)/2-1)} \overbrace{\|\|0\|\| \ldots \|\|0}^{\frac{n+1}{2}-1\ zeroes}\| \, a_{n+1}^{((n+1)/2-1)}].$$

$$NC^{((n+1)/2-1)} = [0||0|| \ldots ||0||t_{(n+1)/2}^{((n+1)/2-1)}||0|| \ldots ||0||t_{n+1}^{((n+1)/2-1)}],$$

$$t_{(n+1)/2}^{((n+1)/2-1)} = \overline{0, m_{(n+1)/2}}; \quad t_{(n+1)/2}^{((n+1)/2-1)} = a_{(n+1)/2}^{((n+1)/2-1)}.$$

$$A^{(Z)} = A^{(n+1)/2} = A^{((n+1)/2-1)} - NC^{((n+1)/2-1)} =$$
$$\left\{ 0||0|| \ldots ||0|| \left[a_{(n+1)/2}^{((n+1)/2-1)} - t_{(n+1)/2}^{((n+1)/2-1)} \right] \bmod m_{(n+1)/2}|| 0 || \ldots 0||0|| \ldots ||0||0|| \right.$$
$$\left. || \left[a_{n+1}^{((n+1)/2-1)} - t_{n+1}^{((n+1)/2-1)} \right] \bmod m_{n+1} \right\} = \left[0||0|| \ldots ||0|| \ldots || 0 || 0 || a_{n+1}^{(n+1)/2} \right],$$

where $\gamma_{n+1} = a_{n+1}^{(n+1)/2}$.

The time T_{Z3} for performing the zeroing procedure for the first (Z3) method of the PNM is defined as

$$T_{Z3} = n \cdot \tau_{add.} \tag{4}$$

When implementing the nullification procedure for the second (Z3) method in the block of nullification constants (NB) of the calculator in the SRC, it is necessary to have $K_{Z3} = \sum_{i=1}^{[\frac{n}{2}]} (m_i \cdot m_{n-i+1} - 1)$ nullification constants. In this case, the number of N_{Z3} double digits of the NB nullification constants is determined by the expression $K_{Z3} = \sum_{i=1}^{[\frac{n}{2}]} (m_i \cdot m_{n-i+1} - 1) \cdot (n - 2i + 1)$.

4   The Method of Successive Subtractions with Preliminary Analysis of the Residual of the Controlled Number

When implementing a NB in a tabular version, we can assume that practically $t_1 = t_2 = \tau_{add}$. In this case, for the SN procedure, the zerofication time equals the value of $T_{Z1} = 2n\tau_{add}$, where: τ_{add} is the subtraction time from the number $A^{(i)}$ of the zeroing constant $NC^{(i)}$; n is the number of information bases of the SRC. In addition, to implement the zeroing procedure by the first method Z1, it is necessary to store $K_{Z1} = \sum_{i+1}^{n} m_i - n$ nullification constants. In this case, the amount of N_{Z1} binary digits of the nullification constants, which indirectly determines the amount of equipment (capacity) of the BNC of CS, is determined by the expression $N_{Z1} = \left(\sum_{i+1}^{n} m_i - 1 \right)(n - i)$. It is obvious that the considered basic procedure of the SN does not exhaust the possibility of increasing the speed of the implementation of the procedure of zeroing numbers, since the execution of the subtraction $A^{(i+1)} = A^{(i)} - NC^{(i)}$ and sampling operation of the next NC is separated in time. This is due to the fact that until the operation of the subtraction is completed, the remainder of the number by which the NC should be selected for the next stage of the nullification procedure is unknown in advance. Consider a nullification procedure that eliminates this disadvantage.

Consider the procedure (Z2) of nullification—the procedure of sequential nullification with the definition of the subsequent remainder (SN DSR). Using this procedure allows you to reduce, compared with the first Z1. The essence of the procedure is that as long as the nullification constant is sampled $NC^{(i)} = [0||0||0||\ldots||0||t_{i+1}^{(i)}||t_{i+2}^{(i)}||\ldots||t_{n-3}^{(i)}||t_{n-2}^{(i)}||t_{n-1}^{(i)}||t_n^{(i)}||t_{n+1}^{(i)}]$ for the number $A^{(i)} = [0||0||0||\ldots||a_{i+1}^{(i)}||a_{i+2}^{(i)}||\ldots||a_{n-3}^{(i)}||a_{n-2}^{(i)}||a_{n-1}^{(i)}||a_n^{(i)}||a_{n+1}^{(i)}]$ of the residual value $a_{i+1}^{(i)}$ on the base m_{i+1}, the residual value $a_{i+2}^{(i+1)}$ can be formed in the computing path (CP) of the CS functioning on the base m_{i+2}, which will be sampled at the next nullification stage $NC^{(i+1)} = [0||0||0||\ldots||0||t_{i+2}^{(i+1)}||t_{i+3}^{(i+1)}||\ldots||t_{n-3}^{(i+1)}||t_{n-2}^{(i+1)}||t_{n-1}^{(i+1)}||t_n^{(i+1)}||t_{n+1}^{(i+1)}]$.

The value Δa_{i+2}, that will be subtracted from the value $a_{i+2}^{(i)}$, to get the value of the residual $a_{i+2}^{(i+1)}$, is determined only by the value of the residual $a_{i+1}^{(i)}$. Analytically this is determined by the following ratio

$$a_{i+2}^{(i+1)} = \left[a_{i+2}^{(i)} - \Delta a_{i+2}\right] \bmod m_{i+2}. \tag{5}$$

In the process of sampling $NC^{(i)}$ by the value of the remainder $a_{i+1}^{(i)}$ of the number $A^{(i)}$, the same remainder will be simultaneously transmitted to the CP of the CS by the base m_{i+2}. In this case, the value $a_{i+2}^{(i+1)}$ is selected from the corresponding, previously compiled two-input tables $F\left\{a_{i+2}^{(i+1)}\right\} = \left[a_{i+1}^{(i)}; a_{i+2}^{(i)}\right]$, by the values $a_{i+1}^{(i)}$ and $a_{i+2}^{(i)}$. The algorithm for performing the procedure is presented in Table 3. The number of additions for the procedure of the SN DSR is equal to n, since the nullification is carried out for all n information bases of the SRC. However, after every two additions, one additional clock cycle is required for the formation of the next address and circulation in the NB. In this regard, for every two such subtractions, there is one clock cycle, free from the subtraction operation from the sample number of the next nullification constant. Thus, the total number of clock cycles, free from the subtraction operation, during which a call is made to the terminal station and the next address is formed, is determined by the value $[n/2]$.

The time T_{Z2} of the procedure for the zeroing of the SN DSR is determined by the value (6)

$$T_{Z2} = \left(\left[\frac{n-1}{2}\right] + n\right) \cdot \tau_{add}. \tag{6}$$

To implement the zerofication procedure by the second method in a NB, it is necessary to have $K_{Z2} = \sum_{i=1}^{n-1}(m_i - 1)$ nullification constants. At the same time, the number of N_{Z2} bits of the BNC of DPS is determined by the expression)

$$N_{Z2} = \sum_{i=1}^{n-1}(m_i - 1) \cdot (n - i). $$

Table 4 presents the calculated data on the relative T_{Zi}/τ_{add} $(i = 1, 2)$ nullification time for the first $(Z1)$ and second $(Z2)$ procedures.

Table 3 SN DSR algorithm

Operation number	Contents of operation	
1	Appeal by the value of the remainder $a_1^{(0)}$ of a number $A^{(0)}$ in BNC_0 for the $NC^{(0)}$	Residue value formation $a_2^{(1)}$ of a number $A^{(1)}$ as $$a_2^{(1)} = t_2^{(1)} = \left[a_2^{(0)} - a_1^{(0)}\right] \bmod m_2$$
2	Performing a subtraction operation $A^{(1)} = A^{(0)} - NC^{(0)}$	Appeal by the value of the remainder $a_2^{(1)}$ of a number $A^{(1)}$ in BNC_1 for the $NC^{(1)}$
3	Performing a subtraction operation $A^{(2)} = A^{(1)} - NC^{(1)}$	Residue value formation $a_3^{(2)}$ of a number $A^{(2)}$ as $$a_3^{(2)} = t_3^{(2)} = \left[a_3^{(1)} - a_2^{(1)}\right] \bmod m_3$$
4	Appeal by the value of the remainder $a_3^{(2)}$ of a number $A^{(2)}$ in BNC_2 for the $NC^{(2)}$	Residue value formation $a_4^{(3)}$ of a number $A^{(3)}$ as $$a_4^{(3)} = t_4^{(3)} = \left[a_4^{(2)} - a_3^{(2)}\right] \bmod m_4$$
5	Performing a subtraction operation $A^{(3)} = A^{(2)} - NC^{(2)}$	Appeal by the value of the remainder $a_4^{(3)}$ of a number $A^{(3)}$ in BNC_3 for the $NC^{(3)}$
⋮	…	…
i	Performing a subtraction operation $A^{(i)} = A^{(i-1)} - NC^{(i-1)}$	Appeal by the value of the remainder $a_{i+1}^{(i)}$ of a number $A^{(i)}$ in BNC_i for the $NC^{(i)}$
$i+1$	Performing a subtraction operation $A^{(i+1)} = A^{(i)} - NC^{(i)}$	Residue value formation $a_{i+2}^{(i+1)}$ of a number $A^{(i+1)}$ as $$a_{i+2}^{(i+1)} = t_{i+2}^{(i+1)} = \left[a_{i+2}^{(i)} - a_{i+1}^{(i)}\right] \bmod m_{i+2}$$
$i+2$	Appeal by the value of the remainder $a_{i+2}^{(i+1)}$ of a number $A^{(i+1)}$ in BNC_{i+1} for the $NC^{(i+1)}$	Residue value formation $a_{i+3}^{(i+2)}$ of a number $A^{(i+2)}$ as $$a_{i+3}^{(i+2)} = t_{i+3}^{(i+2)} = \left[a_{i+3}^{(i+1)} - a_{i+2}^{(i+1)}\right] \bmod m_{i+3}$$
⋮	…	…

(continued)

Table 3 (continued)

Operation number	Contents of operation	
$K-2$	Appeal by the value of the remainder $a_{n-1}^{(n-2)}$ of a number $A^{(n-2)}$ in BNC_{n-2} for the $NC^{(n-2)}$	Residue value formation $a_n^{(n-1)}$ of a number $A^{(n-1)}$ as $a_n^{(n-1)} = t_n^{(n-1)} = \left[a_n^{(n-2)} - a_{n-1}^{(n-2)} \right] \bmod m_n$
$K-1$	Performing a subtraction operation $A^{(n-1)} = A^{(n-2)} - NC^{(n-2)}$	Appeal by the value of the remainder $a_n^{(n-1)}$ of a number $A^{(n-1)}$ in BNC_{n-1} for the $NC^{(n-1)}$
K	Performing a subtraction operation $A^{(n)} = A^{(n-1)} - NC^{(n-1)}$. Getting the nullified number $A^{(Z)} = A^{(n)} = [0\|\|0\| \dots \|\|0\|\| \dots \|\|0\|\|0\|\|(\gamma_{n+1} = a_{n+1}^{(n)})]$	

Table 4 Data of calculation and comparative analysis of the time of nullification of numbers in the SRC

$l(n)$ T_{Zi}	1(4)	2(6)	3(8)	4(10)	8(16)
T_{Z1}	8	12	16	20	32
T_{Z2}	5	8	11	14	23
K_{Z1}	37	30	31	30	28

Table 4 also presents the efficiency ratio $K_{Z1} = \frac{T_{Z1}-T_{Z2}}{T_{Z1}} \cdot 100\%$ of using the second ($Z2$) procedure, relative to the first ($Z1$) procedure for l-byte ($l = 1, 2, 3, 4, 8$) bit CS grids. From Table 4 it can be seen that the use of the $Z2$ nullification procedure by about one third reduces the time for monitoring data in the SRC.

5 Parallel Subtraction Method with Preliminary Analysis of Subsequent Symmetric Residuals of the Controlled Number

A significant drawback of the known methods ($Z1$–$Z3$) of data control in the SRC is the need for time-consuming control, which leads to low control efficiency and considerable unproductive computational costs.

In order to increase the speed of data control, by reducing the time required to implement the nullification procedure, the article proposes the

second (Z4) method for controlling data in the SRC—the parallel nullification method with determining the subsequent residuals (PN DSR). The
proposed control method is based on the procedure of using pair nullification of numbers with the additional operation of preliminary sampling of
residuals (see method Z2). The essence of the control method is that pre-
sampling is performed simultaneously on two residues $a_{i+1}^{(i)}$ and $a_{n-i}^{(i)}$ of a number

$$A^{(i)} = [\overbrace{0||0||\ldots||0||}^{i-\text{zeroes}} a_{i+1}^{(i)}||a_{i+2}^{(i)}||\ldots||a_{n-i-1}^{(i)}||a_{n-i}^{(i)}\overbrace{||0||\ldots||0||0||}^{i-\text{zeroes}} a_{n+1}^{(i)}].$$

Thus, during the implementation of the nullification procedure, the selection
operation is combined in time, according to the residuals $a_{i+1}^{(i)}$ and $a_{n-i}^{(i)}$ of a number
$A^{(i)} = [0||0||\ldots||0||a_{i+1}^{(i)}||a_{i+2}^{(i)}||a_{i+3}^{(i)}||\ldots||a_{n-i-2}^{(i)}||a_{n-i-1}^{(i)}||a_{n-i}^{(i)}||0||\ldots||0||a_{n+1}^{(i)}]$,
the nullification constant $NC^{(i)} = [0||0||\ldots||0||t_{i+1}^{(i)}||t_{i+2}^{(i)}||t_{i+3}^{(i)}||\ldots||t_{n-i-2}^{(i)}||$
$||t_{n-i-1}^{(i)}||t_{n-1}^{(i)}||0||\ldots||0||t_{n+1}^{(i)}]$ and   the   determination   operation,
according to the values of the residuals $a_{i+1}^{(i)}$ and $a_{n-i}^{(i)}$, subsequent
values of the residuals $a_{i+2}^{(i+1)}$ and $a_{n-i-1}^{(i+1)}$ of a number $A^{(i+1)} =$
$[0||0||\ldots||0||0||a_{i+2}^{(i+1)}||a_{i+3}^{(i+1)}||\ldots||a_{n-i-2}^{(i+1)}||a_{n-i-1}^{(i+1)}||0||0||\ldots||0||0||||a_{n+1}^{(i+1)}]$.

Also combined in time is the operation of subtraction $A^{(i+1)} = A^{(i)} -$
$NC^{(i)}$ and the operation of selecting the next nullification constant $NC^{(i+1)} =$
$[0||0||0||\ldots||0||0||||t_{i+2}^{(i+1)}||\ldots||t_{n-i-2}^{(i+1)}||t_{n-i-1}^{(i+1)}||0||0||t_{n+1}^{(i+1)}]$. The nullification algorithm is presented in Table 5.

The values $\Delta a_{i+2} \Delta a_{n-i-1}$, that will be subtracted from the corresponding values
$a_{i+2}^{(i+1)}$ and $a_{n-i-1}^{(i+1)}$, to get the values of the residuals, $a_{i+1}^{(i)}$ and $a_{n-i}^{(i)}$, are determined
only by the values of the corresponding residuals of the number $a_{i+1}^{(i)}$ and $a_{n-i}^{(i)}$.
Analytically this can be represented as the following two expressions

$$a_{i+2}^{(i+1)} = [a_{i+2}^{(i)} - \Delta a_{i+2}] \bmod m_{i+2} \quad \text{and} \quad a_{n-i-1}^{(i+1)} = [a_{n-i-1}^{(i)} - \Delta a_{n-i-1}] \bmod m_{n-i-1}.$$

In the process of sampling $NC^{(i)}$ by the values of the residuals $a_{i+1}^{(i)}$ and $a_{n-i}^{(i)}$ of a
number $A^{(i)}$, the same residuals will be transferred to the calculator on the appropriate
bases m_{i+2} and m_{n-i-1}. From the two-input tables $F_1\left\{a_{i+2}^{(i+1)}\right\} = \left[a_{i+1}^{(i)}; a_{i+2}^{(i)}\right]$ and
$F_2\left\{a_{n-i-1}^{(i+1)}\right\} = \left[a_{n-1}^{(i)}; a_{n-i-1}^{(i)}\right]$ the values $a_{i+2}^{(i+1)}$ and $a_{n-i-1}^{(i+1)}$ are selected (determined).

In this case, the total number of clock cycles that are free from addition, during
which calls are made to the NB and the formation of the next address is equal to
$[(n+1)/2]$, (where [x] is the integer closest to x number, but is not superior to it). At
the same time, nullification is carried out simultaneously on two information bases of
the SRC $a_1, a_n; a_2, a_{n-1}$ etc. After every two subtractions, one more additional time
step is needed to form the next address and refer to the accumulator of nullification
constants. In this regard, for every two cycles of addition ($\tau_{add} = \tau_0$) there is one
cycle free of addition.

Based on the above, the execution time of the nullification operation for the second
Z4 method of operational control will be determined as follows

$$T_{Z4} = \left[\frac{n+1}{2}\right] \cdot \tau_{add} + \left[\frac{\frac{n+1}{2}+1}{2}\right] \cdot \tau_{set}. \tag{7}$$

Considering that $\tau_{add} = \tau_{set}$ we get:

Table 5 PN DSR algorithm

Operation number	Contents of operation	
1	Appeal by the value of the remainders $a_1^{(0)}$ and $a_n^{(0)}$ of a number $A^{(0)}$ in BNC_0 for the $NC^{(0)}$	$a_2^{(1)}$ and $a_{n-1}^{(1)}$ residue values formation of a number $A^{(1)}$ as $a_2^{(1)} = t_2^{(1)} = \left[a_2^{(0)} - a_1^{(0)}\right] \mod m_2$ and $a_{n-1}^{(1)} = t_{n-1}^{(1)} = \left[a_{n-1}^{(0)} - a_n^{(0)}\right] \mod m_{n-1}$
2	Performing a subtraction operation $A^{(1)} = A^{(0)} - NC^{(0)}$	Appeal by the value of the remainders $a_2^{(1)}$ and $a_{n-1}^{(1)}$ of a number $A^{(1)}$ in BNC_1 for the $NC^{(1)}$
3	Performing a subtraction operation $A^{(2)} = A^{(1)} - NC^{(1)}$	$a_3^{(2)}$ and $a_{n-2}^{(2)}$ residue values formation of a number $A^{(2)}$ as $a_3^{(2)} = t_3^{(2)} = \left[a_3^{(1)} - a_2^{(1)}\right] \mod m_3$ and $a_{n-2}^{(2)} = t_{n-2}^{(2)} = \left[a_{n-2}^{(1)} - a_{n-1}^{(1)}\right] \mod m_{n-2}$
⋮	...	...
i	Performing a subtraction operation $A^{(i)} = A^{(i-1)} - NC^{(i-1)}$	Appeal by the value of the remainders $a_{i+1}^{(i)}$ and $a_{n-i}^{(i)}$ of a number $A^{(i)}$ in BNC_i for the $NC^{(i)}$
$i+1$	Performing a subtraction operation $A^{(i+1)} = A^{(i)} - NC^{(i)}$	$a_{i+2}^{(i+1)}$ and $a_{n-i-1}^{(i+1)}$ residue values formation of a number $A^{(i+1)}$ as $a_{i+2}^{(i+1)} = t_{i+2}^{(i+1)} = \left[a_{i+2}^{(i)} - a_{i+1}^{(i)}\right] \mod m_{i+2}$ and $a_{n-i-1}^{(i+1)} = t_{n-i-1}^{(i+1)} = \left[a_{n-i-1}^{(i)} - a_{n-i-2}^{(i)}\right] \mod m_{n-i-1}$

(continued)

Table 5 (continued)

Operation number	Contents of operation	
$i+2$	Appeal by the value of the remainders $a_{i+2}^{(i+1)}$ and $a_{n-i-1}^{(i+1)}$ of a number $A^{(i+1)}$ in BNC_{i+1} for the $NC^{(i+1)}$	$a_{i+3}^{(i+2)}$ and $a_{n-i-2}^{(i+2)}$ residue values formation of a number $A^{(i+2)}$ as $a_{i+3}^{(i+2)} = t_{i+2}^{(i+2)} = \left[a_{i+3}^{(i+1)} - a_{i+2}^{(i+1)} \right] \bmod m_{i+3}$ and $a_{n-i-2}^{(i+2)} = t_{n-i-2}^{(i+2)} = \left[a_{n-i-2}^{(i+1)} - a_{n-i-3}^{(i+1)} \right] \bmod m_{n-i-2}$
$\vdots$	...	...
$k-2$	Appeal by the value of the remainders $a_{n/2-1}^{(n/2-2)}$ and $a_{n/2+2}^{(n/2-2)}$ of a number $A^{(n/2-2)}$ in $BNC_{n/2-2}$ for the $NC^{(n/2-2)}$	$a_{n/2}^{(n/2-1)}$ and $a_{n/2+1}^{(n/2-1)}$ residue values formation of a number $A^{(n/2-1)}$ as $a_{n/2}^{(n/2-1)} = t_{n/2}^{(n/2-1)} = \left[a_{n/2}^{(n/2-2)} - a_{n/2-1}^{(n/2-2)} \right] \bmod m_{n/2}$ and $a_{n/2+1}^{(n/2-1)} = t_{n/2+1}^{(n/2-1)} = \left[a_{n/2+1}^{(n/2-2)} - a_{n/2}^{(n/2-2)} \right] \bmod m_{n/2+1}$
$k-1$	Performing a subtraction operation $A^{(n/2-1)} = A^{(n/2-2)} - NC^{(n/2-2)}$	Appeal by the value of the remainders $a_{n/2}^{(n/2-1)}$ and $a_{n/2+1}^{(n/2-1)}$ of a number $A^{(n/2-1)}$ in $BNC_{n/2-1}$ for the $NC^{(n/2-1)}$
k	Performing a subtraction operation $A^{(n/2)} = A^{(n/2-1)} - KH^{(n/2-1)}$. Getting a nullified $A^{(Z)}$ number $A^{(Z)} = A^{(n/2)} = [0||0|| \ldots ||0|| \ldots ||0||0||\gamma_{n+1} = a_{n+1}^{(n/2)}]$	

$$T_{Z4} = \left(\left[\frac{n+1}{2} \right] + \left[\frac{\frac{n+1}{2}+1}{2} \right] \right) \cdot \tau_{add} \tag{8}$$

When n is even, the expression (3) takes the form:

$$T_{Z4}' = \left(\frac{n}{2} + \left[\frac{\frac{n}{2}+1}{2} \right] \right) \cdot \tau_{add} \tag{9}$$

If $\frac{n}{2}$ is even, then

$$T_{Z4}' = \frac{3}{4}n \cdot \tau_{add}. \tag{10}$$

If $\frac{n}{2}$ is odd, then

$$T'_{Z4} = \left(\frac{3n+2}{4}\right) \cdot \tau_{add}. \tag{11}$$

When n is odd:

$$T''_{Z4} = \left(\frac{n+1}{2} + \left[\frac{\frac{n+1}{2}+1}{2}\right]\right) \cdot \tau_{add}. \tag{12}$$

If $\frac{n+1}{2}$ is even, then

$$T''_{Z4} = \frac{3}{4}(n+1) \cdot \tau_{add}. \tag{13}$$

If $\frac{n+1}{2}$ is odd, then

$$T''_{Z4} = \left(\frac{3n+5}{4}\right) \cdot \tau_{add}. \tag{14}$$

Figure 4 shows the time diagrams of the NB for the method of SN (diagram Z1), for the method of SN DSR (diagram Z2) and also for the first PN (diagram Z3) and the second PN DSR (diagram Z4) of the control methods considered in the article. Where: App $a_i^{(i-1)}$, $a_{n-i+1}^{(i-1)}||$—appealing to the values of digits $a_i^{(i-1)}$ and $a_{n-i+1}^{(i-1)}$ of number

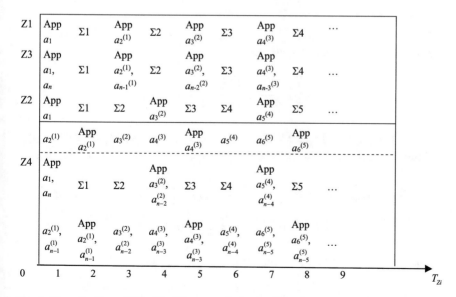

Fig. 4 Time diagrams of NB with different methods of nullification

$A^{(i-1)} = (0|| \ldots ||0||a_i^{(i-1)}||a_{i+1}^{(i-1)}|| \ldots ||a_{n-i}^{(i-1)}||a_{n-i+1}^{(i-1)}||0|| \ldots ||0||a_{n+1}^{(i-1)})$ in NB for

the $NC^{(i)} = (0|| \ldots . |0||t_{i,i}||t_{i+1,i}|| \ldots ||t_{n-i,i}||t_{n-i+1,i}||0|| \ldots ||0||t_{n+1,i})$; $a_{i+1}^{(i+1)}$,

$a_{n-i}^{(i+1)}$ – creation by the values $a_i^{(i)}$ and $a_{n-i+1}^{(i)}$ of a value $A^{(i-1)}$ of the following digits

$a_{i+1}^{(i)}$ and $a_{n-1}^{(i)}$ for the number $A^{(i)} = (0|| \ldots ||0||a_{i+1}^{(i)}|| \ldots ||a_{n-1}^{(i)}||0|| \ldots ||0||a_{n+1}^{(i)})$;

$\sum i$ – operation of subtracting the value of the constant $NC^{(i-1)}$ from a number

$A^{(i-1)}$, i.e. performing operation $A^{(i-1)} - NC^{(i-1)}$.

It is convenient to prove the derived relation (8) by the method of mathematical induction on n.

The first phase of the proof. For the minimum value of $n = 3$ the nullification time is equal to $T_{Z4} = 3 \cdot \tau_{add}$. This is obvious from Fig. 4, Z4 diagram (PN DSR).

Second phase. Suppose that expression (8) is also valid for $n = K$, i.e.

$$T_{Z4} = \left(\left[\frac{K+1}{2} \right] + \left[\frac{\left[\frac{K-1}{2} \right] + 1}{2} \right] \right) \cdot \tau_{add}.$$

The third stage. We prove that expression (7) is also valid for $n = K + 1$, i.e.

$$T_{Z4} = \left(\left[\frac{K+2}{2} \right] + \left[\frac{\left[\frac{K+2}{2} + 1 \right]}{2} \right] \right) \cdot \tau_{add}.$$

When K is even ($K + 1$ is odd) we have:

$$T'_{Z4} = \left(\frac{K}{2} + 1 + \left[\frac{\frac{K}{2} + 2}{2} \right] \right) \cdot \tau_{add}.$$

If $\frac{K}{2}$ is even, then $T'_{Z4} = \left(\frac{3K+8}{4} \right) \cdot \tau_{add}$. If $\frac{K}{2}$ is odd, then $T'_{Z4} = \left(\frac{3K+6}{4} \right) \cdot \tau_{add}$. If K is odd ($K + 1$ is even), we have:

$$T''_{Z4} = \left(\frac{K+1}{2} + \left[\frac{\left[\frac{K+1}{2} \right] + 1}{2} \right] \right) \cdot \tau_{add}.$$

If $\frac{K+1}{2}$ is even, then $T''_{Z4} = \left(\frac{3K+3}{4} \right) \cdot \tau_{add}$; if $\frac{K+1}{2}$ is odd, then $T''_{Z4} = \left(\frac{3K+5}{4} \right) \cdot \tau_{add}$. Then, in accordance with expressions (10), (11), (12) and (14), we write that:

$$\frac{3K+3}{4} \cdot \tau_{add} = \frac{3}{4}(K+1) \cdot \tau_{add}, \quad \frac{3K+5}{4} \cdot \tau_{add} = \left\{ \frac{3(K+1)+2}{4} \right\} \cdot \tau_{add},$$

$$\frac{3K+6}{4} \cdot \tau_{add} = \frac{3}{4}\{(K+1)+1\} \cdot \tau_{add}, \quad \frac{3K+8}{4} \cdot \tau_{add} = \left\{ \frac{3(K+1)+5}{4} \right\} \cdot \tau_{add}.$$

Thus, expression (8) is also valid for $n = K + 1$. $\square$

6 Calculation and Comparative Analysis of the Main Characteristics of the Data Control Methods in the SRC

When choosing a method for controlling data in the SRC, it is necessary to take into account the quantitative values of the indicator characterizing this method. So, to implement the nullification procedure for the second (Z4) method in the NB, it is necessary to have

$$K_{Z4} = \sum_{i=1}^{[\frac{n}{2}]} (m_i \cdot m_{n-i+1} - 2) \tag{15}$$

nullification constants. The number of N_{Z3} binary digits of the nullification constants is determined by the expression

$$K_{Z4} = \sum_{i=1}^{[\frac{n}{2}]} (m_i \cdot m_{n-i+1} - 2) \cdot (n - 2i + 1). \tag{16}$$

The derived expressions (8), (9) and (12) are the working formulas for estimating the performance of the implementation of the nullification procedure depending on the values of n and τ_{add}. In practical calculations, it is recommended to use expressions (10), (11), (13) and (14) to determine the time of data control in the SRC. The summarized main characteristics of all four data control methods are presented in Table 6. Based on the data of Table 7, Table 8 presents the results of calculating the characteristics of the considered control methods for l-byte ($l = \overline{1 \div 8}$) bit grids of the calculator in the SRC.

For simplicity and convenience of conducting a comparative analysis of the effectiveness of the use of the presented control methods, the summarized data of Table 8 should be placed in three (by number of characteristics) separate tables (Tables 9, 10 and 11).

Based on Tables 11 and 12 of the comparative analysis of the effectiveness of the use of the method of operational control of the PN DSR in comparison with the existing control methods in the SRC on the speed of the implementation of the nullification procedure was compiled.

The coefficient K_{Zi} of efficiency of using the method of control of PN DSR, in comparison with existing methods of nullification, is determined by the ratio $K_{Hi} = \frac{T_{Z1} - T_{Zi}}{T_{Z1}} \cdot 100\%$ ($i = \overline{1, 3}$).

From Table 12 one can see the high efficiency of the method of operational control proposed in the article (Z4), based on the implementation of the parallel nullification procedure with the definition of subsequent residuals.

Table 6 The main characteristics of the methods of data control in the SRC

Methods of data control Z_i		Nullification time T_{Zi}	Number of nullification constants K_{Zi}	The number of binary digits of constants N_{Zi}
Z1	Sequential nullification method	$T_{Z1} = 2 \cdot n \cdot \tau_{add}$	$K_{Z1} = \sum_{i=1}^{n} (m_i - 1)$	$N_{Z1} = \sum_{i=1}^{n} (m_i - 1) \cdot (n - i + 1)$
Z2	Successive nullification method with the definition of the subsequent residual	$T_{Z2} = \left(\left[\frac{n-1}{2} \right] + n \right) \times \tau_{add}$	$K_{Z2} = \sum_{i=1}^{n-1} (m_i - 1)$	$N_{Z2} = \sum_{i=1}^{n-1} (m_i - 1) \cdot (n - i)$
Z3	Parallel nullification method	$T_{Z3} = n \cdot \tau_{add}$	$K_{Z3} = \sum_{i=1}^{\left[\frac{n}{2} \right]} (m_i \cdot m_{n-i+1} - 1)$	$N_{Z3} = \sum_{i=1}^{\left[\frac{n}{2} \right]} (m_i \cdot m_{n-i+1} - 1) \cdot (n - 2 \cdot i + 1)$
Z4	Parallel nullification method with the definition of subsequent residues	$T_{Z4} = \left(\left[\frac{n+1}{2} \right] + \left[\frac{n+1}{2} \right] \right) \cdot \tau_{add}$	$K_{Z4} = \sum_{i=1}^{\left[\frac{n}{2} \right]} (m_i \cdot m_{n-i+1} - 2)$	$N_{Z4} = \sum_{i=1}^{\left[\frac{n}{2} \right]} (m_i \cdot m_{n-i+1} - 2) \cdot (n - 2 \cdot i + 1)$

Table 7 The set of bases of the SRC for *l*-byte ($l = \overline{1 \div 8}$) bit grids of the calculator

Bit grid size	SRC information bases	SRC control bases
$l(n)$	$\{m_i\}, i = \overline{1, n}$	m_{n+1}
1(4)	$m_1 = 3, m_2 = 4, m_3 = 5, m_4 = 7$	$m_5 = 11$
2(6)	$m_1 = 2, m_2 = 5, m_3 = 7, m_4 = 9, m_5 = 11, m_6 = 13$	$m_7 = 17$
4(10)	$m_1 = 2, m_2 = 3, m_3 = 5, m_4 = 7, m_5 = 11, m_6 = 13,$ $m_7 = 17, m_8 = 19, m_9 = 23, m_{10} = 29$	$m_{11} = 31$
8(16)	$m_1 = 3, m_2 = 4, m_3 = 5, m_4 = 7, m_5 = 11, m_6 = 13,$ $m_7 = 17, m_8 = 19, m_9 = 23, m_{10} = 29, m_{11} = 31,$ $m_{12} = 37, m_{13} = 41, m_{14} = 43, m_{15} = 47, m_{16} = 53$	$m_{17} = 59$

Table 8 Calculated data of the characteristics of the data control methods in the SRC for *l*-byte ($l = \overline{1 \div 8}$) bit grids

$l(n)$	Z1			Z2			Z3			Z4		
	$\frac{T_{Z1}}{\tau_{add}}$	K_{Z1}	N_{Z1}	$\frac{T_{Z2}}{\tau_{add}}$	K_{Z2}	N_{Z2}	$\frac{T_{Z3}}{\tau_{add}}$	K_{Z3}	N_{Z3}	$\frac{T_{Z4}}{\tau_{add}}$	K_{Z4}	N_{Z4}
1 (4)	8	15	31	5	9	16	4	39	79	3	37	75
2 (6)	12	41	106	8	29	65	6	141	349	4	138	340
3 (8)	16	71	217	11	53	146	8	263	995	6	259	979
4 (10)	20	119	412	14	91	293	10	479	1955	7	474	1930
8 (16)	32	367	1947	23	315	1580	16	2581	16,493	12	2573	16,429

Table 9 Time characteristics of control methods

$l(n)$	T_{Z_i}/τ_{add}			
	Z1	Z2	Z3	Z4
1 (4)	8	5	4	3
2 (6)	12	8	6	4
3 (8)	16	11	8	6
4 (10)	20	14	10	7
8 (16)	32	23	16	12

The minimum values of indicators that characterize the presented methods of nullification are highlighted in bold

7 Conclusion

Based on the analysis of the research results, we can conclude that calculation errors arising due to failures (malfunctions) of binary-bit circuits in an arbitrary computational path of CSCPID are not "multiplied" into adjacent computational paths (remain within the limits of one residual basis of the SRC), which makes it possible to increase the reliability of calculations. It does not matter if there were single or

Table 10 Characteristics of control methods

$l(n)$	K_{Z_i}			
	Z1	**Z2**	Z3	Z4
1 (4)	15	**9**	39	37
2 (6)	41	**29**	141	138
3 (8)	71	**53**	263	259
4 (10)	119	**91**	479	474
8 (16)	367	**315**	2581	2573

The minimum values of indicators that characterize the presented methods of nullification are highlighted in bold

Table 11 Characteristics of control methods

$l(n)$	N_{Z_i}			
	Z1	**Z2**	Z3	Z4
1 (4)	31	**16**	79	75
2 (6)	106	**65**	349	340
3 (8)	217	**146**	995	979
4 (10)	412	**293**	1955	1930
8 (16)	1947	**1580**	16493	16429

The minimum values of indicators that characterize the presented methods of nullification are highlighted in bold

Table 12 Data of comparative analysis of the data control time in the SRC

n	$T_{Z_i} = T/\tau$				Gain in (%)		
	T_{Z1}	T_{Z2}	T_{Z3}	T_{Z4}	K_{Z1}	K_{Z2}	K_{Z3}
4	8	5	4	**3**	62	40	25
6	12	8	6	**4**	66	55	33
8	16	11	8	**6**	62	45	25
10	20	14	10	**7**	65	53	30
16	32	23	16	**12**	62	47	25

The minimum values of indicators that characterize the presented methods of nullification are highlighted in bold

multiple errors or even a packet of errors that is no longer than $[\log 2(m_i - 1) + 1]$ binary digits. he error that occurred in the computational path on the basis of the m_i SRC, is either stored in this path until the end of the computation, or is eliminated in the course of further computations (for example, by multiplying the number by zero).

The use of this feature of SRC allows to create a unique system of monitoring, diagnostics and error correction in the dynamics of the computing process (without stopping the calculations) with the introduction of minimal information redundancy, which is essential for computer systems and components operating in real time.

The developed methods can be effectively used to solve the following problems [10–12]:

- cryptographic and modular transformations;
- signal processing;
- image processing;
- integer data processing of large (hundreds of bits) bit capacity in real time;
- vector and matrix processing of large amounts of information;
- neurocomputer information processing;
- optoelectronic tabular information processing;
- implementation of algorithms for FFT and DFT, etc.

References

1. Ananda Mohan PV (2016) Residue number systems: theory and applications. Springer International Publishing Switzerland, Birkhäuser, Basel
2. Krasnobayev V, Kuznetsov A, Koshman S, Moroz S (2019) Improved method of determining the alternative set of numbers in residue number system. In: Chertov O, Mylovanov T, Kondratenko Y, Kacprzyk J, Kreinovich V, Stefanuk V (eds) Recent developments in data science and intelligent analysis of information. ICDSIAI 2018. Advances in intelligent systems and computing, vol 836. Springer, Cham, 05 Aug 2018, pp 319–328. https://doi.org/10.1007/978-3-319-97885-7_31
3. Akushskii IY, Yuditskii DI (1968) Machine arithmetic in residual classes. Sov Radio. Moscow (in Russian)
4. Krasnobaev V, Popenko V, Kuznetsova T, Kuznetsova K (2019) Examples of usage of method of data errors correction which are presented by the residual classes. In: 2019 IEEE international conference on advanced trends in information theory, pp 45–50. https://doi.org/10.1109/ATIT49449.2019.9030512
5. Krasnobaev V, Koshman S, Kononchenko A, Kuznetsova K, Kuznetsova T (2019) The formulation and solution of the task of the optimum reservation in the system of residual classes. In: 2019 IEEE international conference on advanced trends in information theory, pp 481–487. https://doi.org/10.1109/ATIT49449.2019.9030483
6. Gorbenko I et al (2017) The research of modern stream ciphers. In: 2017 4th international scientific-practical conference problems of infocommunications. Science and technology (PIC S&T). Available at: https://doi.org/10.1109/infocommst.2017.8246381
7. Dubrova E (2013) Fault-tolerant design: an introduction course notes. Royal Institute of Technology, Stockholm, Sweden, 147p. https://pld.ttu.ee/IAF0530/draft.pdf
8. Meyer CH (1978) Ciphertext/plaintext and ciphertext/key dependence vs. number of rounds for the data encryption standard. In: Proceedings of the 1978 National computer conference. AFIPS Press, Montvale, NJ
9. Bajard J, Didier L, Kornerup P (1998) An RNS montgomery modular multiplication algorithm. IEEE Trans Comput 47(7):766–776 (1998). Available: https://doi.org/10.1109/12.709376
10. Zima E, Stewart A (2007) Cunningham numbers in modular arithmetic. In: Program Comput Softw 33(2):80–86. Available: https://doi.org/10.1134/s0361768807020053

11. Karpinski M, Ivasiev S, Yakymenko I, Kasianchuk M, Gancarczyk T (2016) Advanced method of factorization of multi-bit numbers based on Fermat's theorem in the system of residual classes. In: 2016 16th international conference on control, automation and systems (ICCAS), Gyeongju, pp 1484–1486
12. Kuznetsov A et al (2018) Periodic properties of cryptographically strong pseudorandom sequences. In: 2018 international scientific-practical conference problems of infocommunications. Science and technology (PIC S&T). Available at: https://doi.org/10.1109/infocommst.2018.8632021

Traffic Monitoring and Abnormality Detection Methods for Decentralized Distributed Networks

Dmytro Ageyev⬤, Tamara Radivilova⬤, Oksana Mulesa⬤,
Oleg Bondarenko⬤, and Othman Mohammed⬤

Abstract Modern information systems are most often created as distributed, and include a large number of various devices connected by means of networks. Devices included in this distributed information system have different performance and a set of functions, which are often limited. The development of such distributed systems at the present stage has been transformed into the system of the Internet of Things and further into the Internet of Everything. The widespread development of distributed information systems based on public networks, which penetrate many areas of human activity, creates additional threats to information security and makes them extremely relevant. Traffic monitoring is one of the stages of information security systems functioning. Traffic monitoring results are used to solve the problems of detecting traffic anomalies and detecting attacks and intrusions. Modern detections of attacks and other anomalies are not reliable enough and require further re-search. The chapter provides the classification of attacks and intrusions, as well as the classification of anomalies and intrusions. The main attention is paid to intrusion detection systems and a method for detecting network traffic anomalies and intrusions is proposed, which is based on the use of a number of statistical, correlation and informational parameters of network traffic as additional features. The chapter researches a method for detecting traffic anomalies based on machine learning methods. In the end of the chapter, attention is paid to several ways of building and implementing sub-systems for detecting anomalies in distributed networks that include devices with limited performance.

Keywords Traffic abnormalities · Statistical analysis · WSN · IoT · Decentralized distributed networks

D. Ageyev · T. Radivilova (✉) · O. Bondarenko · O. Mohammed
Kharkiv National University of Radio Electronics, 14 Nauki ave., Kharkiv 61166, Ukraine

D. Ageyev
e-mail: dmytro.aheiev@nure.ua

O. Mulesa
Uzhhorod National University, Narodna Sq., 3, Uzhhorod, Ukraine
e-mail: oksana.mulesa@uzhnu.edu.ua

© The Author(s), under exclusive license to Springer Nature Switzerland AG 2022
R. Oliynykov et al. (eds.), *Information Security Technologies in the Decentralized Distributed Networks*, Lecture Notes on Data Engineering and Communications Technologies 115, https://doi.org/10.1007/978-3-030-95161-0_13

1   Introduction

The development of technologies has led to the widespread introduction of distributed information systems, which are based on distributed networks. Industry 4.0 can be considered as an example of a modern information system, which is a modern trend. These systems provide for the use of new technologies in the industrial field and are based on technologies such as big data, cyber-physical systems, IoT, virtualization and cloud computing [1].

Distributed information systems are also widely used in information collection systems and various types of monitoring. Low performance requirements for devices of this class of networks allow them to be made highly autonomous, low cost and low in size. On the other hand, their widespread introduction into many spheres of human life makes them attractive for all sorts of attacks.

Thus, the implementation of information security systems in distributed systems and networks is an urgent task that requires an immediate solution. One of the subsystems of the information protection system is a subsystem for detecting attacks and traffic anomalies.

In these conditions, the low performance of devices does not allow to fully use the accumulated experience in protecting information used in computer networks and large information systems.

The above puts forward requirements for the development of new approaches and solutions to adapt existing technologies for protecting information in distributed networks, which would consider the features of such devices. The application of machine learning methods for detecting computer attacks has been actively discussed in recent years, while an important aspect of the studied subject area is the assessment of the possibility of practical implementation of the developed algorithms. Enough works have been published on this topic, which can serve as the basis for further research.

So, in article [2] the authors consider the problem of detecting low and slow distributed denial of service (DDoS) attacks. A feature of this type of attacks is their significant difference from mass attacks in that they do not affect the traffic intensity on the network. Researchers suggest that slow attacks depend on user behavior. The proposed attack detection method is based on the analysis and prediction of individual user behavior. Dependences of the forecast accuracy on the accumulated statistics of attacks and network traffic are shown.

The studies carried out by us in this work are a development of the previous ones [3, 4], which are based on the use of machine learning methods for detecting anomalies and detecting attacks using an extended set of features by including fractal and information characteristics of traffic and using virtualization methods [5] for their implementation.

The authors of the article [6] are studying machine learning models for detecting anomalies in the network traffic of the Internet of Things. Comparative analysis of such machine learning models as support vector machine, naive Bayesian method, acceleration, K-nearest neighbors, random forest and logistic regression is carried out. According to the results of the study, it was revealed that the time spent on preparing traditional machine learning models is not large, which is explained by the absence of additional costs of computing power and resources.

2 Intrusions and Attack Classification

One of the modern problems is the insufficient security of networks. Malicious influences on the network are also increasing simultaneously with the network infrastructure developing. The last opens opportunities for cybercriminals to access confidential data. Attackers have sufficient interest in obtaining information, as well as in blocking access to these information system resources for legal users.

The deployment of IoT networks and the widespread use of IoT devices with their penetration into the spheres of human activity creates additional possibilities for intrusions. This situation is also complicated by the insufficient resources of IoT devices to implement full-fledged information security solutions on them.

Existing and widely used means of defense such as firewalls, cryptography and access control make it possible to effectively solve the problems of guarding networks from internal and external threats and act as its security perimeter. So, firewalls are designed to protect the edge network access nodes from several attacks and threats. Access control is used for authentication, while cryptography solves communication security problems. The use of these solutions is effective only for providing external security and they become ineffective for providing defense against internal threats and detecting internal attacks. An effective solution to this problem is the use of Intrusion Detection System (IDS), which makes it possible to identify both external and internal attacks.

Automatic identification of attacks on the network and the response to them is provided by IDS, which can have both hardware and software implementation. The IDS sensors, receiving warnings from the IDS, continuously monitor the state of the system to effectively eliminate and identify inappropriate actions or potential incidents [7]. The IDS then takes specific countermeasures to ensure the security of the computing environment. Based on this, it can be argued that the implementation of the notification process is the correct mechanism for checking the optimal response of these systems. In addition, there is a need to implement statistical detection algorithms, as well as to classify attacks depending on their impact on confidentiality, integrity, and availability of data. Thus, if the target of an attack is a network, then the response should be aimed at increasing its performance and the availability of its resources. At that time, if the integrity of the database system is under attack, then the answer must be measures to ensure its integrity.

It is impossible to evaluate the answer without considering the incidents that occur. Based on this, the main goals of the classification of incidents include the following: analysis of potential incidents, identification of the facts of an attack and determination of their goals, as well as the use of appropriate response options that ensure neutralization of the attack. Thus, an incident is an unexpected event on a functioning network. For example, an incident can occur when information resources are accessed during an attack. The classification of attacks on the security system can be represented as shown in Fig. 1.

Unlike passive attacks, when attackers use eavesdropping and do not make any changes to the information circulating in the network, with an active attack, the attackers aim to disrupt the functioning of the network. To do this, they attempt to obtain unauthorized access with the subsequent change or destruction of information. Thus, systems must have the capabilities and functionality of rapid recovery after

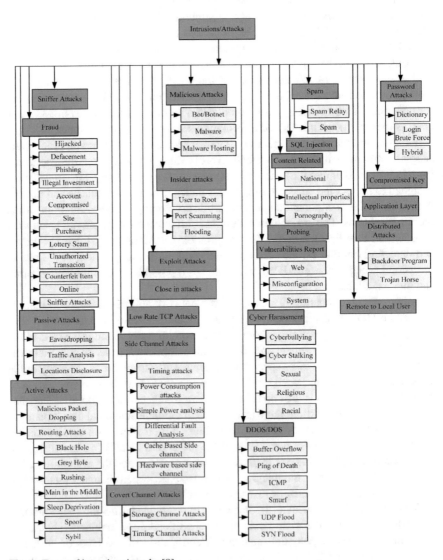

Fig. 1 Types of intrusions/attacks [8]

an active destructive attack. Security incidents can be divided into network incidents and incidents in a node [9]. Network-based incidents contain attacks to networks targeted to influence performance and availability of network. On wired networks, attackers typically infiltrate the network using gateways and firewalls. While in wireless networks, attackers using access from multiple access points usually attack any open node. Host-based attacks that incorporated by application layer can include race condition attacks, MITM attacks, buffer overflow attacks, spamming and mail forgery. Host-based attacks usually against OS performance, web service processes and availability of system [10].

Figure 1 shows the classification of possible attacks aimed at information systems and networks [9, 11]. The presented attacks are divided into distributed, external, internal, passive, active, spoofing, DDoS/DoS and sniffing attacks. These attacks can affect security policies: confidentiality, integrity and availability (CIA).

One of the most active areas in the development of network security is methods of detecting traffic anomalies and attacks and preventing intrusions into information systems and networks.

The solution of these problems is achieved through the use of a number of specialized tools and algorithms that allow detecting already known and unknown attacks, including through the use of signature and behavioral methods, as well as methods for detecting anomalous changes in the behavior of the system and the characteristics of processes occurring in it. The latter are highly effective in detecting zero-day and internal attacks.

For the main classification features can be used next.

1. Type of Source;
2. Origin Place;
3. Method of manifestation;
4. Occurance cause;
5. Sort of changes.

To solve the problem of detecting network attacks, the most important features will be such as the source, the nature of traffic changes and the area of manifestation. The classification of network abnormalities by causes and nature of traffic changes is given in Table 1.

3 Internet of Things Architecture

Until now, there was no architecture standard for the Internet of Things. At the same time, various architectures have already been proposed, differing in the number of levels and functions allocated for each level. The main variants of the architecture standard include, for example, service-oriented architecture; middleware architecture; five-tier, four-tier and three-tier architecture. Let us dwell briefly on these architectures.

Table 1 Description of network traffic abnormalities

Type and cause of network abnormalities	Description	Characteristics of traffic changes
Alpha abnormality	Unusually high point-to-point traffic	Emission in the representation of traffic bytes/s, packets/s on one dominant source–destination stream. Short duration (up to 10 min)
DoS-, DDoS-attack	Distributed denial-of-service attack versus target system	Emission in the traffic view packets/s, streams/s, from multiple sources to a single destination address
Overload	Unusually high demand for one network resource or service	Jump in traffic on streams/s to one dominant IP address and a dominant port. Usually a short-term anomaly
Network/port scanning	Scan the network for specific open ports or scan a single host for all ports to look for vulnerabilities	Jump in traffic on streams/s, with several packets in streams from one dominant IP address
Worm activity	A malicious program that spreads itself over a network and exploits OS vulnerabilities	Discharge in traffic without a dominant destination address, but always with one or more dominant destination ports
Point to Multipoint	Distribution of content from one server to many users	Emission in packets, bytes from the dominant source to several destinations, all to one well-known port
Disconnection	Network problems that cause a drop in traffic between one source–destination pair	The drop in traffic on packets, streams and bytes is usually to zero. Can be long-term and include all source-to-destination streams from or to a single router
Flow switching	Unusual switching of traffic flows from one inbound router to another	Drop in bytes or packets in one traffic stream and release in another. May affect multiple traffic flows

The basic architecture should include a three-tier model, which includes the perception layer, the network layer and the application layer [12, 13]. Considering the four-tier architecture, one can distinguish between the perception, network, middleware and application layers [12, 14]. The intermediate layer includes the roles of data storage, service management, and service composition [15]. At the same time, the proposed five-tier architecture contains such components and layers as object abstraction, service management, application and business layers [13].

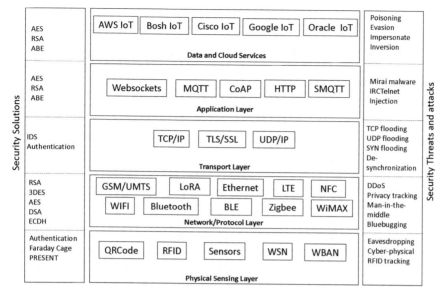

Fig. 2 Extended architecture of the Internet of Things

The architecture of the Internet of Things was further developed in [16], where the authors propose to add to the architecture of the Internet of Things several advanced functions such as Machine Learning components, IoT data and lightweight encryption algorithms. The architecture of the Internet of Things proposed by them is shown in Fig. 2.

This architecture is based on a five-tier model that includes the perception layer; protocol/network layer; transport layer; the application layer and the cloud services and data layer.

The physical layer contains various devices used in the Internet of Things systems such as Wireless Body Area Network, Wireless Sensors Network, different sensors, that used for monitoring object of real-life environment, RFIDs and QR-Codes.

The Protocol and Network Layer encompasses the various wired and wireless network protocols involved in the Internet of Things networks. These protocols include LTE and 5G, Bluetooth, Ethernet, ZigBee, Wi-Fi etc.

Transport layer covers transport layer technologies and protocols of networks such as Transport Layer Security (TLS)/Secure Sockets Layer (SSL), UDP/IP and TCP/IP.

We refer to the functions of application layer components as various application protocols that are specific to IoT systems and that implement its requirements for relatively low device performance and low power consumption. These include protocols and technologies such as Message Queuing Telemetry Transport (MQTT), Constrained Application Protocol (CoAP), and Advanced Message Queuing Protocol (AMQP).

The upper level of the architecture is represented by a layer of cloud services and data that implement the main tasks for which a distributed information system based on the Internet of Things is being created.

In addition, the figure shows the corresponding functional components and protocols that ensure the information security of the Internet of Things.

4 Anomaly-Based Network Intrusion Detection System Generic Architecture

The Anomaly-based network intrusion detection system generic architecture is shown in Fig. 3.

Traffic capture: Network intrusion detection begins with traffic capture. This operation is performed for both inbound and local traffic, because intrusions can be caused by both external and internal attackers. Traffic is captured at different levels of detail, which improves detection algorithms (by increasing the amount of data).

Anomaly detection engine: The main component of the Anomaly-based network intrusion detection system is the anomaly detection mechanism. Network traffic data is pre-processed for this component. This eliminates known attacks using the abuse detection approach and later detects unknown attacks using an anomaly-based approach. Anomaly detection algorithms are divided into controlled, partially

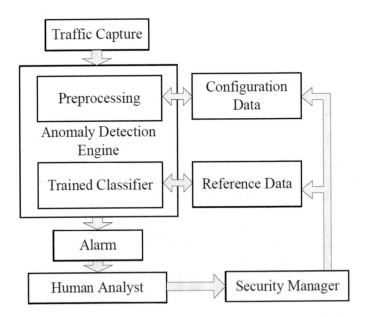

Fig. 3 Anomaly-based network intrusion detection system generic architecture [17]

controlled and uncontrolled training. When training with a teacher, a profile is created in advance with labels of the traffic received in a normally functioning network. The traffic profile is obtained for a specific network environment and is sensitive to its change. The latter can lead to false alarms in the future. It is also worth noting that the data for training we receive from the traffic of the real network and in this set of traffic may contain hidden malicious traffic. This can later lead to the fact that this hidden traffic will later be perceived as legal. This effect can be avoided if going from learning with a teacher to learning without a teacher.

Reference data: This block contains profiles of normal network traffic or information about known attack signatures. When information about new signatures and traffic profiles is detected, processing elements update reference data.

Configuration data: These data are intermediate results, such as intrusion signatures that created partially, for example. Thanks to a combination of intermediate results and previously acquired knowledge, we get relevant results.

Alarm: This unit creates signal of alarm then it obtain signal from the anomaly detector.

Human analyst: This element response for alarm signals analysis and interpretation, and suitable actions.

Security manager: After detection new intrusions, security manager adds new record about intrusion profile to storage.

5   Parameters that Used to Form Features for Traffic Anomality Detection

Important parameters that can be used to detect traffic anomalies are its statistical characteristics. We use the sliding window method to evaluate them. This allows to conduct it while transmitting traffic on the network and receive the values of these parameters in real time. The values obtained in each window are further used to form a space with a sign of IoT network traffic to identify anomalies and detect attacks.

5.1   Statistical Parameters Used for Anomality Detection

The analysis carried out involves calculations for each of such parameters as:

- sample mean calculated as

$$\hat{m}_i = \frac{1}{n} \sum_{j=i}^{i+n} S_j, \tag{1}$$

where S_j is traffic intensity sample value at time moment t_j;
- sample variance

$$\delta_i^2 = \frac{1}{n-1} \sum_{j=i}^{i+n} \left(S_j - \hat{m}_i\right)^2; \tag{2}$$

- asymmetry coefficient

$$\gamma_{i_1} = \frac{1}{n} \frac{\sum_{j=i}^{i+n} \left(S_j - \hat{m}_i\right)^3}{\sigma_i^3}, \tag{3}$$

determines the probability density asymmetry about an axis that passes through its gravity center;
- kurtosis

$$\gamma_{i_2} = \frac{1}{n} \frac{\sum_{j=i}^{i+n} \left(S_j - \hat{m}_i\right)^4}{\sigma_i^4} - 3, \tag{4}$$

this parameter allows to determine the top sharpness degree for probability density in comparison with the Gaussian distribution. A positive value indicates a sharper top than a Gaussian distribution, while a negative value indicates a flatter top of the probability density;
- antikurtosis

$$\gamma_{i_2}' = \frac{1}{\sqrt{\eta}}, \tag{5}$$

where η is kurtosis parameter that can founded as

$$\eta = \gamma_{i_2} = \frac{\mu_4}{\sigma^4}, \tag{6}$$

where μ_4 is sample fourth central moment; σ is standard deviation.

5.2 Correlation Parameters Used for Anomality Detection

The above parameters describe the statistics of the amounts of network traffic, while the time dependences and spectral properties of traffic can be determined using correlation analysis, including such dependences and parameters as correlation function, correlation interval and correlation coefficient.

The correlation function is defined as

$$R_i(k) = \frac{1}{n} \sum_{j=i}^{n+i-k} \left(s_j - \hat{m}_i\right)\left(s_{j+k} - \hat{m}_i\right), \tag{7}$$

where k is time shift or lag.

The correlation interval T_{cor} is defined as autocorrelation function argument value, where this function changes the sign and determines for each window.

Correlation coefficient we defined as normalized correlation function

$$r_j(k) = \left(\frac{R_j(k)}{R_j(0)}\right). \tag{8}$$

5.3 Informational Parameters Used for Anomality Detection

Studies in the field of classification of temporal sequences have shown that informational characteristics can be used as features. For example, they have been successfully used in a model for detecting network traffic anomalies [18]. These parameters can be defined next way.

- Entropy is a key informational parameter; it is a measure of the uncertainty of the collected data and for given dataset X can be calculated as

$$H(X) = \sum_{x \in D} P(x) \log \frac{1}{P(x)}, \tag{9}$$

where $P(x)$ is probability of x in X.
- Entropy of X given that Y is the entropy for $P(x|y)$ named as conditional entropy and can be find as

$$H(X|Y) = \sum_{x,y \in D,Y} P(x, y) \log \frac{1}{P(x|y)}, \tag{10}$$

where $P(x|y)$ the conditional probability of x given y and $P(x, y)$ is the joint probability of x and y.
- Entropy between two distributions $p(x)$ and $q(x)$ for same $x \in X$ named as relative entropy and can be defined as

$$rel_{H(p|q)} = \sum_{x \in X} P(x) \log \frac{p(x)}{p(q)}, \tag{11}$$

- Entropy between two conditional distributions ($p(x|y)$ and $q(x|y)$) for same $x \in X$ and $y \in Y$ named as relative conditional entropy and can be defined as

$$rel_{cond_H(p|q)} = \sum_{x,y \in X,Y} P(x, y) \log \frac{p(x|y)}{q(x|y)}. \tag{12}$$

- Information gain is a measure of the information gain of an attribute or feature A in a dataset X and is

$$Gain(D, A) = H(D) - \sum_{v \in values(A)} \frac{|D_v|}{|D|} H(D_v), \tag{13}$$

where values (A) is the set of possible values of A and D_v the subset of D where A has the value v.

The previously described parameters, together with other indicators of network traffic, are used to train the anomaly symptom detection model.

6 Traffic Abnormality Detection Method

6.1 Features Selection

The set of features that is used in the model has a significant impact on the efficiency of a machine learning model [19]. The selection of features in the formation of the feature space is a mandatory procedure both at the preparatory stage (prior to training) and at the stage of evaluating the results obtained and subsequent adjustment of the training sample and/or model hyperparameters.

At the preliminary stage, the signs associated with time stamps and addresses of the source and destination were excluded. Further, the analysis of the significance of the features is carried out using a method that implements the entropy approach to assessing their importance.

The next step was to further reduce the features from the original dataset using a correlation matrix with linear correlation coefficients (Pearson's correlation coefficients), which was calculated for the twenty most significant features. Using the results of the correlation matrix, we determine the groups (pairs) of features with a high value of the correlation coefficient. In each group, we the sign that is the most informative.

After processing the traffic, we used the following as sets of features.

- "Max Packet Length", the maximum packet length.
- "Mean Packet Size", the mean length of the TCP/IP packet data field.
- "Rate Bytes/s", data traffic rate.
- "Fwd Packet Length Max", the maximum length of a forward transmitted packet.
- "Fwd Header Length", the total length of the packet headers sent in the forward direction.
- "Fwd Packet Length Mean", the average length of forward transmitted packets.

- "Fwd IAT Min", the minimum value of the inter-packet interval in the forward direction.
- "Total Length of Fwd Packets", the total length of forward transmitted packets.
- "Fwd IAT Std", the standard deviation of the value of the interval between packets in the forward direction of packets.
- "Flow IAT Mean", the average value of the inter-packet interval.

The resulting set of features at the next stage was supplemented with additional features, which were statistical, correlation and informational characteristics of the IoT network traffic. For this, for each record of the analyzed dataset, we included the values calculated for the corresponding sliding observation window.

6.2 Traffic Abnormality Detection Method Testing

The quality of the responses of the classifiers (models) was compared using the following metrics:

- part of correct answers (accuracy);
- precision (how much you can trust the classifier);
- recall, (how many objects of the class "there is an attack" is determined by the classifier);
- F1-measure (harmonic mean between accuracy and completeness).

When determining the values of quality metrics, the elements of the confusion matrix are used, which correspond to the number of correct and incorrect answers based on the results of testing the classifier.

TP (True Positive) stands for a true positive answer, TN (True Negative) stands for a true negative answer, FP (False Positive) stands for a false positive answer (false positive, type I error), FN (False Negative) for a false negative response (skip attack, error of the second kind). Taking into account the above designations, the quality metrics used are determined by the following expressions:

$$Accuracy = \frac{TP + TN}{TP + FP + FN + TN}; \tag{14}$$

$$Precision = \frac{TP}{TP + FP}; \tag{15}$$

$$Recall = \frac{TP}{TP + FN}; \tag{16}$$

$$F1 = \frac{2 \cdot Precision \cdot Recall}{Precision + Recall}. \tag{17}$$

Implementation of ML Algorithms: Everything the research were spent by hopping on Python ML libraries (Mat-plotlib, Scikit-learn, NumPy, and Pandas). We tend to unionize the analysis of ML procedures for dataset in 3 stages:

- affecting the projected procedures on separately attack within the dataset disjointedly;
- affecting the procedures on the whole dataset with a group of options joining the simplest features for every attack;
- and applying the procedures on the whole dataset with the seven top features found within the feature choice stage.

For comparative analyses, the following machine learning models (algorithms) were selected.

- k nearest neighbors (KNN) method;
- support vector machine (SVM);
- decision tree (CART);
- random forest (RF);
- an adaptive boosting model over a decision tree (AdaBoost);
- logistic regression (LR);
- Bayesian classifier (NB).

Table 2 shows the obtained values of the quality metrics.

The best results are demonstrated by the KNN, CART, RF, AdaBoost, LR models. Given the minimum execution time, applying a RF model to the task at hand is a reasonable choice.

Table 2 Results of evaluation of classifiers quality

Model	Accuracy	Precision	Recall	F1
KNN	0.968	0.953	0.589	0.973
SVM	0.734	0.687	0.042	0.598
CART	0.974	0.969	0.953	0.972
RF	0.973	0.974	0.949	0.965
AdaBoost	0.984	0.961	0.962	0.973
LR	0.954	0.941	0.926	0.962
NB	0.737	0.529	0.966	0.772

7 Traffic Monitoring and Abnormality Detection Methods for Distributed IoT Networks Using Cloud ML-Based IDS Approach

The proposed architecture consists of two key layers that interact over the IoT (see Fig. 4): the layer of cloud service, which in control of managing ML-based IDS models, and the layer of device, where the the devices are placed in managed, organized assemblies, each operating the same IDS model [20]. Next, we will apply the layers of architecture and their collaborations.

Represents that cloud portion in Cloud MaaS that uses present SaaS (Software as a Service) cloud services to design and train models for every collection of equipment. Cases of attacks and anomalies from participating devices will also be collected by the service. An enormous number of peripherals with heterogeneous designs and roles are connected to the IoT network, creating the enormous workload required to maintain such a service and requiring the use of huge cloud computing resources.

The role of Cloud MaaS can be allocated to one or several hosts in the cloud. It depends on the utilization of nodes for the assigned device groups and the number of devices contained in these groups. Therefore, numerous nodes can divide the equipment sets from one another for load balancing. MaaS cloud nodes contain a number of functional elements described in [20] and shown in Fig. 5.

Fig. 4 Cloud-based IDS architecture

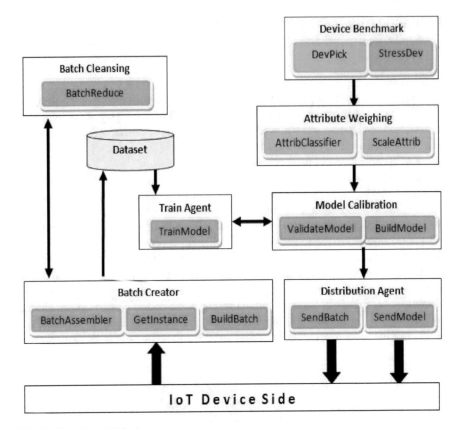

Fig. 5 Cloud-based IDS elements

Device Benchmark: Specifies existing resources (such as memory and CPU) and mean traffic rate for IDS model working on edge devices. Given element installs a test agent on the selected device element from each set of devices and collects agent measurements to perform the attribute weighting component.

Attribute weighting: assigns a weight to all significant features participating in building the model. These weights are indicative of the feature mining capacity at the peripherals and have to impact model structure. Weighted functions are presented to the model calibration module.

Calibration module: Revises model constraints created on active weights of critical properties and may ignore detected critical properties that overwork appropriate edge devices and initiate packet check error.

Batch Generator: assembles related samples from devices and combines given samples to batches for additive training.

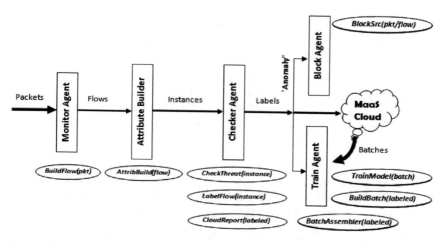

Fig. 6 IoT devices IDS componets [20]

Batch cleaning: adjust created batches for every defined device. It involves eliminating reported versions of device from their respective bundles to avoid duplication on the device side.

Delivery agent: this element distributes template to devices that participating in specific set of devices, when new template is created for this set.

Training Agent: prepares new models with a relevant assembled and modernized set of data, else it will gradually train the existing model with new batches.

At the IoT device level, edge devices are grouped into groups with the same roles, software, and hardware. The edge device is endowed with the functions of exchanging and updating model parameters from the cloud service. The model downloaded from the cloud service is used to detect attacks and anomalies. If an attack instance and an anomaly are detected, the edge device notifies the cloud service. The IDS structure for IoT devices was proposed and described in [20] (Fig. 6).

The IDS that runed on IoT edge devices contains five basic elements can operate in 2 modes: flow-based or packet-based.

Flow-based mode involves combining packets transmitted in one direction and belonging to one connection or session into a single object for analysis. Packet mode involves packet-by-packet analysis of network traffic, based on which the prediction of any anomalies in the packet is made. Both modes require a training dataset supporting them. The functional elements are the same for both modes.

Monitor agent—responsible for intercepting incoming packets, transmitting them to the first analyzer element, the property generator, either as a single packet or aggregated into sets of streams.

Attribute builder—responsible for extracting attribute data from individual packages or sets of streams and generating what is called "instance data" for processing in machine learning models.

Checker Agent—Predict anomalies by passing data from each instance to a machine learning model. When an anomaly is detected in the device, an alarm is triggered, and the source of the incoming packet/stream is blocked while the packet (in burst mode) is dropped. The instance that triggered the alarm is pushed to the cloud node for deployment to other devices.

Train Agent—Trains the model step-by-step using locally found instances marked as "anomalies", but first assembles them into packages for application to the model when the device is not in use. This agent also receives flagged instance packages (except device-reported examples) from cloud node to train model.

Blocking Agent—Blocks incoming traffic from unhealthy sources. This can be sent to the firewall for appropriate action or done on the device itself.

8 Conclusions

The rapid development of information technologies in the modern world, the creation and deployment of IoT networks is inevitably accompanied by an increase in the number of threats and factors leading to the disruption of the functioning of network information systems. The paper considers the issues of building a model for detecting network traffic anomalies and attacks based on machine learning methods.

The proposed method for detecting network traffic anomalies based on a reduced feature space and the use of statistical, correlation and information features has shown its efficiency and high efficiency. Among the studied machine learning models, the KNN, CART, RF, AdaBoost, LR models showed the best results.

The result of this research is a cloud solution for IoT devices with limited resources, which is based on machine learning. The cloud component of the IDS system is used to offload IoT devices from the need to perform complex ML calculations. Downstream devices receive updated IDS parameter information from cloud MaaS. In this case, the functions of lower-level devices are limited to extracting signs of network traffic and detecting anomalies. The devices inform the cloud service about the detected anomalies and transmit their parameters for their analysis and model training.

References

1. Mabkhot M, Al-Ahmari A, Salah B, Alkhalefah H (2018) Requirements of the smart factory system: a survey and perspective. Machines 6(2):23. https://doi.org/10.3390/machines6020023

2. Savchenko V (2020) Detection of slow DDoS attacks based on user's behavior forecasting. Int J Emerg Trends Eng Res 8(5):2019–2025. https://doi.org/10.30534/ijeter/2020/90852020
3. Ageyev D, Radivilova T, Mohammed O (2020) Traffic monitoring and abnormality detection methods analysis. In: 2020 IEEE international conference on problems of infocommunications. Sci. Technol. (PIC S&T), pp 823–826. https://doi.org/10.1109/PICST51311.2020.9468103
4. Ageyev D, Radivilova T (2021) Traffic monitoring and abnormality detection methods for decentralized distributed networks. CEUR Worksh Proc 2923:283–288
5. Ageyev D, Bondarenko O, Radivilova T, Alfroukh W (2018) Classification of existing virtual-ization methods used in telecommunication networks. In: 2018 IEEE 9th international conference on dependable systems, services and technologies (DESSERT), pp 83–86. https://doi.org/10.1109/DESSERT.2018.8409104
6. Istratova E,Grif M, Dostovalov D (2021) Application of traditional machine learning models to detect abnormal traffic in the internet of things networks. In: Trawiński B, Nguyen NT, Iliadis L, Maglogiannis I (eds) Computational collective intelligence. ICCCI 2021. Lecture notes in computer science, vol 12876. Springer, Cham, pp 735–744
7. Scarfone K, Mell P (2007) Guide to intrusion detection and prevention systems (IDPS). Natl Inst Stand Technol 800–894:127 [Online]. Available: http://csrc.ncsl.nist.gov/publications/nistpubs/800-94/SP800-94.pdf
8. Anwar S et al (2017) From intrusion detection to an intrusion response system: fundamentals, requirements, and future directions. Algorithms 10(2):39. https://doi.org/10.3390/a10020039
9. Neela K, Kavitha V (2013) A survey on security Issues and vulnerabilities on cloud computing. Int J Comput Sci Eng Technol 4(7) (2013)
10. Wu Z, Xu Z, Wang H (2012). Whispers in the hyper-space: high-speed covert channel attacks in the cloud. In: Proceedings of the 21st USENIX security symposium, pp 159–173
11. Yarom Y, Falkner K (2014) FLUSH+RELOAD: a high resolution, low noise, L3 cache side-channel attack. In: Proceedings of the 23rd USENIX security symposium, pp 719–732. https://doi.org/10.5555/2671225.2671271
12. Ray PP (2018) A survey on Internet of Things architectures. J King Saud Univ Comput Inf Sci 30(3):291–319. https://doi.org/10.1016/j.jksuci.2016.10.003
13. Al-Fuqaha A, Guizani M, Mohammadi M, Aledhari M, Ayyash M (2015) Internet of Things: a survey on enabling technologies, protocols, and applications. IEEE Commun Surv Tutorials 17(4):2347–2376. https://doi.org/10.1109/COMST.2015.2444095
14. Da Xu L, He W, Li S (2014) Internet of things in industries: a survey. IEEE Trans Industr Inf 10(4):2233–2243. https://doi.org/10.1109/TII.2014.2300753
15. H. Suo, J. Wan, C. Zou, J. Liu, Security in the internet of things: a review. In: Proceedings—2012 international conference on computer science and electronics engineering. ICCSEE 2012, vol 3, pp 648–651 (2012). https://doi.org/10.1109/ICCSEE.2012.373
16. Mrabet H, Belgulth S, Alhomoud A, Jemai A (2020) A survey of IoT security based on a layered architecture of sensing and data analysis. Sensors 20(13):3625. https://doi.org/10.3390/s20133625
17. He D, Chan S, Ni X, Guizani M (2017) Software-defined-networking-enabled traffic anomaly detection and mitigation. IEEE Internet Things J 4(6):1890–1898. https://doi.org/10.1109/JIOT.2017.2694702
18. Ahmed M, Naser Mahmood A, Hu J (2016) A survey of network anomaly detection techniques. J Netw Comput Appl 60, 19–31 (2016). https://doi.org/10.1016/j.jnca.2015.11.016
19. Leskovec J, Rajaraman A, Ullman JD (2014) Mining of massive datasets. Cambridge University Press, Cambridge
20. Alsharif M, Rawat DB (2021) Study of machine learning for cloud assisted IoT security as a service. Sensors 21(4):1034. https://doi.org/10.3390/s21041034

Printed in the United States
by Baker & Taylor Publisher Services